Protein purification methods

a practical approach

TITLES PUBLISHED IN
THE
PRACTICAL APPROACH
SERIES

Series editors:

Dr D Rickwood
Department of Biology, University of Essex
Wivenhoe Park, Colchester, Essex CO4 3SQ, UK

Dr B D Hames
Department of Biochemistry and Molecular Biology, University of Leeds
Leeds LS2 9JT, UK

Affinity chromatography
Animal cell culture
Antibodies I & II
Biochemical toxicology
Biological membranes
Carbohydrate analysis
Cell growth and division
Centrifugation (2nd Edition)
Computers in microbiology
DNA cloning I, II & III
Drosophila
Electron microscopy
in molecular biology
Fermentation
Gel electrophoresis of nucleic acids
Gel electrophoresis of proteins
Genome analysis
HPLC of small molecules
HPLC of macromolecules
Human cytogenetics
Human genetic diseases
Immobilised cells and enzymes
Iodinated density gradient media
Light microscopy in biology
Liposomes
Lymphocytes
Lymphokines and interferons
Mammalian development
Medical bacteriology
Medical mycology
Microcomputers in biology
Microcomputers in physiology
Mitochondria
Mutagenicity testing
Neurochemistry
Nucleic acid and
protein sequence analysis
Nucleic acid hybridisation
Nucleic acids sequencing
Oligonucleotide synthesis
Photosynthesis:
energy transduction
Plant cell culture
Plant molecular biology
Plasmids
Prostaglandins
and related substances
Protein function
Protein purification applications
Protein purification methods
Protein sequencing
Protein structure
Proteolytic enzymes
Ribosomes and protein synthesis
Solid phase peptide synthesis
Spectrophotometry
and spectrofluorimetry
Steroid hormones
Teratocarcinomas
and embryonic stem cells
Transcription and translation
Virology
Yeast

Protein purification methods

a practical approach

Edited by

E L V Harris and S Angal

Celltech Ltd, 216 Bath Road,
Slough SL1 4EN, UK

IRL PRESS
——at——
OXFORD UNIVERSITY PRESS
Oxford New York Tokyo

Oxford University Press, Walton Street, Oxford OX2 6DP

Oxford   New York
Athens   Auckland   Bangkok   Bombay
Calcutta   Cape Town   Dar es Salaam   Delhi
Florence   Hong Kong   Istanbul   Karachi
Kuala Lumpur   Madras   Madrid   Melbourne
Mexico City   Nairobi   Paris   Singapore
Taipei   Tokyo   Toronto

and associated companies in
Berlin   Ibadan

Oxford is a trade mark of Oxford University Press

A Practical Approach 🔾 *is a registered trade mark*
of the Chancellor, Masters, and Scholars of the University of Oxford
trading as Oxford University Press

Published in the United States
by Oxford University Press Inc., New York

A catalogue record for this book is available from the British Library

Library of Congress Cataloging in Publication Data

Protein purification methods: a practical approach / edited by
E.L.V.Harris and S.Angal
p.   cm.— (The Practical Approach series)
Includes bibliographies and index.
1. Proteins—Purification.   I. Harris,E.L.V.   II. Angal,S.
III. Series.
QP551.P69752   1989
574.19'296—dc19

ISBN 0 19 963002 X (Hbk)
ISBN 0 19 963003 8 (Pbk)

Previously announced as:
ISBN 1 85221 113 X (hardbound)
ISBN 1 85221 112 1 (softbound)

Printed by Information Press Ltd, Oxford, England.

Preface

Protein purification is key to many areas of biochemical research. With the recent advances in gene cloning and expression it is not only protein biochemists who need to be able to purify proteins. Despite its key role protein purification is not a subject studied in depth in many undergraduate courses. Our experience of training graduates and post-graduates has led to the realization of the often limited knowledge and experience in this field. So when the subject of editing a 'Practical Approach' book on protein purification came up there was an immediate attraction. As we began the detailed planning of the book it became apparent that to cover this broad field adequately would entail division of the contents into two volumes, one on basic methods and one on applications. We have aimed to cover both the fundamentals and the more recent advances in the field. We therefore anticipate that both volumes will be useful as bench manuals to a wide range of biochemists and molecular biologists, from novices to experienced protein purifiers.

In this, the first volume, *Protein purification methods*, we have included all the major techniques. Basic principles are explained together with detailed protocols and advice on troubleshooting. The first chapter covers the initial planning stages including design of a purification process. Sections on protein assays, electrophoretic analysis and prevention of proteolysis, a problem frequently encountered, are also included. The following chapters then take the reader through the techniques used in protein purification starting in Chapter 2 with making a clarified protein extract and moving on to separations exploiting the biological and physical properties of the protein. Chapter 3 covers methods for concentrating the protein extract, and crude separations often used in the early stages of the purification process by precipitation and by aqueous two-phase partitioning. Chromatographic methods exploiting differences in structural properties such as charge or hydrophobicity and size are described in Chapters 4 and 6 respectively. Methods dependent on biospecific activity are the subject of Chapter 5 and are also covered in more detail in another book in this series (*Affinity Chromatography: A Practical Approach*, edited by P.D.G.Dean, F.A.Middle and W.S.Johnson).

We are indebted to Margaret Turner for her expert secretarial assistance without which the book would not have reached publication. We are also grateful to Martin and Shantanu for their moral support over the two years it has taken to complete the book. Finally we wish to dedicate this volume to Dr Andrew Doherty, who was to have been an author but unfortunately was tragically killed in a road accident.

E.L.V.Harris
S.Angal

Contributors

B.A.Andrews
Department of Biochemical Engineering, University of Reading, PO Box 226, Reading RG6 2AP, UK

S.Angal
Celltech Ltd, 216 Bath Road, Slough SL1 4EN, UK

J.A.Asenjo
Department of Biochemical Engineering, University of Reading, PO Box 226, Reading RG6 2AP, UK

R.J.Beynon
Department of Biochemistry, University of Liverpool, Liverpool L69 3BX, UK

M.Cusack
Department of Biochemistry, University of Liverpool, Liverpool L69 3BX, UK

P.D.G.Dean
Cambridge Life Sciences plc, Cambridge Science Park, Milton Road, Cambridge CB4 4GN, UK

M.J.Dunn
National Heart and Lung Institute, Dovehouse Street, London SW3 6LY, UK

G.H.Goodwin
Chester Beatty Research Institute, Fulham Road, London, UK

D.Griffiths
Department of Biological Sciences, Hatfield Polytechnic, College Lane, Hatfield, Herts AL10 9AB, UK

E.L.V.Harris
Celltech Ltd, 216 Bath Road, Slough SL1 4EN, UK

M.R.Hartley
Department of Biological Sciences, University of Warwick, Coventry CV4 7AL, UK

C.R.Hill
Celltech Ltd, 216 Bath Road, Slough SL1 4EN, UK

A.C.Kenney
Oros Systems Ltd, Albion Close, Petersfield Avenue, Slough SL2 5DV, UK

N.Lambert
AFRC Institute of Food Research, Colney Lane, Norwich NR4 7UA, UK

A.Z.Preneta
Celltech Ltd, 216 Bath Road, Slough SL1 4EN, UK

S.Roe
Biotechnology Group, Biotechnological Separations Club, Building 353, Harwell Laboratory, Oxfordshire OX11 0RA, UK

T.Salusbury
Biotechnology and Separation Division, Warren Spring Laboratories, Gunnels Wood Road, Stevenage, Herts SG1 2BK, UK

C.Sutton
Celltech Ltd, 216 Bath Road, Slough SL1 4EN, UK

L.G.Thompson
Celltech Ltd, 216 Bath Road, Slough SL1 4EN, UK

P.N.Whittington
Biotechnology and Separations Division, Warren Spring Laboratories, Gunnels Wood Road, Stevenage, Herts SG1 2BK, UK

Contents

4. SEPARATION BASED ON STRUCTURE   175
S.Roe

Abbreviations

Ab	antibody
Ag	antigen
AK	adenylate kinase
AP	alkaline phosphatase
4-BA	4-aminobenzamidine
BCA	bicinchoninic acid
Bis	N,N'-methylene bisacrylamide
BSA	bovine serum albumin
CBB	Coomassie Brilliant Blue
CDI	N,N'-carbonyldiimidazole
CM	carboxymethyl
ConA	concanavalin A
CNBr	cyanogen bromide
CPG	controlled pore glass
CZE	continuous zone electrophoresis
DCI	dichloroisocoumarin
DipF	diisopropylphosphofluoridate
DMSO	dimethylsulphoxide
DTT	dithiothreitol
EDTA	ethylenediamine tetra-acetic acid
EGTA	ethylenebis (oxyethylenenitrilo)-tetra-acetic acid
ELISA	enzyme-linked immunosorbent assay
ER	endoplasmic reticulum
ETP	electron-transport particles
FMP	2-fluoro-1-methylpyridinium toluene-4-sulphonate
FPLC	fast protein liquid chromatography
GPC	gel permeation chromatography
HA	hydroxylapatite
Hepes	N-2-hydroxyethylpiperazine-N'-2-ethanesulphonic acid
HIC	hydrophobic interaction chromatography
HPLC	high performance liquid chromatography
IEF	isoelectric focusing
Ig	immunoglobulin
IL-2	interleukin 2
IPG	immobilized pH gradient
LCA	lentil lectin
LDH	lactate dehydrogenase
LPLC	low pressure liquid chromatography
LPS	lipopolysaccharides
MAb	monoclonal antibody
MAO	monoamine oxidase
Mes	2-(N-morpholino) ethanesulphonic acid
MPLC	medium pressure liquid chromatography
MZE	multiphasic zone electrophoresis
NCCR	NADPH cytochrome c reductase
NDPase	nucleoside diphosphatase
NMWC	nominal molecular weight cut-off

PAGE	polyacrylamide gel electrophoresis
PBA	phenyl boronate
PBS	phosphate buffered saline
PDI	protein disulphide-isomerase
PEG	polyethylene glycol
PMSF	phenylmethylsulphonyl fluoride
PVDF	polyvinylidene difluoride
RIA	radioimmunoassay
RPC	reverse-phase chromatography
SCP	single cell protein
SDS−PAGE	sodium dodecyl sulphate−polyacrylamide gel electrophoresis
STI	soybean trypsin inhibitor
TCA	trichloroacetic acid
TEMED	N,N,N',N'-tetramethylethylenediamine
TFA	trifluoroacetic acid
WGA	wheatgerm agglutinin

CHAPTER 1

Initial planning

1. INTRODUCTION—E.L.V.Harris

Before embarking on the purification of a protein a few key questions should be addressed first. Thus: what is the protein required for? Which would be the most suitable source? What is known about the protein? How should the protein be assayed? By answering these questions the aims of the purification and criteria for success will be defined, and the background knowledge required to plan a suitable strategy will be acquired.

1.1 What is the protein required for?

The purification of a protein is frequently not the end point itself, but is the means to obtain a pure protein for further studies. These studies may be on the activity of the protein, on its structure, or on its structure—function relationships. The requirements of these studies will define how much of the purified protein is required, whether loss of activity can be tolerated, how pure it should be, and the time and cost of purifying it. Thus, for studies on activity relatively small amounts of active protein will be required. High purity will probably not be essential provided any interfering activities are removed. Cost will probably not be very important, but speed will be important to minimize activity losses. In contrast, structural studies will require larger amounts of highly pure protein. Cost and time will be of secondary importance, except for structure—function studies where activity is required and therefore speed will probably be important. For economic reasons cost and time become much more important in the manufacture of proteins. The amounts required and purity levels will depend on the end use of the manufactured protein. Thus, if the protein is to be sold for research use (e.g. those supplied by Sigma, BCL or Amersham International) the quantities required are small, whilst in terms of purity the removal of interfering activities will be essential. For industrial applications, such as in the food industry (e.g. enzymes produced by Novo and Sturge for degrading starch into glucose and maltose), or for use in domestic detergents (e.g. Ariel, Bold), large quantities are required, and purity is usually of secondary importance to cost. In contrast, for therapeutic applications (e.g. tissue plasminogen activator for treating myocardial infarction, or monoclonal antibodies for cancer imaging) purity is of the utmost importance and quantities required are relatively small.

   The amount of purified protein will not only depend on the amount of starting material, but will also depend on yield. Protein is lost at each step of a purification procedure, therefore to maximize yield the minimum number of steps should be used. However by minimizing the number of steps, the final purity of the protein will be lower. Some methods of purification inherently give higher yields than others, thus by judicious choice of the individual steps yield can be maximized with a minimal loss in purity (see Section 6).

Purity of the protein of interest may simply be defined in terms of the percentage of total protein. Alternatively, other types of contaminants may also be important. Thus, for enzymic studies purity is usually defined in terms of total protein. Typically about 80−90% is adequate, although the presence of active proteases or other enzymes which interfere with activity is not acceptable. However in some cases other types of contaminant may interfere with the activity of the enzyme and must therefore be removed (e.g. DNA, metal ions). For structural studies the protein is typically required to be ≥95% pure protein. For therapeutic use, all types of contaminant must be considered. The final product must therefore be assayed for self-aggregates, contaminating proteins, DNA, endotoxins, and any in-process additives (e.g. anti-foam, affinity ligands). Specific steps may have to be introduced to remove these contaminants.

Additional purification steps to increase purity not only result in additional loss in yield, but also increase cost and time taken. For research use, the scale is small and therefore the cost of an additional step may not be important, whereas in the manufacture of proteins introduction of an additional step may make a purification process uneconomic. Removal of the final few percent contaminants is often considerably harder than the initial purification steps, and may therefore take several more steps with a resultant decrease in yield and increase in cost and time.

In most cases, except for some structural studies such as protein sequencing, the purified protein should be as 'native' and active as possible. Steps to minimize denaturation and proteolysis should therefore be taken. Thus, typically harsh conditions, such as extremes of pH or chaotrophic solutions, should be avoided (see Section 6). This requirement may place constraints on the methods of purification used, for example, elution from affinity columns frequently requires harsh conditions. Proteolysis can be minimized by working quickly, and at low temperatures (typically 4°C). In addition, protease inhibitors should be added (see Section 5). Frequently the protein may become more susceptible to proteolysis as it is purified, since there is less other protein around to 'mop up' the protease activity.

1.2 **What source should be used?**

Frequently the source is defined by the final application. When there is a choice, the source where the protein of interest is most stable and most abundant should be selected. In addition, the availability and quantity of the source should be considered. Thus, mouse livers may be suitable for purifying small quantities of an enzyme, but bovine liver would be more suitable for larger quantities. Also, animals which can be kept in the laboratory are usually preferable to animals from the wild. Cells which can be grown in culture, such as *Escherichia coli, Saccharomyces cerevisiae*, or mammalian cells, are preferable to animals. Not only can large quantities of these be obtained but the culture conditions can be carefully controlled giving more batch to batch reproducibility. In addition, by using cultured cells, the need to kill animals is minimized.

With the advent of gene cloning techniques proteins can now be expressed in high amounts in cells which can be grown in culture. In addition to the above-mentioned advantages the percentage of the protein of interest is usually higher than in its native source, thus making purification easier. The host cell should be chosen with care since each has its advantages and disadvantages. *E.coli* is the most commonly used host due to ease of handling, however, proteins are often not secreted, and in addition are often produced in an insoluble form (known as inclusion bodies). Secreted proteins are usually easier to purify since there are fewer contaminating proteins present. Inclusion bodies can be relatively easily purified and consist mainly of the expressed protein, however, the protein must be denatured and refolded to obtain a soluble, active form. *Bacillus subtilis* may become a favoured host, since proteins are naturally secreted. However, the genetics of this organism are currently not as well understood as those of *E.coli*, and in addition highly active proteases are usually also secreted, resulting in degradation of the cloned protein. The lack of a protein glycosylation pathway in the prokaryotes is a major disadvantage for producing many eukaryote proteins. In order to overcome this, *S.cerevisiae* is often used, but this organism will only secrete small proteins and peptides, and the glycosylation is not typical of the higher eukaryotes. Due to the above, mammalian cells (e.g. Chinese hamster ovary cells) are gaining popularity. Even large, multimeric glycosylated proteins, such as antibodies, can be cloned, expressed and secreted. However, culture of mammalian cells is technically more complex than the other potential host cells and since the cell densities achieved are lower the concentrations of protein are lower. Cloned proteins can also be modified to ease purification. Thus, a basic tail may be added to aid purification by ion-exchange chromatography, or part of another protein may be added and exploited for affinity purification (1,2). After purification, the foreign part of the protein is removed by chemical or enzymatic cleavage.

The levels of protein in a particular source are a balance between synthesis and breakdown. These activities can often be influenced by hormones or chemicals in whole animals but more easily in cultured cells.

1.3 What is known about the protein?

Knowledge of at least some of the chemical and physical properties of the protein and its cellular localization will aid the design of a purification process. If the protein has been previously isolated from a different source much of the knowledge can be applied to the protein from the new source. Thus, it's localization is likely to remain the same, as is whether it is a glycoprotein or lipoprotein. Size will often remain similar, whereas pI and hydrophobicity may vary more. Knowledge of the intracellular localization may be gained by assaying subcellular fractionations or by microscopic examination using a label specific for the protein (e.g. a suitably labelled ligand or antibody). Alternatively, the activity of the protein may identify its localization, for example, enzymes involved in transcription will be located in the nucleus, whilst receptors for extracellular growth factors will be found in the plasma membrane.

If the protein is known to be a glycoprotein or lipoprotein its properties will be markedly different from most other contaminating proteins. These properties can be exploited during purification, for example, glycoproteins can be purified by lectin affinity chromatography (Chapter 5).

Whether the protein is intracellular or extracellular, soluble, insoluble or membrane-bound, or located in a subcellular organelle, will influence the choice of extraction method and buffer components used. Thus, for extracellular proteins a high degree of purification is achieved by removal of the cells. Membrane-bound proteins will require detergents or organic solvents to solubilize them. Isolation of the subcellular organelle prior to extraction of the protein will give a high degree of purification (but usually a low yield).

If the protein to be purified is an enzyme or receptor it may be possible to exploit its activity by affinity purification on a substrate or ligand, or an analogue. Knowledge of the size and pI of the protein will indicate suitable matrices and conditions for gel filtration and ion-exchange chromatography. A prior knowledge of the stability of the protein and its sensitivity to temperature, extremes of pH, proteases, air and metal ions will also aid the design of a purification procedure.

2. BASIC PREPARATIONS—E.L.V.Harris

2.1 **Equipment**

In addition to the usual equipment available in most laboratories, such as pH meters, balances and magnetic stirrers, the protein purification laboratory requires other, more specialized equipment, as discussed below.

In order to minimize denaturation and proteolysis of proteins many of the purification steps must be carried out at about 4°C. A cold cabinet (e.g. Gallenkamp) or cold room suitable for performing dialysis, precipitation and column chromatography steps is required. In addition large quantities of ice from an ice-making machine will usually be necessary for short-term storage of samples, for example, prior to assay, and will also allow some manipulation to be carried out at the bench whilst still maintaining the solutions at about 4°C. Fridges (~4−8°C) and freezers (−15 to −20°C) are required for storage of samples and buffers. In addition a −70°C freezer may be necessary to preserve activity in stored samples.

A spectrophotometer covering the range 205−850 nm is essential for measuring total protein. In addition the spectrophotometer can often be used to assay enzyme activity, and one fitted with a chart recorder is particularly useful for this purpose. Other equipment to allow assay of the specific protein may also be required, for example, a scintillation counter, an ELISA plate reader or an HPLC. Electrophoresis is often used to assess protein purity; suitable power packs and tank assemblies are available from several suppliers (e.g. Bio-Rad, BRL and Pharmacia-LKB).

To extract the protein from its source, equipment for disruption may be necessary. Typically a domestic blender, Waring blender or Polytron blender will be suitable for most laboratory-scale processes. These and other types of equipment used for disruption are discussed in detail in Chapter 2. To clarify the extract or other protein solutions one or more centrifuges will be required. These should ideally be refrigerated to allow use at 4°C, capable of giving centrifugal forces of up to 15 000 g and should cover the range of 100 ml to a few litres. In addition a small-bench top centrifuge for handling approximately 1.5 ml tubes (e.g. Eppendorf) may be useful.

For column chromatography a fraction collector, UV monitor and recorder, peristaltic pump and a range of sizes of columns are required. Although columns can be run by gravity and fractions collected by hand and measured for absorbance in a spectrophotometer this does not usually make efficient, economical use of the laboratory worker's time. The range of column sizes should cover those suitable for ion-exchange and affinity chromatography (i.e. relatively short, fat columns) and those for gel permeation chromatography (i.e. relatively long, thin columns). The volumes of the columns will depend on the scale of the purification; for typical laboratory-scale work the range 10 ml to a few hundred ml will usually be appropriate. In addition to the above equipment some means of generating a buffer gradient is often required. Suitable gradient mixers are available from several suppliers (e.g. BRL, or Pharmacia-LKB). Suitable chromatography equipment is covered in more detail in Chapter 4.

Additional, more specialized, equipment may be necessary for specific applications, for example, an amino acid analyser and a protein sequencer for primary structural studies. Further equipment may speed up the process of developing a purification procedure, or make life easier for the protein purifier, but is usually not essential. For example, the Phast system (Pharmacia-LKB) for electrophoresis uses pre-cast gels and allows rapid screening (typically < 1 h) of samples. The microprocessor controlled medium pressure chromatography systems (e.g. FPLC, Pharmacia-LKB), used with gel matrices which can be scaled up easily, are extremely useful for rapid screening of conditions for optimum chromatography.

2.2 Buffers

At all stages of the purification it is necessary to control the pH of the solution to avoid denaturation or inactivation of the protein. This is achieved by using buffers which absorb or release protons by a shift in the equilibrium:

$$HA \rightleftharpoons H^+ + A^-$$

or

$$HB^+ \rightleftharpoons H^+ + B$$

The buffering capacity is optimal at the pK_a of the buffer where the concentration of the acidic and basic forms are equal. Buffers only work over a limited pH range, in practice approximately 1.0 pH unit on either side of the pK_a. *Table 1* shows the working range for many of the commonly used buffers.

The concentration of buffer required will depend on the application, thus if a process is known to release protons a higher buffer concentration is required. Also, if the buffer is being used at the limit of its buffering range higher buffer concentrations will be necessary. Typically buffer concentrations of $20-50$ mM are adequate. Unfortunately references rarely give full details of how buffers are made. Thus for example a 50 mM sodium acetate buffer may be made by mixing 50 mM sodium acetate and 50 mM acetic acid in appropriate proportions to obtain the desired pH, alternatively sodium hydroxide can be added to 50 mM acetic acid to obtain the desired pH. Frequently the pH of the buffer is more critical than the ionic strength, therefore in most instances either of the above two methods, which would result in different ionic strengths, would be acceptable.

This also means that it is not necessary to weigh buffer constituents to more than one decimal place, and graduated cylinders rather than volumetric flasks are adequate for adjusting to the final volume. Many buffers, such as Tris and sodium acetate, can be made by dissolving the buffering salt and adjusting the pH with HCl or NaOH, the final volume is then adjusted to give the desired concentration. Alternatively standard recipes can be used, for example, sodium phosphate buffer is made by mixing disodium hydrogen phosphate and sodium dihydrogen phosphate in appropriate proportions (see

Table 1. pH range of buffers.
A. Commonly used buffers

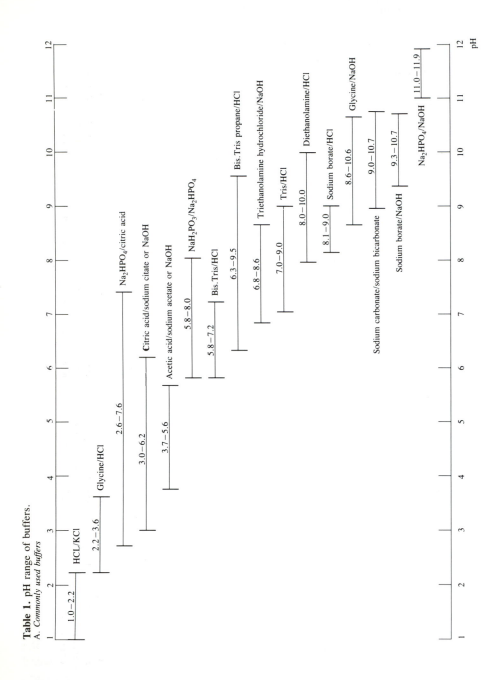

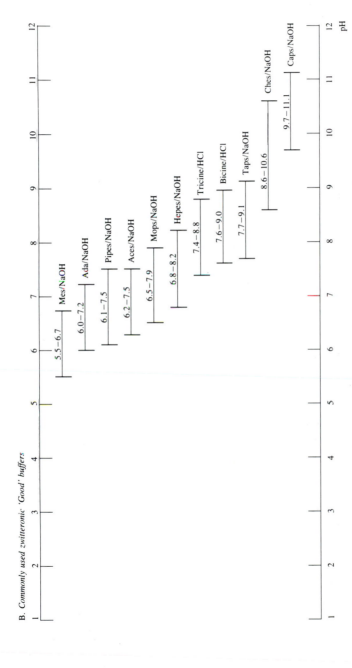

B. *Commonly used zwitteronic 'Good' buffers*

Mes, 2-(N-morpholino) ethanesulphonic acid; Ada, N-(2-acetamido) iminodiacetic acid; Pipes, Piperazine-NN'-bis-2-ethanesulphonic acid; Aces, N-(2-acetamido)-2-aminoethanesulphonic acid; Mops, 3-(N-morpholino) propanesulphonic acid; Hepes, N-2-hydroxyethylpiperazine-N'-2-ethane-sulphonic acid; Tricine, N-Tris (hydroxymethyl) methylglycine; Bicine, NN-bis (2-hydroxyethyl) glycine; Taps, N-Tris (hydroxymethyl) methyl-3-aminopropanesulphonic acid; Ches, 2-(cyclohexylamino) ethanesulphonic acid; Caps, 3-(cyclohexylamino)-1-propanesulphonic acid.

Appendix and references 3—5). In this case the final pH should be always checked and adjusted if necessary to allow for variability in the water content of the buffer components and to check for errors. This is particularly important if other components, such as EDTA, or 2-mercaptoethanol, are present in the solution.

The pK_a of a buffer is affected by the temperature. The pH of some buffers therefore can change significantly with temperature, for example, the pK_a of Tris at 4°C is 8.8 whereas at 20°C it is 8.3. For this reason buffers should be equilibrated at the temperature at which they will be used prior to adjustment of the pH. The temperature compensation on the pH meter should also be adjusted to that of the solution to be measured.

For convenience stock buffer solutions of 10- or 100-fold higher concentrations than they will be used at are made. However, with some buffers (e.g. phosphate and citrate) the pH is significantly affected by the ionic strength, therefore the pH should be checked after dilution.

Factors which influence the choice of buffer are as follows.

(i)   Desired pH.
(ii)   Anionic or cationic buffer species. Thus, for anionic exchange chromatography a cationic buffer should be used (e.g. Tris not phosphate).
(iii)   Variation of pH with ionic strength or temperature.
(iv)   Chemical reactivity. For example, primary amines such as Tris can interfere with analysis of proteins (e.g. by Lowry method or amino acid analysis) or can inhibit some enzymes.
(v)   Biological activity. For example, phosphate is often a participant in biological reaction and may therefore inhibit or activate an enzyme.
(vi)   Interaction with other components. For example, phosphate complexes with di- and polyvalent metal ions, thus inhibiting metal ion-dependent enzymes. Borate complexes with many organics and hydroxyl groups, especially those on carbo-hydrates.
(vii)   Penetration of biological membranes (e.g. Tris).
(viii) Toxicity. Barbitone and cacodylate buffers have been largely replaced due to their toxicity.
(ix)   Absorption at 280 nm or less (e.g. maleate).
(x)   Expense, if used on a large scale.
(xi)   Solubility.

A series of buffers which overcome many of the above potential problems commonly observed with traditional buffers were developed by Good and his colleagues (6,7) (*Table 1B*). These 'Good' buffers are biologically and chemically non-reactive, non-toxic, do not absorb in the UV region, and their pK_a's show minimum dependence on temperature or ionic strength. They are however more expensive than some of the traditional buffers, such as acetate or citrate.

Care and proper use of the pH meter and probe are essential for accurate pH measurements. Standard buffer solutions available from several suppliers (e.g. Russell, Sigma and BDH) should be used to adjust the pH meter. Unless the buffer pH is to be the same as that of the standard buffer at least two standard buffers should be used and the slope adjusted to ensure the pH meter reads both buffers correctly.

Specially designed pH probes should be used with Tris solutions, since the Tris ions can cross the probe membrane and thereby lead to false values. Such probes (available from Sigma and Russell) are more expensive than standard probes. Frequent measuring of solutions containing high concentrations of protein can also shorten the useful life of a probe by adsorption of protein onto the membrane (specially treated probes to prevent this are available from Russell). Probes must not be allowed to dry out, and should therefore be stored in liquid, ideally saturated potassium chloride.

2.3 Assays

After each stage of the purification, assays for the protein of interest and its purity are required to assess the efficiency of the purification. Specific assays for the protein of interest are used to identify which fractions contain the protein prior to pooling for the next step. In addition the yield for each step can be determined.

Results from total protein assays when combined with those from specific assays provide information on the degree of purification achieved by each step and the specific activity of the protein of interest.

$$\text{Specific activity} = \frac{\text{Protein of interest (mg or units)}}{\text{Total protein (mg)}}$$

$$\text{Degree of purification} = \frac{\text{Specific activity at step 2}}{\text{Specific activity at step 1.}}$$

Thus, each step can be assessed by yield and degree of purification for its efficiency. Methods for total protein assays are covered in Section 3. It is beyond the scope of this book to detail methods for assaying specific proteins, since there will be as many methods as there are proteins. However, the requirements of an assay and the relative merits of different assay types will be discussed.

Ideally an assay should be simple, highly specific and rapid, allowing many fractions to be screened for activity prior to the next stage of the purification. Rapid assays will minimize storage times between steps, thereby minimizing proteolysis. Many enzymes can be rapidly assayed (< 10 min) spectrophotometrically, though problems may be encountered with other components in impure samples interfering with the assay, and screening of multiple samples may be time-consuming. Immunological assays can be used for many proteins provided an appropriate antibody is available. The most frequently used immunoassays are radioimmunoassays (RIA) or enzyme-linked assays (ELISA). These are not usually as rapid as enzyme activity assays, typically taking at least one day. Since only the immunological activity is necessary for detection, these methods will often also detect denatured protein and thus loss of activity of the protein may not be detected. However, these types of assays are usually highly specific and multiple samples can be screened relatively easily. Assays of biological activity using either animals or cultured cells frequently take a long time, even up to a few weeks! Often they are notoriously variable since the metabolic state of the animal or cells may profoundly influence the assay sensitivity. Many assays may be inhibited by compounds present in the crude protein extract. When such a component is removed by a purification step the apparent yield of the protein of interest may be more than 100%. Conversely removal of a factor required for activity of the protein will result in a 0% yield. In such cases activity may be restored by mixing fractions containing the factor with the protein, often a trial and error process!

Purity is often assessed by electrophoresis usually under denaturing conditions (Section 4). A single band is indicative of a pure protein, although potentially another protein may co-migrate and therefore an additional alternative method should ideally be tried. Heterologous multimeric proteins will not give a single band, therefore the definition of purity is when further purification steps do not remove any of the bands. HPLC may be used in addition to electrophoresis; in this case an alternative property of the protein should be used to achieve the separation (i.e. if the method of electrophoresis used separates on the basis of size, gel permeation HPLC should not be used). Alternatively the ratio of specific protein activity to total protein (specific activity) can be measured at each step; purity is achieved when subsequent purification steps no longer increase the specific activity.

3. DETERMINATION OF TOTAL PROTEIN CONCENTRATION—M.J.Dunn

It is useful, but often not essential, to determine the total amount of protein present at each stage of a fractionation procedure. In practice, the biological or biochemical activity of the fractions is usually more important. However, it is essential to estimate protein content under certain circumstances, for example to estimate the degree of purification obtained by a particular fractionation scheme, or to determine the specific activity of the final product. In addition, certain fractionation procedures are critically sensitive to protein concentration.

Over the years many indirect and direct techniques have been described for the measurement of protein concentration (for review, see 8). Of the former procedures, the most important technique is the determination of nitrogen by the Kjeldahl method. The latter group includes procedures such as infrared spectrophotometry, turbidimetry, fluorimetry, refractometry and polarography. If the primary sequence of the protein under investigation has been established, then analysis of amino acid composition can yield accurate quantitative data. It should be remembered that certain amino acids, especially tryptophan, cystine and cysteine, are particularly sensitive to acid hydrolysis, making data on these amino acids unreliable. In addition, glutamine and asparagine are converted to their respective acids by acid hydrolysis from which they are therefore indistinguishable. In practice, of course, the requisite amino acid sequence data will often not be available to the purification scientist. The methods most commonly used today for the measurement of protein concentration are based either on UV absorption of the protein solution or visible region spectrophotometry after reaction of the protein with a chemical to generate chromophores, and only these methods will be discussed here.

It is essential to remember that all methods for the determination of protein concentration, with the exception of amino acid analysis and dry matter content determined by drying at constant temperature to constant weight (8), are at best only semiquantitative unless the procedure has been standardized using the same purified protein measured by dry weight estimation. It is also important to realize that the response of proteins, such as colour development in a spectrophotometric assay, varies with the amino acid composition of the particular protein and can be influenced by the presence of non-protein prosthetic groups such as carbohydrate. Thus, the almost universal use of bovine serum albumin (BSA) as a standard protein in these procedures is not entirely satisfactory and it is much preferable, where possible, to calibrate the assay using a purified sample of the protein under investigation or of a protein with a similar colour yield.

3.1 Ultraviolet spectrophotometry

3.1.1 *Measurement at 280 nm*

Most proteins exhibit an absorption maximum at 280 nm which is attributable to the phenolic group of tyrosine and the indolic group of tryptophan. The extinction coefficient, usually expressed either as $E_{280}^{1\%}$ or $E_{280}^{1\,mg/ml}$, varies significantly from protein to protein depending on the precise amino acid composition. Values of $E_{280}^{1\,mg/ml}$ for most proteins lie in the range $0.4 - 1.5$, but extremes include some parvalbumins and related Ca^{2+}-binding proteins (0.0) and lysozyme (2.65). The method is consequently rather inaccurate unless the protein is relatively pure and its extinction coefficient is known and calibrated against dry weight. This technique is relatively insensitive and requires between 0.05 and 2 mg of protein. Moreover, it is very sensitive to interference from other compounds which also absorb at 280 nm. This is especially true if compounds containing purine and pyrimidine rings (nucleic acids, nucleotides) are present, in which case the absorbance of the sample is measured at both 280 nm and 260 nm and a correction formula applied (9):

$$\text{protein (mg ml}^{-1}) = 1.55\,A_{280} - 0.76\,A_{260}$$

If the solvent itself absorbs at 280 nm, an accurate blank must be taken. The very real advantage of this approach, however, is that it is rapid and non-destructive, so that the assayed sample can be used for subsequent processing. Quartz cuvettes are essential for measuring absorbance at this wavelength and below.

3.1.2 *Measurement in the far-UV*

The peptide bonds in proteins and peptides exhibit a well-defined absorption maximum at $191 - 194$ nm, which is more sensitive $(0.01 - 0.05$ mg protein) and less dependent on amino acid composition than measurements at 280 nm. Routine spectrophotometers cannot be used for measurements at these wavelengths due to UV absorption by oxygen. However, useful measurements can be made at 205 nm, particularly if a correction is made for the tyrosine and tryptophan content by measurement at 280 nm (10). The formula used is:

$$E_{205}^{1\,mg/ml} = 27.0 + 120 \times \frac{A_{280}}{A_{205}}$$

The success of this approach depends on the use of clean quartz cuvettes and buffers with minimal UV absorption at this wavelength.

Method Table 1. Biuret method.

---

1. Prepare Biuret reagent by dissolving 1.5 g of copper sulphate ($CuSO_4 \cdot 5H_2O$) and 6 g of sodium potassium tartrate in 500 ml of water. Add 300 ml of 10% w/v sodium hydroxide and make up to 1 litre with water.
2. Store the reagent in a plastic container in the dark. It will keep indefinitely if 1 g of potassium iodide is added to inhibit reduction of copper.
3. To 0.5 ml samples containing up to 3 mg of protein, add 2.5 ml of Biuret reagent.
4. Stand for 30 min.
5. Measure absorbance at 540 nm against a blank containing 0.5 ml of sample buffer plus 2.5 ml of Biuret reagent.

Method Table 2. Lowry method.

---

1. Stock reagents: A, 1% w/v copper sulphate ($CuSO_4 \cdot 5H_2O$); B, 2% w/v sodium potassium tartrate; C, 0.2 M sodium hydroxide; D, 4% w/v sodium carbonate. These reagents are stable at room temperature.
2. To 49 ml of reagent C, add 49 ml of reagent D. Then add 1 ml of reagent A, followed by 1 ml of reagent B. This is the copper−alkali solution (reagent E) which must be prepared fresh when needed.
3. To 10 ml of Folin−Ciocalteau reagent, add 10 ml of water. This is reagent F.
4. To 0.5 ml samples containing up to 0.5 mg protein, add 2.5 ml of reagent E.
5. Mix and stand for 10 min.
6. Add 0.25 ml of reagent F.
7. Mix and stand for 30 min.
8. Measure absorbance at 750 nm against a blank of 0.5 ml of sample buffer processed through steps 4−7.

3.2 Biuret method

Under alkaline conditions the copper (II) ion is bound to peptide nitrogen of proteins and peptides to produce a purple colour with an absorption maximum at 540−560 nm. There is no interference from free amino acids and there is little dependence on amino acid composition as the copper reagent reacts with the peptide chain itself rather than with side groups. A recommended procedure (11) is given in *Method Table 1*. The main disadvantage of this method which severely limits its applicability is its low sensitivity (range 1−6 mg protein/ml). The sensitivity can be increased either (i) by separation of the copper−protein complex by gel filtration followed by release of copper from the complex and its colorimetric determination (12), or (ii) by measurement of absorbance of the standard Biuret reaction in the near-UV at 310 nm (13). Some buffers, particularly those based on the use of Tris, can interfere with the Biuret reaction, as does ammonia which complicates the measurement of fractions obtained using ammonium sulphate precipitation.

3.3 Lowry procedure

The most commonly used method of protein determination is the Folin−Lowry method.

Indeed, this has made the original paper of Lowry *et al.* (14) a *citation classic* as it is one of the most frequently cited articles in the life sciences. The basis of the method is the Biuret reaction of proteins with copper under alkaline conditions and the Folin−Ciocalteau phosphomolybdicphosphotungstic acid reduction to heteropoly-molybdenum blue by the copper-catalysed oxidation of aromatic amino acids. The reaction results in a strong blue colour and is more sensitive (range 0.10−1 mg protein/ml) than the Biuret method. A recommended procedure is given in *Method Table 2*. It should be noted that the reaction is very dependent on pH which should be maintained between pH 10 and 10.5. Due to batch-to-batch variations in the Folin−Ciocalteau reagent, it is essential to establish the dilution of any particular batch of this reagent required to obtain the desired final pH. In addition, the instability of this reagent in an alkaline environment necessitates that precise timings of both reagent addition and mixing with sample must be used if accurate and reproducible results are to be obtained.

The reaction shows a moderate variation of response to protein amino acid composition, but more importantly a diverse range of substances have been found to interfere with this procedure. These compounds include amino derivatives, amino acids, buffers, detergents, drugs, lipids, sugars, nucleic acids, salts and sulphydryl reagents. An extensive list of substances which can interfere with the Lowry procedure is given in (15) together with the concentrations tested, the tolerable concentration limit, and the effect on the reagent blank and the colour produced by the standard protein. The effect of a selection of some of these compounds is shown in *Table 2*. Of particular importance is the inability of the Lowry procedure to tolerate a variety of different types of detergent (anionic, zwitterionic and non-ionic) since these reagents are often used to solubilize proteins. This problem has resulted in the development of modified procedures designed to overcome interference by specific substances and this subject is discussed in detail in reference 15. However, the high sensitivity of the assay is of advantage here, as often in practice interfering substances can be diluted out to minimize their deleterious effects. An additional disadvantage of this procedure, but one which is often overlooked, is that the standard curve is non-linear at higher concentrations. Modified procedures and curve-fitting techniques have been described to overcome this problem (15). Finally, the two-step nature of the method complicates the assay and presents problems for its automation.

3.4 Bicinchoninic acid procedure

Many of the problems associated with the Lowry procedure discussed in Section 3.3 are due to the properties of the detection reagent (Folin−Ciocalteau) which is used. Recently, a procedure based on an alternative detection reagent, bicinchoninic acid (BCA), has been developed (16). The reaction results in an intense purple colour with an absorbance maximum at 562 nm. The method has a similar sensitivity (range 0.1−1.2 mg protein ml^{-1}) to the Lowry procedure. The recommended procedure is given in *Method Table 3*. The reagents are commercially available from Pierce. A more sensitive (range 0.5−10 μg protein ml^{-1}) microassay procedure has also been developed (16) and is given in *Method Table 4*.

Since the mechanism of the BCA assay is closely related to that of the Lowry

Table 2. Effect of selected potential interfering compounds on the Lowry and bicinchoninic acid assay methods. Reproduced with permission from (16).

Sample (50 μg BSA) in	BCA assay (μg BSA found)		Lowry assay (μg BSA found)	
	Water blank corrected	Interference blank corrected	Water blank corrected	Interference blank corrected
Water (reference)	50.00	–	50.00	–
0.1 M HCl	50.70	50.80	44.20	43.80
0.1M NaOH	49.00	49.40	50.60	50.60
0.2% Sodium azide	51.10	50.90	49.20	49.00
0.02% Sodium azide	51.10	51.00	49.50	49.60
1 M Sodium chloride	51.30	51.10	50.20	50.10
100 mM EDTA (4 Na)	No colour		138.50	5.10
50 mM EDTA (4 Na)	28.00	29.40	96.70	6.80
10 mM EDTA (4 Na)	48.80	49.10	33.60	12.70
50 mM EDTA, pH 11.25	31.50	32.80	72.30	5.00
4 M Guanidine−HCl	48.30	46.90	Precipitated	
3 M Urea	51.30	50.10	53.20	45.00
1% Triton X-100	50.20	49.80	Precipitated	
1% SDS	49.20	48.90	Precipitated	
1% Brij 35	51.00	50.90	Precipitated	
1% Lubrol	50.70	50.70	Precipitated	
1% Chaps	49.90	49.50	Precipitated	
1% Chapso	51.80	51.00	Precipitated	
1% Octyl glucoside	50.90	50.80	Precipitated	
40% Sucrose	55.40	48.70	4.90	28.90
10% Sucrose	52.50	50.50	42.90	41.10
1% Sucrose	51.30	51.20	48.40	48.10
100 mM Glucose	245.00	57.10	68.10	61.70
50 mM Glucose	144.00	47.70	62.70	58.40
10 mM Glucose	70.00	49.10	52.60	51.20
0.2 M Sorbitol, pH 11.25	40.70	36.20	68.60	26.60
1 M Glycine	No colour		7.30	7.70
1 M Glycine, pH 11.25	50.70	48.90	32.50	27.90
0.5 M Tris	36.20	32.90	10.20	8.80
0.25 M Tris	46.60	44.00	27.90	28.10
0.1 M Tris	50.80	49.60	38.90	38.90
0.25 M Tris, pH 11.25	52.00	50.30	40.80	40.80
20% Ammonium sulphate	5.60	1.20	Precipitated	
10% Ammonium sulphate	16.00	12.00	Precipitated	
3% Ammonium sulphate	44.90	42.00	21.20	21.40
10% Ammonium sulphate, pH 11.25	48.10	45.20	32.60	32.80
2 M Sodium acetate, pH 5.5	35.50	34.50	5.40	3.30
0.2 M Sodium acetate, pH 5.5	50.80	50.40	47.50	47.60
1 M Sodium phosphate	37.10	36.20	7.30	5.30
0.1 M Sodium phosphate	50.80	50.40	46.60	46.60
0.1 M Caesium bicarbonate	49.50	49.70	Precipitated	

procedure, it would be expected that the two methods would show a similar variation of response to protein amino acid composition. This has been shown to be the case using a series of test proteins with the curious exception of avidin (16). A distinct advantage of the BCA method is that it is generally more tolerant to the presence of

Method Table 3. Standard BCA assay.

---

1. Stock reagents: A, 1% w/v BCA-Na$_2$, 2% w/v sodium carbonate, 0.16% w/v sodium tartrate, 0.4% w/v sodium hydroxide and 0.95% sodium bicarbonate. If required, 50% w/v sodium hydroxide or solid sodium bicarbonate is added to adjust the pH to 11.25. B, 4% w/v copper sulphate (CuSO$_4 \cdot$5H$_2$O). Reagents A and B are stable indefinitely at room temperature.
2. Standard working reagent (reagent C): prepared by mixing 100 vol of reagent A with 2 vol of reagent B. Reagent C is apple green in colour and is stable for 1 week at room temperature.
3. To 100 μl samples containing 10−120 μg protein (0.1−1.2 mg protein ml^{-1}) add 2 ml of reagent C.
4. Mix and stand for 30 min at 37°C.
5. Measure absorbance at 562 nm against a blank of 100 μl sample buffer processed through steps 3 and 4.

Method Table 4. BCA microassay procedure.

---

1. Stock reagents: A, 8% w/v sodium carbonate, 1.6% w/v sodium hydroxide, 1.5% w/v sodium tartrate, and sufficient solid sodium bicarbonate to adjust the pH to 11.25. B, 4% w/v BCA-Na$_2$. Reagents A and B are stable indefinitely at room temperature. C, 4 vol of 4% w/v copper sulphate (CuSO$_4 \cdot$5H$_2$O) mixed with 100 vol of reagent B. This reagent should be prepared freshly as needed.
2. Stock working reagent (reagent D): prepared when required by mixing 1 vol of reagent C with 1 vol of reagent A.
3. To 1 vol of sample (0.5−10 μg protein ml^{-1}) add 1 vol of reagent D.
4. Mix and stand for 60 min at 60°C.
5. Cool to room temperature.
6. Measure absorbance at 562 nm against a blank of sample buffer processed through steps 3−5.

---

compounds which have been found to interfere with the Lowry procedure. The effects of a selection of these compounds on the BCA assay compared with their effect on the Lowry procedure are shown in *Table 2*. Importantly, the BCA assay is more tolerant to a variety of detergents (anionic, non-ionic and zwitterionic) which are frequently used to solubilize proteins. Denaturing agents such as guanidine−HCl and urea are also well tolerated, but the BCA assay is more sensitive to interference from reducing sugars. Of course, copper-chelating reagents such as EDTA cause severe problems for both methods, as do buffer solutions which shift the pH of the BCA or Lowry working reagents away from their pH optima.

3.5 Dye-binding procedure
Under appropriate conditions, the acidic and basic groups of proteins interact with the dissociated groups of organic dyes to form coloured precipitates. This dye-binding

Method Table 5. Dye-binding method.

---

1. Dye reagent: use vigorous homogenization or agitation to dissolve 100 mg of Coomassie Brilliant Blue G-250 in 50 ml of 95% ethanol. This solution is mixed with 100 ml of 85% w/v phosphoric acid, diluted with water to 1 litre and filtered. The reagent is stable at room temperature for at least 2 weeks.
2. Standard assay: add 5 ml of dye reagent to 0.1 ml protein samples containing $20-140$ μg of protein ($0.2-1.4$ mg protein ml^{-1}).
or Microassay: add 0.2 ml of dye reagent to 0.8 ml samples containing $1-20$ μg of protein ($5-100$ μg protein ml^{-1}).
3. Mix and stand for $5-30$ min.
4. Measure absorbance at 595 nm against a blank prepared from 0.1 ml (0.8 ml for the microassay) of sample buffer and 5 ml (0.2 ml for the microassay) of dye reagent.

---

phenomenon can be readily exploited for quantitative analysis but many dyes (e.g. Orange G) are insensitive whilst others (e.g. Amido Black 10-B) have other disadvantages. This approach only became popular with the development by Bradford [17] of a method using Coomassie Brilliant Blue G-250. The binding of the dye to protein causes a shift in the absorption maximum of the dye from 465 nm (red form) to 595 nm (blue form). The dye is prepared as a stock solution in either phosphoric [17,18] or perchloric [19] acid, the latter reagent being more stable. Commercial preparations of this reagent are available from Bio-Rad and Pierce. The method (*Method Table 5*) is a rapid, simple one-step procedure in which the reagent is added to the sample and the absorbance measured at 595 nm. A difficulty observed in performing the assay is the tendency of the protein–dye complex to bind to glass surfaces. Disposable cuvettes can, therefore, be used to advantage, but it is relatively easy to remove the dye from glass cuvettes by soaking in 0.1 M HCl or by washing with concentrated detergent followed by water and acetone.

The procedure is very sensitive, working in the range $0.2-1.4$ mg protein ml^{-1} for the standard assay and $5-100$ μg protein ml^{-1} for the microassay procedure. It exhibits a significant dependence on protein amino acid composition and this has recently been shown [20] to be a consequence of the dye binding primarily to basic (especially arginine) and aromatic amino acid residues. The assay is also adversely affected, particularly the microassay procedure, by a range of common laboratory chemicals as shown in *Table 3*. However, it is somewhat better than the Lowry procedure in this respect. Additionally, particular protein samples may be relatively insoluble in this acidic system.

3.6 Silver-binding procedure

The methods described so far for the determination of protein concentration are sufficiently sensitive for use in the majority of protein purification procedures. However, there is currently considerable interest in proteins which can mediate cellular processes such as proliferation and differentiation and which are biologically active in the $10^{-9}-10^{-13}$ M range. This high potency facilitates bioassay, but they are generally

Table 3. Effect of various laboratory reagents on the Bradford procedure. Reproduced with permission from ref. 17.

Substance	Change in OD_{595} (nm)	Equivalent BSA (μg)
1 M Potassium chloride	0.000	0.00
5 M Sodium chloride	0.000	0.00
1 M Magnesium chloride	0.000	0.00
2 M Tris	0.026	2.34
0.1 M EDTA	0.004	0.36
1 M Ammonium sulphate	0.000	0.00
99% Glycerol	0.012	1.08
1 M 2-Mercaptoethanol	0.004	0.36
1 M Sucrose	0.013	1.17
95% Ethanol	0.000	0.00
Acetone	0.069	6.21
5% Phenol	0.046	4.14
0.1% Triton X-100	0.013	1.17
1% Triton X-100	0.590	53.10
0.1% SDS	0.011	0.99
1% SDS	0.495	44.55
0.1% Haemosol	0.004	0.36
1% Haemosol	0.108	9.72

The above values were obtained when 0.1 ml of each substance was assayed in the standard assay.

purified in vanishingly small amounts. For example, 150 litres of conditioned medium will yield only $2-10$ μg of purified interleukin 3. The standard assay procedures would consume too much of such pure materials and, therefore, there is a need for more sensitive protein assay procedures. Recently, a highly sensitive method has been developed (21) based on the ability of proteins to bind silver, a property already effectively exploited for the sensitive detection of proteins separated by polyacrylamide gel electrophoresis.

The useful range of the assay is claimed (21) to lie between 15 ng and 2 μg (150 ng to 20 μg protein ml^{-1}), representing an approximately 100-fold increase in sensitivity over the dye-binding procedure. The recommended method is detailed in *Method Table 6*. The assay shows a marked variation of response to protein amino acid composition comparable to that of the Coomassie dye-binding assay (21). A serious problem with such a microassay is the loss of protein due to adsorption to glass and plastic surfaces. This can be minimized by the addition of the detergent, Tween 20, to the samples prior to their assay. Another major problem is interference caused by many compounds commonly present in protein samples and buffers, including anions (e.g. Cl^-) which form insoluble silver salts, EDTA, SDS in excess of 0.01%, and the reducing agents dithiothreitol and 2-mercaptoethanol (21). It is, therefore, usually necessary to remove any potentially interfering substances by subjecting samples to gel filtration chromatography on small Bio-Gel P2 columns. It should be noted that silver-treated samples left in cuvettes for more than a few minutes will cause silver deposition on the glass walls. This can be easily removed with nitric acid, or plastic disposable cuvettes can be used. This assay promises very high sensitivity, but its use should be considered only if insufficient material is available to make an alternative assay procedure practicable.

Method Table 6. Silver-binding method.

---

1. Stock reagents: A, 7.5% w/v Tween 20, 100 mM Tris, 100 mM sodium carbonate. B, 2.5% solution of glutaraldehyde prepared fresh daily from a 25% stock glutaraldehyde solution stored at 4°C. C, Ammoniacal silver solution prepared fresh daily by adding 1.4 ml of 20% w/v sodium hydroxide and 0.2 ml of concentrated ammonium hydroxide (29%) to 18.2 ml of distilled water, followed by the dropwise addition of 0.2 ml of 20% w/v silver nitrate. D, 3% w/v sodium thiosulphate prepared fresh daily.
2. Add 11 μl of reagent A to 100 μl of protein sample containing from 15 ng to 2 μg protein (150 ng to 20 μg protein ml^{-1}).
3. Centrifuge at 450 g for 5 min through a 2-ml Bio-Gel P-2 column pre-equilibrated in 10× diluted reagent A and then drained of void volume.
4. Add 0.9 ml of distilled water to make the sample volume 1 ml.
5. Add 20 μl of reagent B and vortex for 2 sec.
6. Add 200 μl of reagent C and vortex for 2 sec.
7. Stand for precisely 10 min at room temperature.
8. Add 40 μl of reagent D.
9. Measure absorbance at 420 nm against a blank prepared from 100 μl of sample buffer and processed through steps 2−8.

---

4. ELECTROPHORETIC ANALYSIS METHODS—M.J.Dunn

Proteins are charged at a pH other than their isoelectric point (pI) and thus will migrate in an electric field in a manner dependent on their charge density. If the sample is initially present as a narrow zone, proteins of different mobilities will travel as discrete zones and thus separate during electrophoresis. Electrophoresis is, therefore, an ideal analytical technique to resolve and separate the individual components of protein mixtures. Such separations are best carried out in a support medium to counteract the effects of convection and diffusion that occur during electrophoresis, and to facilitate the immobilization of the separated proteins. A variety of matrices including starch, agarose and cellulose acetate can be used for electrophoresis. However, the high resolution capacity of polyacrylamide gel electrophoresis (PAGE) makes this the method of choice for most applications.

A range of electrophoretic techniques are available which can separate proteins on the basis of one or a combination of their three major properties: i.e. size, net charge and relative hydrophobicity. Electrophoresis under native conditions is often used to analyse soluble proteins with the advantage of the retention of their biological and enzymatic properties. In contrast, more vigorous, and often denaturing conditions, are used for the analysis of less soluble proteins, with the consequent loss of biological and biochemical activity. A particularly important procedure is PAGE in the presence of the anionic detergent, sodium dodecyl sulphate (SDS), by which proteins can be characterized in terms of the molecular size of their constituent polypeptides. Another procedure of importance to the purification scientist is isoelectric focusing (IEF) which can be used to characterize purified proteins in terms of their pI. In the field of protein purification, PAGE has become an almost mandatory analytical procedure for the characterization of protein purity. It is also routinely used for monitoring the progress of a purification procedure and for the identification of fractions containing the protein of interest. In addition, electrophoretic methods can be used to characterize purified proteins in terms of microheterogeneity, degradation and subunit structure.

A considerable disadvantage of electrophoretic techniques is that they are often very time-consuming compared with many of the other techniques used in protein purification procedures. Using standard electrophoretic techniques, it can often take from many hours to several days to obtain the results of an electrophoretic separation. One approach to this problem that is becoming increasing popular is to use miniaturized electrophoresis systems in which the separation distance is significantly reduced (typically 2- to 4-fold). When used in combination with an efficient cooling system to allow the application of high field strengths, such equipment is capable of producing high-speed, high-resolution protein separations. Apparatus of this type is available from several commercial suppliers (e.g. Hoefer, Bio-Rad). This approach has been further extended with the development of a microprocessor-controlled, integrated, miniaturized electrophoresis system known as the PhastSystem (Pharmacia-LKB). This system uses small (43 × 50 mm), ultra-thin gels pre-cast on plastic supports and is compatible with most of the major gel electrophoresis procedures (native PAGE, SDS−PAGE, IEF, titration curves, immunoelectrophoresis). An additional feature of this equipment is that protein detection by staining is also performed automatically under programmable control. This combination of speed, simplicity of operation, reliability and increased reproducibility must make such automated electrophoresis equipment very attractive to the protein purification scientist.

In this section the advantages and problems are discussed of those PAGE procedures of most importance to the purification scientist. For additional details of PAGE techniques the interested reader is referred to a previous volume in this series (22) and to some other excellent current texts (23−26).

4.1 Properties of polyacrylamide gels

4.1.1 *Basic properties*

Polyacrylamide gels are formed by the copolymerization of acrylamide monomers with a cross-linking agent to form a three-dimensional lattice. The best and most commonly used cross-linking agent for the majority of PAGE applications is N,N'-methylene bisacrylamide (Bis). Gel polymerization is usually initiated with ammonium persulphate and N,N,N',N'-tetramethylethylenediamine (TEMED). This reaction is more efficient when compared with the alternative of UV-activated polymerization with riboflavin, but careful control of polymerization conditions is essential if consistent results are to be obtained. The size of the pores within the gel matrix is dependent both on the total monomer concentration (%T) and on the concentration of the cross-linking agent (%C). It should be noted that %T is expressed as total gel concentration (acrylamide plus Bis as g per 100 ml), while %C is expressed as a percentage of the total gel concentration (i.e. g Bis $\%T^{-1}$). Pore size can be progressively increased by reducing %T at a fixed %C, but very dilute gels are mechanically unstable and pore sizes in excess of 80 mm cannot be obtained. The alternative approach is progressively to increase %C at a fixed %T, where the increase in pore size is thought to be due to the formation of bead-like

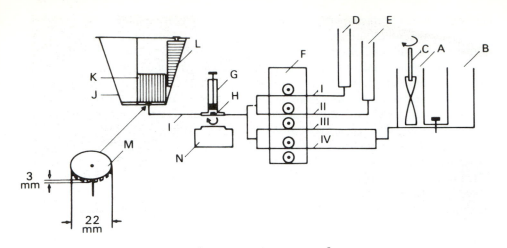

Figure 1. Diagram of an apparatus based on a two-chamber gradient former, used to prepare simultaneously several gradient polyacrylamide slab gels [reproduced from (28) with permission]. **A** and **B**, two-chamber gradient mixer (inner diameter = 41 mm, height = 100 mm); **C**, stirrer; **D**, reservoir for ammonium persulphate solution; **E**, reservoir for TEMED solution; **F**, proportioning pump; **G**, modified disposable syringe, used as a small volume mixing chamber; **H**, magnetic flea; **I**, connecting tubing; **J**, gel casting apparatus; **K**, gel casettes; **L**, perspex wedge; **M**, distributor; **N**, magnetic stirrer; **I** and **II**, Tygon tubing (0.64 ml min^{-1}); **III** and **IV**, Tygon tubing (6.80 ml min^{-1}).

structures within the gel rather than a three-dimensional network. Stable gels of high pore size (>200−250 nm) can be formed in this way, but at concentrations of Bis in excess of 30%C the gels become hydrophobic and prone to collapse.

4.1.2 *Gradient gels*

Theoretically there is a gel concentration which is optimal for the resolution of a given pair of proteins. However, there is no single gel concentration which will give maximum separation of the components of a more complex protein mixture. Thus, in practice it is often better to use gels containing a linear or non-linear concentration gradient of polyacrylamide. The average pore radius of these gels decreases with increasing gel concentration, so that there is an effective band-sharpening effect during electrophoresis and proteins with a wider range of molecular masses can be separated. It is sometimes referred to as 'pore limit electrophoresis' as the migration rates decrease until each protein reaches its pore limit, after which further migration occurs at a slow rate proportional to time (27). The resulting separation profiles are consequently stable and highly resolved.

Most routine separations are usually carried out using linear polyacrylamide gradients as their preparation is comparatively straightforward. However, it should be remembered that for a particularly difficult separation problem it may be necessary to resort to the use of convex, concave or more complicated (e.g. flattened) gradient shapes. Various procedures have been developed for the preparation of gradient polyacrylamide gels. The simplest approach is to use a two-chamber gradient-forming apparatus of fixed shape (28) as shown in *Figure 1*. Equipment of this type is commercially available (Pharmacia-LKB, Hoefer, Bio-Rad), reproducible, and easy to use, but it lacks the

Figure 2. Computer-controlled stepmotor-driven burette system for the preparation of polyacrylamide gel gradients [reproduced from (30) with permission]. From **left to right**: printer (3), floppy disk unit (2), computer (1), four burettes (4), mixing chamber (5) on magnetic stirrer (6), polymerization stand and gel cassettes (7).

flexibility required for the generation of gradients of different shapes. A greater degree of flexibility can be obtained using a two-chamber system of variable complimentary shape (29) or an electromechanical gradient-forming device such as the Ultrograd (Pharmacia-LKB). Alternatively, a microcomputer-controlled stepmotor-driven burette system (*Figure 2*) has been developed (30) which is very versatile and provides full documentation of the gradient produced (*Figure 3*). The high precision stepmotor-driven burettes used in this apparatus are expensive, but a commercial version of this equipment, using less expensive proportional pumps, is available from Desaga.

4.2 Electrophoresis under denaturing conditions

4.2.1 SDS−PAGE

The anionic detergent, SDS, is a very effective solubilizing agent for a wide range of proteins. Moreover, the majority of proteins bind 1.4 g SDS per 1 g protein (31), effectively masking the intrinsic charge of the polypeptide chains, so that net charge per unit mass becomes approximately constant. Subsequent electrophoretic separation is dependent only on the effective molecular radius (M_r), which roughly approximates to molecular size, and occurs solely as a result of molecular sieving through the gel matrix. The polyacrylamide gel concentration used determines the effective separation range of SDS−PAGE. For example, a 5%T gel will separate proteins in the range 20−350 kd, while a 10%T gel is suitable for proteins in the range 15−200 kd. If the sample to be analysed contains proteins with a wide range of molecular weights it is then advantageous to use a gradient SDS−PAGE system. For example, a 3−30%T gradient gel can separate proteins in the range 10 kd to nearly 500 kd. However, small polypeptides and peptides (< 10 kd) cannot be separated by the SDS−PAGE system. This failure results from the fact that small molecules form SDS−protein complexes of the same dimension and charge, which then migrate together during SDS−PAGE. This problem can be overcome by adding a solute such as 8 M urea that decreases the

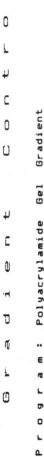

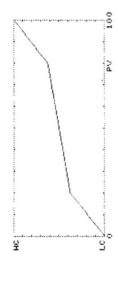

G r a d i e n t   C o n t r o l

P r o g r a m :   Polyacrylamide Gel Gradient

D a t a :

TV= 63 mL;   BV= 10 mL;   MV= 0.2 mL;   TV/BV= 6.3 ;   Speed: 65 %
LC= 8.5 Units;   HC= 22 Units;
H1= 3 % ;   L1= 7 % ;   H2= 2 % ;   L2= 10 % ;

ENDPOINTS OF LINES

EP No.	PV(%)	C(U)	EP No.	PV(%)	C(U)	EP No.	PV(%)	C(U)	EP No.	PV(%)	C(U)	EP No.	PV(%)	C(U)	EP No.	PV(%)	C(U)
0	0	8.5	1	20	13.5	2	80	17	3	100	22						

(EP=End Point; PV(%)=Partial Vol.; C(U)=Conc. in Units; TV=Total Vol.; BV=Burette Vol.; MV=Mixing Vol.; LC=Low Conc.; HC=High Conc. ;
L1,H1,L2,H2 = Proportions of catalyst solutions 1 and 2 admixed to solutions with low and high conc. of gradient constituent)

Figure 3. Final document of a polyacrylamide gel gradient produced by the system shown in *Figure 2* including all data except the recipes for preparing the stock solutions for the reproduction of the gradient in any laboratory (reproduced from ref. 30 with permission).

size of the detergent micelles (32,33). These modified SDS−PAGE systems are capable of separating polypeptides and peptides in the range 1−45 kd.

The technique of SDS−PAGE is presently the most commonly used electrophoretic separation technique for the majority of protein samples. It is a relatively simple and reproducible technique, yet is capable of very high resolution as it can separate proteins differing in mobility by as little as 1% (equivalent to a difference in molecular weight of ~1 kd for an 'average' 100 kd protein). For optimal reaction with SDS, samples should be boiled for 3−5 min in the presence of 2−5% w/v SDS and reagents such as 0.1 M 2-mercaptoethanol or 20 mM dithiothreitol to cleave disulphide bonds.

The most commonly used buffer system for SDS−PAGE is that devised by Laemmli (34) based on the discontinuous buffer system of Ornstein and Davis (35,36) with the addition of 0.1% w/v SDS. The advantage of this buffer system compared with a homogeneous buffer system is that it results in the concentration of sample proteins into a narrow starting zone prior to separation, thereby resulting in very sharp protein zones during the separation process. This effect is due to the moving boundary formed between a rapidly migrating (leading) and a slowly migrating (trailing) species. The lower gel phase contains the leading constituent whilst the trailing constituent is present in the upper buffer phase. If these constituents are chosen so that the sample proteins have intermediate mobilities, the latter will be concentrated in the moving boundary ('stack'). In the Ornstein−Davis buffer system (35,36) the leading and trailing ions are chloride and glycinate respectively. The moving boundary is formed in a low concentration, 'stacking' gel and there is an increase in the operational pH in the

Method Table 7. Stock solutions for SDS−PAGE.

---

1. Acrylamide solution (30%T, 2.7%C): dissolve 146 g of acrylamide and 4 g of Bis acrylamide and make up to 500 ml. This solution is stable for at least 1 month at 4°C.
2. Separating gel buffer: 1.5 M Tris−HCl, pH 8.8. This solution is stable for at least 1 month at 4°C.
3. Stacking gel buffer: 1 M Tris−HCl, pH 6.8. This solution is stable for at least 1 month at 4°C.
4. Electrode reservoir solution: 0.192 M glycine (free acid), 0.025 M Tris base, 0.1% w/v SDS. This should be prepared fresh for each electrophoretic run.
5. SDS solution: 10% w/v SDS in distilled water. This solution should be prepared fresh weekly.
6. Ammonium persulphate: 10% v/v in distilled water. This solution should be prepared fresh daily.
7. Sample solubilization solution (double strength): mix 1 g of SDS, 2 ml of glycerol, 2 ml of Bromophenol Blue tracking dye (0.1% w/v solution in distilled water), 1.25 ml of 1 M Tris−HCl, pH 6.8, 2 ml of 2-mercaptoethanol and make up to 10 ml with distilled water. When this solution is diluted to single strength, samples will contain 5% w/v SDS, 10% v/v glycerol, 5% v/v 2-mercaptoethanol and 0.0625 M Tris−HCl, pH 6.8. This solution should be prepared fresh each week and stored at 4°C.

Method Table 8. Method for the preparation and electrophoresis of 10%T SDS gels.

---

1. Assemble the gel cassettes using clean, dry glass plates according to the manufacturer's instructions for the particular electrophoresis apparatus being used.

2. Prepare a sufficient volume of separating gel mixture (e.g. 100 ml) by mixing 33.3 ml of stock acrylamide solution, 25 ml of 1.5 M Tris−HCl, pH 8.8, 40 ml of distilled water and 150 μl of 10% ammonium persulphate. Degas this solution using a vacuum pump to remove oxygen which inhibits polymerization. Then add 1 ml of 10% SDS and 50 μl of TEMED. Mix gently and pour into the gel cassettes to about 1 cm below the level which will be occupied by the well-forming combs when they are in position.

3. Overlay the gels carefully with distilled water (or isobutanol) and allow to polymerize for at least 2 h (or overnight) at room temperature. The separating gels will then contain 10%T, 2.7%C acrylamide, 0.1% w/v SDS, 0.375 M Tris−HCl, pH 8.8.

4. Prepare the required volume of stacking gel mixture (e.g. 30 ml) by mixing 3 ml of stock acrylamide solution, 3.75 ml of 1 M Tris−HCl, pH 6.8, 22.75 ml of distilled water and 150 μl of 10% ammonium persulphate. Degas this solution using a vacuum pump. Then add 0.3 ml of 10% SDS and 50 μl of TEMED. Mix gently and use immediately. The stacking gels will then contain 3%T, 2.7%C acrylamide, 0.1% SDS, 0.125 M Tris−HCl, pH 6.8.

5. Pour off the water/isobutanol from the top of the polymerized separating gels and fill the cassettes with the stacking gel mixture. Insert the sample well-forming combs and leave to polymerize for least 2 h at room temperature.

6. When the stacking gels have polymerized, remove the sample combs from the gels and install the gels in the electrophoresis apparatus according to the manufacturer's instructions.

7. Fill the electrode chambers with the reservoir buffer solution. This should be circulated (if this is possible with the apparatus being used) between the cathode and anode chambers to equalize the pH conditions.

8. Prepare dry samples by dissolving in single strength (i.e. diluted 1:1 with distilled water) sample solubilization solution. Liquid samples should be diluted to a suitable protein concentration and added to an equal volume of double strength sample solubilization solution. The total sample volume should be less than the volume of the sample wells in the gel. Heat the samples in a boiling water bath for 3 min.

9. Load the samples into the sample wells using a suitable microsyringe.

10. Run the gels under constant current conditions. A 1.5 mm thick, 15 cm long 10%T gel can conveniently be run overnight at 15 mA per gel, by which time the Bromophenol Blue tracking dye reaches the bottom of the gels. Higher field strengths can be used to accelerate electrophoresis, but care must be taken to minimize heating effects within the gels. This can be accomplished using a refrigerated recirculating water bath system to cool the electrode buffer solution.

11. At the completion of electrophoresis, remove the gels from the cassettes and fix them in a solution of 20% w/v TCA.

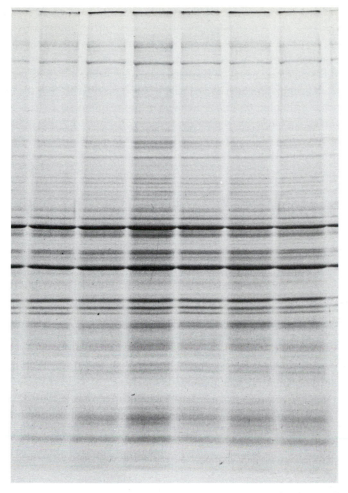

Figure 4. Separation of human skin fibroblast total cell proteins by 5−20%T gradient SDS−PAGE (reproduced from ref. 65 with permission). The Laemmli buffer system was used and the gel was stained wtih Coomassie Brilliant Blue R-250.

restrictive separating gel. This causes further dissociation of glycinate, increasing its mobility so that it moves just behind the chloride ion. This effect, together with the decrease in gel pore size causes the proteins to become 'unstacked' so that they are subsequently separated on the basis of their size. It should be noted that the nature of stacking is somewhat modified in the presence of SDS. Since SDS-coated proteins have a constant charge to mass ratio (that is all proteins and their subunits have a uniform charge), they will migrate with the same mobility and thus will automatically stack. Moreover, as SDS−protein complexes are not titratable between pH 7 and pH 10, mobility is not variable within this range. It is, therefore, not strictly necessary to have a discontinuity in pH and unstacking of the proteins will occur by the change in gel concentration (37). In addition, SDS can act as a separate ion boundary to Cl⁻ ions (38).

A recommended procedure for SDS−PAGE in homogeneous 10%T gels is given in *Method Tables 7* and *8* and a typical protein separation obtained by SDS−PAGE is shown in *Figure 4*. Despite the almost universal adoption of SDS−PAGE for the characterization of protein samples, the technique does have certain limitations. Obviously, proteins of the same or similar molecular size will not be resolved by this procedure. In addition, although the majority of proteins bind SDS in the expected ratio (1.4 g SDS per 1 g protein), proteins containing non-protein moieties (e.g. glycoproteins, phosphoproteins, lipoproteins, nucleoproteins) can bind varying amounts of SDS resulting in anomalous mobility and separation artefacts during SDS−PAGE. For glycoproteins, this problem can be alleviated using borate buffer, since borate ions bind to *cis*-hydroxyl groups of sugars (39). Perhaps the most serious disadvantage of this procedure is that after electrophoresis it is difficult, if not impossible, to characterize the separated proteins in terms of their enzymic or biological properties.

4.2.2 *Gels containing urea*

SDS−PAGE separates proteins on the basis of size alone and provides no information on their charge properties. However, gel electrophoresis techniques are available for the characterization of proteins in terms of both their charge and size, and like SDS−PAGE these are also suitable for the analysis of proteins of limited solubility or where the sample components are liable to precipitate during electrophoresis. The simplest approach is to add urea at a concentration of 6−8 M to the sample prior to electrophoresis. Urea should also be added to both the separation gel and the stacking gel (if used), but it is not necessary to include urea in the electrode buffer solutions. It is sometimes possible to reduce the urea level in the gel to around 4−5 M provided that the sample has been dissociated with 8 M urea. Uncharged disulphide-bond disrupting reagents such as 0.1 M 2-mercaptoethanol or 20 mM dithiothreitol are often also added to samples to cleave inter- and intra-chain disulphide bonds. In this case, multimeric disulphide-bonded proteins will be dissociated into their constituent polypeptides. It is also commmon practice to add 0.5% w/v non-ionic detergent such as Triton X-100 to gels containing urea. Another variation on this technique which has been applied particularly to basic proteins such as histones is to use an acid−urea gel system containing 2.5 M urea and 0.9 M acetic acid (pH 2.7) (40). Separation in this system occurs essentially on the basis of charge differences between the proteins. Low concentrations (e.g. 0.5% w/v) of non-ionic detergents such as Triton X-100, Triton DF-16 or Lubrol-WX can also be added to form an acid−urea−detergent gel system (41,42). When proteins are electrophoresed as cations at low pH in the presence of non-ionic detergents their separation depends on their ability both to bind the detergent and to form mixed micelles between the detergent and the hydrophobic domains of the polypeptide chains (41). Thus, mobility depends on protein hydrophobicity in addition to size and charge. A potential problem with this technique is that oxidation, especially of methionine residues, can result in altered mobilities leading to artefactual protein heterogeneity (43).

When using gels containing urea it must be remembered that at alkaline pH urea progressively breaks down to form cyanate ions which can readily react with protein amino groups to form stable carbamylated derivatives. These derivatives will have altered charge properties which can result in anomalous electrophoretic behaviour and lead to artefactual separation profiles. Urea solutions should, therefore, be treated with a suitable cationic or mixed-bed ion-exchange resin to minimize contamination with cyanate ions. Samples containing urea should not be stored for extended periods and heating should be avoided in order to minimize carbamylation effects. It can also be useful to subject gels containing urea to a pre-electrophoresis step to remove cyanate ions prior to sample application.

4.3 Electrophoresis under native conditions

Techniques for polyacrylamide gel electrophoresis of proteins under denaturing conditions such as SDS−PAGE and acid−urea−detergent gels are excellent techniques for characterizing proteins in terms of their physicochemical properties (i.e. size, charge, hydrophobicity). However, these procedures suffer from the distinct disadvantage that they generally result in the complete loss of biological and biochemical activity of the proteins being separated. It can, therefore, be beneficial to use electrophoresis under non-denaturing, native conditions in circumstances where it is desired to characterize the activity of the proteins after electrophoretic separation (i.e. enzyme staining, antibody binding, receptor activity). Unfortunately, native electrophoresis techniques can only be applied to protein samples which are soluble and which will not precipitate or aggregate during electrophoresis.

4.3.1 *Continuous zone electrophoresis (CZE)*

The simplest form of PAGE involves the use of gels of a uniform polyacrylamide concentration in conjunction with a single homogeneous buffer system. In such a system, separation occurs on the basis of both the charge and the size properties of the proteins being analysed. The former property of the proteins can be exploited to optimize the separation by the careful choice of the buffer pH used for electrophoresis. In addition, the appropriate gel concentration should be selected for optimal separation of the proteins on the basis of their size. A rough guide for the selection of the appropriate gel concentration is given in *Table 4*. The buffer and pH used for the separation can be freely chosen, but the concentration of the buffer should not exceed 0.01 M to minimize the effects of Joule heating. The most commonly used buffer systems operate in the slightly alkaline region (pH 8−9) as most proteins are negatively charged under these conditions. In this case the samples are applied at the cathode and migrate towards the anode during electrophoresis. However, it is important to realize that in such a system any basic proteins present in the sample being analysed will migrate in the other direction and

Table 4. Separation ranges of acrylamide gels of various concentrations (%T) (reproduced with permission from ref. 26).

%T	Optimum molecular weight range
3−5	>100 000
5−12	20 000−150 000
10−15	10 000−80 000
15+	<15 000

Table 5. Commonly used homogeneous buffer systems for PAGE (reproduced with permission from ref. 26).

Approximate pH range	Primary buffer constituent	pH adjusted to the desired value with
2.4−6.0	0.1 M Citric acid	1 M NaOH
2.8−3.8	0.05 M Formic acid	1 M NaOH
4.0−5.5	0.05 M Acetic acid	1 M NaOH or Tris
5.2−7.0	0.05 M Maleic acid	1 M NaOH or Tris
6.0−8.0	0.05 M KH_2PO_4 or NaH_2PO_4	1 M NaOH
7.0−8.5	0.05 M Na diethyl-barbiturate (veronal)	1 M HCl
7.2−9.0	0.05 M Tris	1 M HCl or glycine
8.5−10.0	0.015 M $Na_2B_4O_7$	1 M HCl or NaOH
9.0−10.5	0.05 M Glycine	1 M NaOH
9.0−11.0	0.025 M $NaHCO_3$	1 M NaOH

be lost from the gel. If the basic proteins are of importance it is possible to perform the electrophoresis under acidic conditions where most of the proteins will exist as cations. Of course, it is then necessary to apply the samples at the anodic end of the gels. Some commonly used buffer systems are given in *Table 5*.

The main advantage of CZE procedures is that gel preparation and the electrophoresis procedure are both rapid and simple. In addition, the buffer composition and separation pH are precisely established which can be an important consideration if the stability of the proteins being analysed is sensitive to pH and/or buffer composition. The main disadvantage of these techniques is that they have an inherently rather low resolution capacity. The method is not suitable for the analysis of dilute sample solutions since there is no concentration effect in the initial stage of the separation. Thus, large sample volumes result in broad protein zones in the final separation pattern. Acceptable and useful separations can be achieved, however, if the sample is available in a concentrated state (>1 mg ml^{-1}) so that it can be applied as a narrow starting zone.

4.3.2 Multiphasic zone electrophoresis (MZE)

Despite their simplicity, PAGE techniques based on CZE are seldom used nowadays and they have been largely replaced by methods using discontinuous (multiphasic) buffer systems. These techniques are generally known as 'disc electrophoresis', the term 'disc' referring to the nature of the buffer system and not to the fact that this methodology was developed using cylindrical rod gels rather than the slab gels generally used today. The popularity of this group of PAGE techniques is due to their ability to concentrate the sample proteins into a narrow starting zone or stack. The mechanism of this stacking process has already been described in relation to SDS−PAGE (see Section 4.2.1). The major difference is that, in the absence of SDS, when the proteins are unstacked on entering the separating gel their subsequent separation depends on both their size and charge properties. These procedures have an inherently high resolution capacity due to their ability to separate proteins into very sharp protein bands. The method is, therefore, applicable to quite dilute sample solutions. However, one should be aware that during the stacking phase the concentration of the sample proteins can be so great as to induce concentration-dependent artefacts, such as irreversible protein—protein interactions. Care should also be taken to ensure that the different pH conditions operative during stacking and separation are not such as to affect adversely the chemical, physical or biological properties of the sample proteins.

The majority of separations are carried out using the Ornstein—Davis buffer system (35,36). However, it should be pointed out that this was developed specifically for the separation of serum proteins and that there is no universal buffer system suitable for all types of sample. Thus, if the Ornstein—Davis buffer system produces unsatisfactory results or if a particular separation is to be optimized, it is necessary to select the appropriate buffer system. Computational treatments have been developed to establish the constituents of moving boundary systems with the desired properties. The most powerful of these is that of Jovin (44) which generated in excess of 4000 buffer systems, forming the so-called 'extensive buffer system output'. A selection of 19 of the most useful of these buffer systems has been published to aid the investigator to select the most appropriate system for his particular separation problem (45).

4.3.3 *Non-ionic detergents*

Electrophoresis under native conditions is usually used for the analysis of samples of soluble proteins. However, the method can sometimes be applied successfully to less soluble proteins, such as some membrane proteins and enzymes, simply by the addition of non-ionic detergent (e.g. 0.5% w/v Triton X-100, Tween 80, Brij 35) to the gel. These conditions can often be sufficiently mild as to be essentially non-denaturing, therefore retaining the enzymic or biological properties of the proteins being separated.

4.4 **Isoelectric focusing**

IEF is a high resolution method in which proteins are separated in the presence of a continuous pH gradient. Under these conditions proteins migrate according to their charges until they reach the pH values at which they have no charge (i.e. their iso-electric points, pI). The proteins will, therefore, attain a steady state of zero migration and will be concentrated or focused into narrow zones. As IEF separates proteins on the basis of their charge properties, it is essential to use conditions which minimize molecular sieving effects. This necessitates the use of gels of low acrylamide concentration (3—5%T). The technique is best performed using slab gels run on a horizontal flat-bed apparatus with an efficient cooling platten to cope with the Joule heating caused by the high field strengths generally employed. It is also beneficial to use thin or ultra-thin (0.5—0.02 mm) gels, which can be cast on thin plastic supports (e.g. GelBond PAG, Pharmacia-LKB) to improve heat dissipation and ease of handling. IEF is not generally used for monitoring the progress of purification procedures, but is commonly used as a method for the characterization of the final purified protein product. When used in conjunction with a parallel analysis of a sample by SDS—PAGE, these two techniques form a powerful method for assessing protein purity with IEF detecting charge heterogeneity and SDS—PAGE detecting size heterogeneity. Only a brief outline of IEF technology can be given here and for a fuller account the reader is referred to refs 23, 24 and 46.

4.4.1 *IEF using synthetic carrier ampholytes*

The most popular method for generating pH gradients for IEF is the incorporation of low molecular weight carrier ampholytes in the polyacrylamide gel matrix. A variety of procedures have been developed for the synthesis of these carrier ampholytes and several of these preparations covering the range pH 2.5−11 are available commercially, for example Pharmalyte (Pharmacia-LKB), Ampholine (Pharmacia-LKB), Servalyte (Serva), Resolyte (BDH), and Biolyte (Bio-Rad). As a result of the variety of synthetic procedures used, different carrier ampholyte preparations are often found to give better resolution in different portions of IEF gels. For example, we have found (47) that for human skin fibroblast proteins, Ampholine yielded superior cathodic resolution, Pharmalyte gave better resolution in the mid-range, and Servalyte gave the best resolution of acidic proteins. It is possible by blending of these different full-range ampholyte preparations to generate a mixture containing a greater diversity of charge varieties and so incorporate the advantages of each preparation and enhanced resolution. In addition to full range ampholytes, narrow range preparations are also available to produce gradients spanning as little as 1−2 pH units and capable of resolving proteins differing in pI by as little as 0.01 pH units. These ampholytes can also be added to a wide-range ampholyte mixture to flatten the region of the pH gradient where the proteins of interest are distributed. A range of ready-made, wet IEF gels bound to plastic supports is commercially available (Ampholine PAGplates from Pharmacia-LKB), covering both broad (pH 3.5−9.5) and narrow (pH 4−5, 4−6.5, 5−6.5 and 5.5−8.5) pH intervals. These ready-made gels are simple to use and give reproducible separations. However, they are expensive and restricted to use under native conditions since it is difficult to incorporate detergents and/or urea into the system. No simple guide can be given to the choice of the pH gradient interval, ampholyte preparation or mixture best suited for the separation of any particular protein sample. For complex mixtures containing proteins with a wide range of pI values, a wide-range pH gradient is obviously indicated. If a mixture of ampholytes is to be used, the best blend can only be established empirically to optimize the separation of the proteins being analysed. Narrow-range pH gradients are often used in circumstances where the pI of the protein(s) of interest has already been established. However, it is worth remembering that if such a narrow pH gradient is used to assess the purity of a protein preparation, any impurities with pI values outside the range of the pH gradient will be lost from the gel.

IEF can be performed

(i)   under native conditions where the gels contain only the added carrier ampholytes,

(ii)   in the presence of non-ionic detergents (0.5−5% w/v) such as Triton X-100 to increase sample solubility, or

(iii)   under denaturing conditions in the presence of high levels (typically 8 M) of urea (either in the absence or in the presence of non-ionic detergent).

For gels containing urea, it is essential to minimize the potential of cyanate ions formed from the breakdown of urea to carbamylate the sample proteins (see Section 4.2.2) and so cause artefacts in the IEF separation. It should be noted that the use of charged detergents such as SDS is incompatible with IEF, although it is sometimes possible to solubilize proteins with SDS and subsequently replace the SDS in the sample with non-ionic or zwitterionic (e.g. Chaps) detergent. Thus, samples which require stringent conditions for their solubilization are often not amenable to analysis by IEF. It should be noted that the salt concentration of samples must be minimized (e.g. by gel filtration, ultrafiltration, or dialysis) if band distortion, extended focusing times and excessive heating effects are to be avoided during IEF.

The main disadvantages of IEF are concerned largely with the properties of the synthetic carrier ampholytes used to generate the pH gradients. In theory IEF is a stable, equilibrium technique, but in practice the electroendosmotic properties of IEF gels result in long-term instability of pH gradients. This phenomenon results in cathodic drift so that with time the pH gradient and the proteins within it migrate towards the cathode (anodic drift can also occur under some circumstances), thus causing decay of the pH gradient and loss of proteins from the gel. The effects of this drift can be minimized, but not totally overcome. Another serious problem, particularly if the technique is to be used as a test of protein purity, is that IEF is exquisitely sensitive for the detection of protein microheterogeneity. This protein microheterogeneity can be the result of the presence of stable conformers, different oligomeric combinations of the same subunits, varying degrees of substitution (e.g. phosphorylation, acetylation, methylation, glycosylation), proteolytic damage, partial deamidation of asparagine and glutamine residues and to tightly bound cofactors or inhibitors. Additionally, there are classes of compounds, generally highly charged species such as heparin, polyanions, polycations and certain dyes, which have been shown to display artefactual heterogeneity when separated by IEF. This effect has been ascribed to interaction of these compounds with the synthetic carrier ampholytes themselves, resulting in the formation of a complex with an altered pI. Thus, in a practical situation using IEF to assess protein purity it can be difficult, if not impossible, to determine whether the presence of minor bands represents impurities in the sample, or is the result of real or artefactual microheterogeneity.

4.4.2 *IEF using immobilized pH gradients*

An important recent innovation in IEF has been the development of Immobiline reagents (Pharmacia-LKB) for the preparation of polyacrylamide gels containing immobilized pH gradients (IPGs). For a detailed review of this topic the interested reader is referred to ref. 46. The Immobiline reagents are a series of seven acrylamide derivatives containing either a carboxyl (acidic Immobilines) or a tertiary amino (basic Immobilines) group, forming a series of buffers with different pK values. IPG IEF gels are prepared, generally as slab gels on plastic supports, by generating a gradient using the appropriate Immobiline solutions. A range of ready-made, rehydratable IPG IEF gels with different pH gradients (Immobiline DryPlates) can be obtained from Pharmacia-LKB. Unfortunately, the widest ready-made IPG currently available spans pH 4−7 and no alkaline (>pH 7) ranges are available. Hopefully, the available range of pre-cast IPG IEF gels will be extended in the near future. These ready-made, rehydratable gels simplify the use of IPG technology and are compatible with additives such as urea and detergents since they can be reswollen in a solution containing the desired additives. In addition to their convenience and versatility, these gels should also benefit from the greater gel-to-gel reproducibility that can be gained from the preparation of large batches of gels.

The main advantage of the IPG system is that during polymerization the buffering

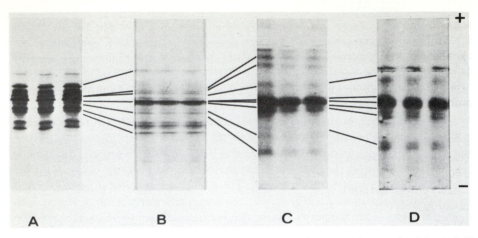

Figure 5. Separation of ovalbumin by IEF using synthetic carrier ampholytes and IPGs (reproduced from ref. 66 with permission). (**A**) IEF gel containing Ampholine pH 4−6 (0.2 pH cm^{-1}, 140 V cm^{-1}); (**B**) IPG gel spanning 1 pH unit (0.1 pH cm^{-1}, 250 V cm^{-1}); (**C**) IPG gel spanning 0.2 pH unit (0.02 pH cm^{-1}, 500 V cm^{-1}); (**D**) IPG gel spanning 0.1 pH unit (0.01 pH cm^{-1}, 1000 V cm^{-1}). The dark, central band which appears as a single, homogeneous component in (**A**) is clearly split into two components in (**C**). The pI difference between these two bands is estimated to be 0.002 pH units.

groups forming the pH gradient are covalently attached and immobilized via vinyl bonds to the polyacrylamide backbone. This results in pH gradients which are effectively infinitely stable, thereby eliminating the deleterious effects of pH gradient drift. However, this does not completely eliminate electroendosmosis, which can still cause problems in the extreme pH ranges (i.e. below pH 5 and above pH 9). The stability of the pH gradients combined with the defined chemical nature of Immobiline reagents should ensure greater reproducibility of protein separations than can be achieved using conventional IEF technology.

It is possible to generate narrow and ultra-narrow IPGs spanning from as little as 0.1 to 1 pH units (46). Such narrow and ultra-narrow pH gradients have an extremely high resolving power and are claimed to be able to resolve a difference in pI of as little as 0.001 pH units as illustrated in *Figure 5*. IPG IEF is, therefore, an excellent technique for assessing protein purity as it should be able to separate components with nearly identical pI values. Extended pH gradients (> 1 pH unit) can also be generated using the Immobiline reagents. However, some problems have been found to be associated with the use of wide-range IPGs, including slow entry of sample proteins, lateral band spreading, prolonged focusing times and increased electroendosmosis. These problems have been attributed to the inherently low conductivity of the Immobiline system (48). Using a technique known as hybrid or mixed-bed IEF, improved separations can be obtained by the addition of low concentrations (e.g. 0.5% w/v) of synthetic carrier ampholytes of the appropriate range to the IPG IEF gels (48).

An additional advantage of IPGs is that they are not particularly sensitive to disturbance by the presence of salt. Sample composition is, therefore, not so critical as for IEF using synthetic carrier ampholytes. Moreover, separations can be performed under conditions of controlled ionic strength and buffering power. The elimination of synthetic carrier ampholytes from the IPG system (unless hybrid IEF is being used) avoids potential artefactual protein microheterogeneity due to protein – ampholyte reactions (see Section 4.4.1). However, certain classes of proteins, including histones, high-mobility group chromatin proteins, albumin and ferritin can interact with IPG matrices and can in some cases form insoluble complexes (49). This interaction probably involves both ionic and hybrophobic interactions, and there is some evidence (50) that this may be a more general problem leading to loss of protein and increased background staining in IPG IEF gels.

4.5 Two-dimensional electrophoresis

Two-dimensional polyacrylamide gel electrophoresis (2-D PAGE), where proteins are separated in each dimension on the basis of independent physicochemical properties (usually charge and size), is extensively used for the characterization of complex protein mixtures. This method is routinely capable of resolving between 1000 and 2000 proteins and specially modified procedures are claimed to resolve as many as 5000 to 10 000 species. The method most frequently used today is based on the procedure developed by O'Farrell (51) using a combination of IEF under denaturing conditions in gels containing 8 M urea and non-ionic detergent (e.g. 0.5% w/v NP-40) in the first dimension with gradient SDS – PAGE in the second dimension. This technique has rather limited applicability in the field of protein purification since protein samples are unlikely to be of sufficient complexity and can generally be analysed adequately using one-dimensional electrophoretic techniques. Moreover, 2-D PAGE is a technically demanding, time-consuming and labour-intensive procedure. For a detailed discussion of this technology the interested reader is referred to ref. 52.

4.6 Analysis of electrophoretic profiles

4.6.1 *Fixation*

Gels are usually fixed after electrophoresis to precipitate and immobilize the separated proteins within the gel and to remove non-protein components which might subsequently interfere with staining procedures. The best general purpose fixative is 20% w/v tri-chloroacetic acid (TCA), whilst sulphosalicylic acid (10 – 20% w/v) or mixtures of TCA and sulphosalicylic acid (10% w/v of each) are less efficient. Methanolic solutions of acetic acid (e.g. 45% v/v methanol, 45% v/v distilled water, 10% v/v acetic acid) are very popular for gel fixation, but it should be remembered that low molecular weight species, basic proteins and glycoproteins may not be adequately fixed by this procedure. Aqueous solutions of reagents such as 5% w/v formaldehyde or 2% w/v glutaraldehyde which covalently cross-link proteins to the gel matrix can be used, but they are not generally popular.

4.6.2 *Staining procedures*

The most popular general protein staining methods are based on the use of the non-polar, sulphated triphenylamine dye Coomassie Brilliant Blue R-250 (CBB R-250). Staining is usually carried out using 0.1% w/v CBB R-250 in 45% v/v methanol, 45% v/v distilled water and 10% v/v acetic acid. The time required for staining depends on the thickness of the gel and its polyacrylamide concentration. A 10%T gel of 0.5 – 1 mm thickness should be stained for about 2 h, whilst the staining time should

Method Table 9. Silver staining procedure.

A. *Materials*

1. All water used must be distilled and deionized. All solutions should be prepared fresh. Gloves should be worn at all times since the stain will detect skin and sweat proteins from the fingers.
2. Gel fixation solution: 20% w/v TCA.
3. Silver diamine solution: add 21 ml of 0.36% w/v NaOH to 1.4 ml of 35% w/v ammonia and then add 4 ml of 20% w/v silver nitrate dropwise with stirring. If the mixture fails to clear with the formation of a brown precipitate, add the minimum amount of ammonia required to dissolve the precipitate. Make the solution up to 100 ml with water. The solution is unstable and should be used within 5 min.
4. Reducing solution: 2.5 ml of 1% w/v citric acid, 0.26 ml of 36% w/v formaldehyde made up to 500 ml with water.
5. Farmer's reducer: 0.3% w/v potassium ferricyanide, 0.6% v/v sodium thiosulphate, 0.1% w/v sodium carbonate.
6. Clearing solution: 100 ml of 1.5 M sodium thiosulphate mixed with an equal volume of a solution of 0.15 M copper sulphate, 0.6 M sodium chloride, and 0.9 M ammonium hydroxide.

B. *Method*

1. After electrophoresis fix the gel for at least 2 h (1.5 mm thick gel) or longer for thicker gels (overnight soaking can be used).
2. Wash the gel twice for 30 min in 200 ml of 50% w/v methanol with agitation.
3. Wash the gel twice for at least 20 min in a large volume of water to rehydrate the gel and to remove methanol.
4. Incubate the gel in the silver diamine solution for 15 min with constant agitation. Caution should be exercised in disposal of the ammoniacal silver reagent since it decomposes on standing and may become explosive. The ammoniacal silver reagent should be treated with dilute hydrochloride acid (1 M) prior to diposal.
5. Wash the gel twice for 5 min in water.
6. Incubate the gel in reducing solution. Proteins are visualized within 10 min, after which the background will gradually increase. It is important to note that the reaction displays inertia, and that staining will continue for 2−4 min after removal of the gel from the reducing solution. The background generally becomes unacceptable after 15 min.
7. Wash the gel in water for 2 min.
8. Stop further development of the stain by immersing the gel in 45% v/v methanol, 10% v/v acetic acid.
9. If staining is of low sensitivity it can be further intensified by recycling. The stained gel is washed in water for 5 min, then returned to 50% v/v methanol and steps 2−7 repeated.

10. Selective destaining is used for the controlled removal of background staining that obscures the protein pattern. The gel is washed in water for 5 min to remove the stop solution. The gel is then placed in Farmer's reducer for a time dependent on the intensity of the background. Destaining is stopped by returning the gel to the stop solution.

11. Complete destaining for a gel which is so overstained that it is unusable can be achieved by immersing it in clearing solution. Once the gel has cleared, wash it extensively in water (3×30 min) prior to return to 50% v/v methanol. Then recycle the gel through steps $2-7$.

---

be increased for thicker and/or more concentrated gels. Destaining is accomplished by gentle agitation in the same acid $-$ methanol solution but in the absence of dye. The time required for destaining is also dependent on gel thickness and polyacrylamide concentration. The process can take up to 24 h but can be accelerated by using several changes of the destaining solution. The proteins will be visualized as intense blue bands on a colourless background. Stained gels can be stored in 7% v/v acetic acid. The dimethylated form of the dye, CBB G-250, can be used as a colloidal dispersion in TCA (0.25% w/v CBB G-250 in a 50% v/v aqueous methanol solution containing 12.5% w/v TCA). The advantage of this approach is that the dye does not penetrate the gel matrix, thus giving rapid staining of the proteins (only a few minutes at 60°C) without the development of background staining. Gels are destained in 5% w/v TCA for $10-60$ min and stored in 7% v/v acetic acid. The main disadvantage of staining procedures using CBB is their relative insensitivity, being capable of detecting about 0.5 μg protein cm^{-2} of gel. Recently in a test of over 600 modifications of the CBB staining procedure, Neuhoff and his coworkers (53) developed an optimal staining technique using 0.1% w/v CBB G-250 in 2% w/v phosphoric acid containing 6% w/v ammonium sulphate. This procedure is claimed to have a significantly increased sensitivity (0.7 ng protein mm^{-2} of gel) compared with other CBB staining techniques.

Recently, numerous staining techniques based on the use of silver have been developed for the detection of proteins separated by PAGE. These procedures are based either on techniques developed for histology (diamine silver stains) or on photographic protocols (non-diamine stains, photodevelopment stains) and the various methods are reviewed in detail in ref. 54. A recommended procedure in routine use in our laboratory is given in *Method Table 9*. A typical SDS $-$ PAGE separation visualized by silver staining is shown in *Figure 6*. Silver staining techniques are claimed to be between 20 and 200 times more sensitive than CBB R-250 and can detect as little as $0.05-0.1$ ng protein mm^{-2} of gel. Silver staining is, therefore, an ideal method for the detection of trace impurities during protein purification. It is compatible with most forms of PAGE, IEF and 2-D PAGE techniques. However, various chemicals commonly used in these procedures will inhibit staining (e.g. acetic acid), whilst others can produce artefacts (agarose, 2-mercaptoethanol) or increased background (e.g. Tris, glycine). These substances must be efficiently washed out of the gel before staining is attempted.

The main disadvantages of the silver staining technique are that it is relatively expensive and time-consuming. In addition, reproducibility can be difficult to control,

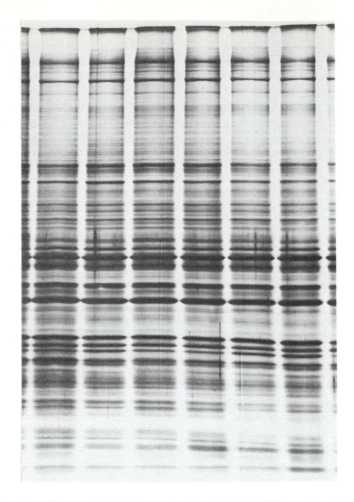

Figure 6. Separation of human skin fibroblast total cell proteins by 5−20%T gradient SDS−PAGE. The protein separation was visualized by silver staining.

although the availability of commercial silver staining kits (Amersham, Pierce, Bio-Rad) may go some way to alleviate this problem. Silver staining is often criticized on the basis that plots of the optical density of silver-stained proteins against their concentration produce different slopes for each of the proteins studied. However, it should be pointed out that such protein-specific staining slopes have also been observed with most of the organic dyes including CBB (54). Nevertheless, protein patterns visualized by silver staining should be interpreted with caution since certain proteins (e.g. calmodulin, immunoglobulin light chains) stain very poorly by this procedure. Finally, the absolute sensitivity of the technique can be an embarrassment as nearly all samples can show the presence of minor impurities using this method!

4.6.3 *Specific enzyme staining*

Techniques have been developed for the detection of specific proteins on the basis of their enzymic activity. They can be applied to the visualization of electrophoretic protein profiles by two types of procedure. In the first method, the gel is immersed in the required reagents and the enzyme activity is visualized directly within the gel matrix. This method is quite satisfactory if the final reaction product is insoluble, but a soluble reaction product present at the gel surface will rapidly diffuse away from the site of enzyme activity. The alternative approach is partially to immobilize the reagents by incorporating them into an agarose gel or by adsorbing them onto a substrate such as cellulose acetate or filter paper. The agarose or filter support is then placed in direct contact with the gel surface with resultant initiation of the enzymic reaction. This type of 'print' method can work quite satisfactorily even for soluble reaction products as their rate of diffusion is significantly reduced.

The majority of enzyme staining methods currently used are based on electron transfer dyes such as methyl thiazolyl tetrazolium, which is reduced by electron donors to form a dark blue insoluble formazan. This reaction is catalysed by phenazine methosulphate and can be used to detect enzymes which lead to the production of NADH or NADPH. This approach can be extended to other enzymes if they can be coupled by intermediate reactions to the reduction of NAD or NADP. A large number of enzyme staining procedures have been developed for use with a variety of gel systems (55).

The main disadvantage of this group of procedures is that they are totally dependent on the retention of enzymatic activity by the separated proteins. The technique is, therefore, limited to use with gels run under native conditions and is incompatible with methods involving the use of urea or SDS. Gels cannot be fixed prior to visualization of enzyme activity, so that resolution can be degraded if the separated proteins diffuse appreciably during the time taken to carry out the reaction.

4.6.4 *Blotting techniques*

The high specificity and affinity of antibodies and other ligands such as lectins makes these reagents powerful tools for the identification and characterization of proteins separated by electrophoretic procedures. Antibodies can be used to precipitate proteins directly in the gel matrix (immunofixation), but it is generally much better to use the technique known as Western blotting to transfer the separated proteins from the gel onto the surface of a thin matrix such as nitrocellulose. In this way the proteins are immobilized on the surface of the substrate and are readily accessible by a variety of techniques but the most efficient and rapid transfer is achieved by electrophoretically transferring the proteins from the gel onto the immobilizing matrix by application of an electric field perpendicular to the plane of the gel. This technique, pioneered by Towbin (56), is known as electroblotting. Good reviews on this topic are refs 57 and 58.

There are various designs of electroblotting apparatus, many of which are commercially available. These are either of the vertical type with a large buffer reservoir and platinum electrodes (e.g. Hoefer, Pharmacia-LKB, Bio-Rad) or of the horizontal, semi-dry type with graphite plate electrodes (59) (e.g. Pharmacia-LKB, Dako, Sartorius, Fisher). In the vertical type of apparatus the electrodes are usually several cm apart so that the maximum voltage gradient that can be applied is limited (typically 5 V cm^{-1}), even if an efficient cooling system is available to deal with Joule heating. Consequently, rather prolonged transfer times (12−18 h) have to be used, although this can be reduced (2−6 h) if only low molecular weight proteins (<10 kd) are to be transferred or if gels of a low polyacrylamide concentration (e.g. IEF) have been used. Higher field strengths (10 V cm^{-1}) can be used with the semi-dry, horizontal type of apparatus due to the short inter-electrode distance so that very rapid transfer (1 h or less) can often be achieved.

The most commonly used buffer for electroblotting of both native and SDS gels is that developed by Towbin (56) (25 mM Tris base, 129 mM glycine, pH 8.3). Methanol (10−20% v/v) is often added to this buffer in the case of SDS gels since it increases the binding capacity of nitrocellulose filters for SDS−proteins. However, it should be remembered that methanol acts as a fixative and reduces the efficiency of protein elution so that extended transfer times (up to 24 h) must be used. This effect is worse for high molecular weight proteins, so that methanol is best avoided if proteins >100 kd are to be transferred. A discontinuous buffer system has been recommended for use with the semi-dry type of apparatus (59), but it is claimed to work as efficiently using standard electroblotting buffers (60). If basic proteins are to be transferred or if acid−urea gels have been used, a transfer solution of 0.7% v/v acetic acid is recommended and the proteins are transferred as cations.

Nitrocellulose (pore size 0.45 μm for proteins >15 kd, 0.2 μm for proteins <15 kd) remains the most popular matrix for Western blotting. Positively charged nylon membranes have the potential advantage of higher binding capacity (480 μg cm^{-2} compared with 80 μg cm^{-2} for nitrocellulose), but suffer from problems with subsequent protein visualization. A new polyvinylidene difluoride (PVDF) membrane (Immobilon from Millipore) is claimed to have high mechanical strength and to be compatible with all immunostaining protocols. Although both nitrocellulose and nylon membranes have higher initial binding capacities, PVDF membranes provide superior retention of bound proteins after treatment with commonly used blocking and washing reagents, resulting in higher sensitivity with immunostaining.

Stains such as Amido black (0.1% w/v in 45% v/v methanol, 7% v/v acetic acid) can be used to visualize rapidly (staining time, 10−15 min; destaining time, 5−10 min) the transferred proteins. CBB staining can be used, but generally results in a rather high background which is difficult to reduce by destaining. Sensitive detection can be achieved by overnight staining with India ink [0.1% v/v in phosphate buffered saline (PBS) containing 0.3% v/v Tween 20] (61) or with colloidal gold particles (Aurodye from Janssen) (62). Nylon membranes are more difficult to stain, but sensitive detection is now possible on this type of matrix using a cationic cacodylate iron colloid followed by treatment with potassium ferrocyanide (Ferridye from Janssen) (63). The sensitivities of some of these staining methods is shown in *Table 6*. Individually coloured (Amersham) and pre-stained (Bio-Rad) protein molecular weight markers are now available and these can be used to provide a visual check of the efficiency of protein transfer.

Before probing with a specific ligand, all unoccupied binding sites on the filter must be blocked. This is often achieved with a solution of a protein such as BSA, ovalbumin, haemoglobin or gelatin. Membranes should be incubated for 1 h in PBS containing 3% w/v BSA (for nitrocellulose) or for 12 h in PBS containing 10% w/v BSA (for nylon). However, the proteins used in the blocking step are not always unreactive during subsequent probing resulting in a high background. A solution of powdered non-fat milk (5% w/v for nitrocellulose, 10% w/v for nylon) in PBS has recently become popular

Table 6. General protein stains for blot transfers arranged in order of increasing sensitivity.

Detection reagent	Approximate sensitivity	Matrix
Fast Green FC	—	NC
Ponceau S	—	NC
Coomassie Brilliant Blue R-250[a]	1.5 μg	NC
Amido Black 10B	1.5 μg	NC
India ink	100 ng	NC
Colloidal iron	30 ng	NC + nylon
In situ biotinylation + HPR−avidin	30 ng	NC + nylon
Colloidal gold	4 ng	NC

[a]Results in a high background.
NC, nitrocellulose.

Table 7. Specific detection methods for blot transfers arranged in order of increasing sensitivity.

Method	Approximate sensitivity (ng mm^{-2})
Peroxidase−protein A	2.0
Peroxidase−second antibody	1.5
Gold−second antibody	1.5
^{125}I−second antibody	1.0
Peroxidase double sandwich	0.8
Peroxidase−antiperoxidase	0.5
Avidin−biotin−peroxidase (ABC) complex	0.5
Gold−second antibody with silver enhancement	0.1

for use as a blocking reagent and usually results in a very low background staining. Another popular alternative to proteins for the blocking step is the non-ionic detergent, polyethylene sorbitan monolaurate (Tween 20) (64), which should be used at a concentration of 0.5% w/v in PBS for 1 h. However, there is a risk of displacing proteins from the matrix by use of the detergent (58).

The most important step in the procedure is the reaction with the specific antibody or ligand and visualization of the proteins present in the sample with which it reacts. Indirect sandwich techniques using second or third ligands are generally used for immunodetection due to the significant increase in sensitivity which can be achieved. The detection systems used involve either fluorescence (e.g. fluorescein isothiocyanate), radiolabels (usually ^{125}I), or enzyme reactions (e.g. peroxidase, alkaline phosphatase, β-galactosidase, glucose oxidase). The latter approach using reagents conjugated with one of these enzymes is becoming increasingly popular. Recently, even more sensitive detection systems using gold-labelled ligands or the avidin−biotin system have been developed to extend further the limits of detection of specific proteins on blot transfers. A guide to the relative sensitivities of some of these specific detection procedures is given in *Table 7*.

4.6.5 *Quantitative analysis*

The methods of electrophoresis and visualization of the separated protein profiles which have been described should enable a thorough assessment of the purity of any fraction from a particular purification protocol. However, these methods as described are purely subjective and give no direct quantitative assessment of the levels of impurities present in any given sample. Quantitative analysis is best carried out by scanning densitometry where a graphical representation of protein zones against migration distance is produced and from which areas can be calculated to yield quantitative data. Many commercial scanning densitometers are available (e.g. Joyce Loebel, Bio-Rad, Pharmacia-LKB, Hoefer) for the analysis of one-dimensional electrophoretic separations and several of these are provided with extensive suites of computer analysis software to provide flexible and thorough quantitative analysis.

One caveat that should be made concerning quantitation of electrophoretic patterns is that at best one is obtaining semiquantitative data since individual proteins can respond quite differently to different visualization procedures. Moreover, there is only a limited range of protein concentration over which there is a linear relationship between absorbance and concentration. The same, of course, is also true for assays of total protein. Thus, colour development in a staining procedure or spectrophotometric assay can vary dramatically with the amino acid composition of the particular protein and can be significantly influenced by the presence of non-protein groups such as carbohydrate. It is a delusion to believe that all proteins stain equally with any given visualization procedure and that absolute quantitative data can be obtained. Caution should, therefore, be exercised in the interpretation of quantitative estimates of protein purity based solely upon scanning densitometry.

5. PREVENTION OF UNCONTROLLED PROTEOLYSIS—R.J.Beynon

There is growing awareness that every cellular compartment is pervaded by multiple proteases, although the roles of most of these enzymes remain a mystery (67). Perhaps therefore it is surprising that so many proteins escape from rather than succumb to adventitious proteolysis. It is inappropriate in this section to catalogue the large number of reports of proteolytic artefacts (see for example ref. 68) and my intention is to provide general guidelines to assist (i) in recognition of proteolysis as a source of artefacts, and (ii) to eliminate those problems as far as possible. This section builds upon two very readable accounts by Pringle (69,70); other aspects have also been covered in a recent publication (71).

Ultimately, all proteins are degraded, either intracellularly or extracellularly and thus, under appropriate conditions, can be completely and effectively hydrolysed to amino acids by proteolytic enzymes or systems (72). Although protein degradation must be progressive *in vivo* and involve partially proteolysed derivatives of a protein, these intermediates are difficult to detect, from which we may infer that inside the cell, the commitment of a protein to the degradative machinery brings about its rapid and complete hydrolysis. However, when a tissue or cell is disrupted two new conditions can ensue. First, the proteins may be brought into contact with new proteolytic enzymes from a different intracellular compartment, extracellular space or cell. Secondly, the intracellular degradative machinery is disassembled and diluted, such that limited proteolytic events occur without complete hydrolysis of the products.

In both instances, the result may be endopeptidase or exopeptidase attack on the protein of interest. Proteolysis may be limited, either in extent or in the number of bonds that are attacked. For the purposes of this volume, proteolytic artefacts are defined as undesirable proteolytic events that occur after tissue disruption. These events must be distinguished from proteolysis that occurs *in vivo* and which may be a consequence of normal biological activity. A protein or peptide may also exhibit structural heterogeneity as a consequence of proteolysis *in vivo*; no modification to purification procedures will be able to prevent this heterogeneity.

Losses of protein or biological activity are often attributed to proteolysis *in vitro*. At the outset it is important to realize that (i) proteolysis is not the only cause of losses of material during purification procedures and (ii) even if proteolysis can be established as the cause, there is rarely a simple solution to the problem. Effective control of proteolysis *in vitro* will require at least some information about the protease(s) that are responsible. Some strategies are discussed below that may be of value in controlling adventitious proteolysis during protein purification procedures. It would be inappropriate and even impossible to give specific guidelines for different proteins or cells. Rather an approach is presented that requires some knowledge of the proteolytic systems that are responsible.

5.1 **Proteolytic susceptibility of native proteins**

Proteins differ greatly in their intrinsic susceptibility to proteolytic attack, but at present there is no way to predict this property. It is far from clear why many proteins containing several hundred peptide bonds appear to be refractory to the action of a wide range of endopeptidases and exopeptidases, yet other proteins display an equally unusual fragility and tendency to be digested by proteases. Since denatured proteins are uniformly susceptible to proteolysis it follows that resistance or vulnerability to proteolysis must be a function of higher order structure. Flexible regions of the higher order structure might be able to fold into the active site of a protease and thus undergo preferential attack. By contrast, a fragment of polypeptide chain involved in secondary structure would be highly constrained and offer a sterically constrained and thus inefficient site of proteolysis. Disulphide bridges may offer enhanced resistance to proteolysis, a general property of extracellular proteins. A reasonable guideline derives from many observations on the relationship between susceptibility to proteases and other thermodynamic properties; if the protein is vulnerable to thermal inactivation it may also be a good target for proteases.

A valuable approach to the study of proteolysis of native proteins is based upon a two stage model (73). The first stage comprises hydrolysis of the peptide bonds that are most vulnerable to proteolysis; whereas the second stage generalizes all subsequent proteolysis leading to the formation of limit peptides, in which there are no further peptide bonds vulnerable to the action of the protease. The behaviour of the overall process will depend greatly upon the relative rates of the two stages (k_1 and k_2, respectively). When k_1 is much larger than k_2 a partially proteolysed form of the protein will accumulate; this is the classical 'proteolytic artefact'. By contrast, if k_2 is much larger than k_1 then the net result is complete hydrolysis of the protein, controlled by the rate-limiting k_1. Such a system would be manifest as a time-dependent loss of activity and detectable protein. Finally, when k_1 and k_2 are similar in magnitude, the result is the transient generation of intermediate partially proteolysed forms of the protein; the consequence of which would be an ever changing protein population.

The rate of proteolysis of a protein can be influenced by alteration of the solvent conditions, or often by inclusion of ligands of the protein of interest. In my experience, native proteins are often resistant to proteolysis in high salt concentrations (at least 0.1 ionic strength) or in organic co-solvents that are protein stabilizers, such as glycerol. Ligands may help to stabilize the protein, and indeed, a large number of protein purification protocols routinely include a substrate, inhibitor or cofactor in the buffers.

Over-generalization may be dangerous as there are many examples of ligand-induced *destabilization* of a protein (74), but protection of a protein by such devices is worthy of investigation.

One set of circumstances should be singled out for special mention. During sample preparation for SDS−PAGE, the protein solution is heated in a powerfully denaturing solution. In general, many proteases are relatively resistant to the physicochemical denaturation and may survive in the detergent solution for longer than the substrate protein, a brief death-throe during which they may wreak havoc. Under such circumstances, proteolytic artefacts may be suspected when they are not present in the original sample. There is a strong case for including protease inhibitors in sample preparation buffer if such behaviour is suspected.

5.2 Identification of proteolysis as a problem

Although apparently self-evident, an essential step in controlling proteolytic artefacts is to establish that proteolysis is indeed the source of the problem. In my experience, there is a tendency to blame proteolysis, without good evidence, for losses of material or activity that occur in protein purification schemes. Other reasons may often be invoked. First, a multimeric protein may have been resolved into components that possess no biological activity when separated. This phenomenon will often manifest itself as a dramatic loss of activity when a high resolution purification step is employed. Complementation experiments will be needed to identify active fractions. Secondly, a protein may have been separated from an essential low molecular weight ligand (e.g. a metal ion or cofactor). Such behaviour may be observed when the sample is dialysed or subject to gel permeation chromatography. Reconstitution experiments may indicate such behaviour. Finally, the protein may be intrinsically vulnerable to physicochemical denaturation; this is difficult to establish with certainty and even more difficult to prevent.

However, there are a number of indicators that implicate proteolysis as a source of artefacts. Limited proteolysis can lead to structural heterogeneity in the protein. If the protein can be purified then paradoxically, heterogeneity in subunit size, amino or carboxy terminal amino acids, catalytic properties or isoelectric point may be ascertained. More difficult is the problem of establishing which of those changes represent true heterogeneity *in vivo*, and which can be attributed to post-homogenization proteolysis.

Structural heterogeneity may be more difficult to detect in crude, or semi-purified preparations. Such heterogeneity may be manifest as changes in isoenzyme patterns; in turn this requires some method (immunological or enzymic) to detect the isoenzyme patterns that change during the purification scheme. If an antibody is available that cross-reacts with the protein of interest (not necessarily raised to the particular protein under study; limited cross-reactivity may suffice in these studies), immunoblotting of SDS−polyacrylamide gels can indicate molecular weight changes even in crude

Method Table 10. Assay for proteases using azocasein.

1. Prepare an azocasein (e.g. Sigma) stock solution by dissolving the substrate in a suitable assay buffer[a], at a final concentration of 11 mg ml^{-1}. The buffer should be appropriate for the conditions used, but the following guidelines are apposite. Azocasein is not soluble below pH 5−6. The buffer should not contain any reduced thiols or chelators (which inhibit metallo-proteases and activate cysteine proteases) unless these are essential components of the purification buffers. The buffer should ideally have an ionic strength of 0.1−0.2. Azocasein has an absorbance maximum at 440 nm in alkaline solution; consult the manufacturer's specifications for the $E_{440}^{1\%}$ of the batch being used (Sigma typically quote figures of 35−40). Initially, prepare the azocasein solution at ∼ 15 mg ml^{-1} in the buffer, adjust the pH to the correct value if necessary and determine the concentration of this initial solution by diluting 10 μl of the azocasein solution in 1.0 ml of 0.1 M NaOH ($E_{440}^{1\%}$ at 440 nm is typically 0.5). Dilute the solution to 1 mg ml^{-1} with assay buffer.

2. Prepare assay incubations by pre-equilibrating 1.0 ml of the buffered azocasein at a suitable temperature.

3. Start the digestion by adding a protease sample in a final volume of 0.1 ml[b]. Begin a blank incubation by adding 0.1 ml of buffer to a second incubation.

4. Immediately remove 0.25 ml of the incubations and add each to 1.0 ml of 5% (w/v) TCA in a 1.5 ml capped Eppendorf tube. The red/orange azocasein will precipitate as a yellow flocculate; the colour change is because the dye has a pK_a between neutral and acidic pH values.

5. At suitable times (typically 30, 60 and 90 min)[c] remove further 0.25 ml samples and add to TCA as in step 4.

6. Centrifuge each Eppendorf tube in a bench centrifuge (10 000 g for 2 min) to compact the undigested substrate. Remove the supernatant and measure the absorbance at 340−360 nm (the absorbance maximum at low pH values)[d].

7. The increase in absorbance of the supernatant from sample incubations is a measure of the rate of hydrolysis of azocasein. The slope of the blank should typically be less than 0.00005 per min; values much higher might indicate contamination of the azocasein solution by a protease. The initial absorbance should be ∼ 0.05−0.1. The time course of azocasein digestion becomes non-linear when the absorbance (at 340 nm) exceeds ∼ 0.5. In this assay trypsin (1 μg) gives a slope of 0.02 per min.

[a]Little guidance can be given as to the choice of buffer, as requirements in the protein purification scheme may be of over-riding importance.
[b]Start by using as much sample as can be accommodated in the assay (0.1 ml) and take the first sample after a short incubation period (e.g. 15 min). A very high proteolytic activity may then be detected very quickly and the assays redone with lower amounts of sample.
[c]As above, little guidance can be given over time intervals. It may be wise to conduct early assays over expanding time intervals (e.g. 15 min, 60 min, 5 h and overnight); this sampling 'window' will give some indication of the absolute level of proteolytic activity.
[d]For the assay of large numbers of samples (such as an elution profile) it is feasible to set up incubations in an analogous fashion to step 2 and 3, but incubate for a fixed time period before proceeding to steps 4 and 6. The single time point is sufficient to define the elution profile of the proteolytic activity. Include appropriate buffer blanks, from the beginning and end of an elution profile if a desorbing solute was used (e.g. affinity eluent or salt).

Method Table 11. Preparation of radioiodinated insulin B-chain.

---

1. Unlike many iodination reactions, the goal is not to produce a peptide of very high specific activity but to generate substantial amounts of substrate labelled to a workable specific radioactivity. The iodination is performed at high protein concentration and in high volumes.

2. Dissolve 2−3 mg of insulin B-chain in 0.25 ml of 0.5 M potassium phosphate buffer, pH 7.5. The insulin B-chain is not particularly soluble at this pH and it may be preferable to make the solution slightly alkaline to dissolve the peptide, then readjust the pH to 7.5 using narrow range pH paper.

3. Add 5 μl of 0.15 M potassium iodide (25 mg ml^{-1}) to lower the specific radioactivity of the label.

3. All subsequent stages to step 8 should be performed behind ≥ 3 mm lead shields in a fume hood using suitable precautions for the handling of ^{125}I. The specific radioactivity of stock solutions is so high that microaerosols can introduce significant amounts of radioactivity into the atmosphere. Wear two pairs of disposable gloves when handling the stock Na^{125}I and discard one pair immediately after you have finished with the stock. Check gloves frequently for contamination with a hand monitor and replace as necessary.

4. Add ~200 μCi (7.4 MBq) of Na^{125}I in a final volume of 2−20 μl as convenient.

5. Add 50 μl of 0.44 M chloramine T (100 mg ml^{-1} in water, dissolved just before use). Mix well and allow reaction to proceed for 15 min.

6. Add 20 μl to 0.32 M sodium metabisulphite (60 mg ml^{-1} in water, dissolved just before use) to terminate the reaction.

7. Add 20 μl of 0.15 M potassium iodide to act as a carrier for the free iodine during the gel filtration. If the solution shows signs of yellow/brown coloration due to elemental iodine, add a further 20 μl of 0.32 M sodium metabisulphite. If the insulin B-chain appears to have precipitated, add a few drops of 1.0 M NaOH to raise the pH and dissolve the peptide.

8. Apply the whole reaction mixture to a Sephadex G-25 (medium or fine grade) column of 10−15 ml total volume, with a diameter of approximately 1 cm and a length of 12−20 cm. Elute the column with 6 mM potassium iodide, dissolved in water adjusted to pH 9−10 with NaOH. Collect 20−25 fractions of ~0.7 ml.

9. For each fraction, remove 5 μl and transfer to a second tube containing 1.0 ml of water. Determine the radioactivity in 100 μl of each dilution by γ counting. This is the total radioactivity in 0.5 μl of each original fraction.

10. Remove 100 μl of each dilution and add to 100 μl of 2% (w/v) casein as a co-precipitant. Precipitate the sample by adding 200 μl of 25% (w/v) TCA and sediment the precipitated material by centrifugation in a bench centrifuge at 10 000 g for 2 min. Determine the radioactivity in 100 μl of the supernatant; this is equivalent to the acid-soluble radioactivity in 0.125 μl of each fraction (see *Figure 7*).

11. Pool fractions that contain the high molecular weight radioactivity that is >95% precipitated. Specific activities are usually high (100 000 d.p.m. μg^{-1}).

12. Determine the peptide concentration of the pooled fractions by absorbance at 280 nm (no dilution should be necessary for this determination). The $E_{280}^{1\%}$ of insulin B-chain at this wavelength is 9.92. If possible, use an acrylic cuvette that is reasonably transparent at 280 nm and which can be discarded as radioactive waste after use. Take care not to contaminate the spectrophotometer.

13. Adjust the pH of the pooled fractions to near neutrality and dilute the pooled fractions to a final concentration of 0.5 mg ml^{-1}. Aliquot and store the radiolabelled substrate at $-20°C$ in a lead-lined box.

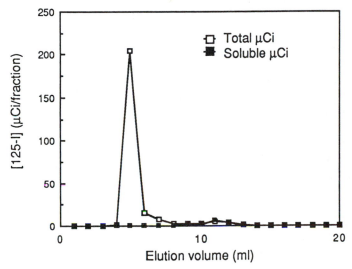

Figure 7. Gel permeation chromatography of iodinated insulin B-chain. Insulin B-chain (2 mg) was iodinated using the chloramine T method before separation on Sephadex G-25. The total radioactivity and acid-soluble radioactivity were determined in a small part of each fraction. Highly labelled fractions with less than 5% soluble counts were pooled and used as substrate.

preparations, although slight decrease in size consequential to exopeptidase attack may not be detectable at the resolution normally obtained with this technique.

Some indication of proteolysis-induced heterogeneity may come from studies of the protein purified from alternative tissues or cells. For example, liver and kidney express far higher levels of proteolytic enzymes than skeletal muscle. The artefacts may be less apparent in protease-defective mutants of bacteria or yeast. Tissues that are permeated by a vascular bed are perfused by plasma that is a rich source of proteolytic enzymes (75); perfusion of such tissues to remove blood contamination may eliminate a proteolytic artefact (note also the converse; plasma is also a rich source of protease inhibitors, see ref. 76).

Measurement of protease activity in protein solutions may be valuable as an indicator of overall proteolytic potential. Protease assays vary from the non-specific, capable of detecting a large proportion of proteases, to the highly specific, utilizing substrates that are hydrolysed by a very restricted set of enzymes (77). For assessing total proteolytic activity, general protease assays are preferable. Predominantly, these consist

45

Method Table 12. Assay for proteases using iodinated insulin B-chain.

---

1. Dilute the stock radiolabelled insulin B-chain (0.5 mg ml^{-1} in water) with 10 volumes of assay buffer[a]. The same considerations of suitability of buffer apply here as for the azocasein assay (*Method Table 10*, step 1). For *n* assays, dilute $25 \times n$ μl of substrate into $250 \times n$ μl of buffer.
2. Pre-equilibrate the buffered substrate at the assay temperature.
3. To 275 μl of buffered substrate add the protease preparation in a final volume of 225 μl. The final assay volume should be 0.5 ml and the assay buffer should be twice the concentration needed in the final assay.
4. Immediately remove 100 μl of the incubation mixture and add to 100 μl of 2% (w/v) casein in an Eppendorf tube. Add to this mixture 200 μl of 25% (w/v) TCA.
5. At suitable times (typically 10, 20 and 30 min)[b] repeat step 4 with further samples of the incubation mixture[c].
6. At the end of the reaction, when up to four samples have been removed, take a further 50 μl of the reaction mixture and count directly, without the precipitation step. This gives the specific radioactivity of the substrate in the assay and permits calculation of the value corresponding to 100% solubilization of the substrate.
7. Centrifuge each Eppendorf tube in a bench centrifuge (10 000 g for 2 min). Determine the radioactivity in 200 μl of the supernatant.
8. A plot of percentage of substrate solubilized should be linear with respect to time, up to $\sim$25% solubilization. If the reaction shows signs of non-linearity the assay should be repeated with less enzyme preparation.

[a,b]As in *Method Table 10*, little guidance can be given as to assay buffer and incubation times.
[c]For assay of large numbers of samples (such as an elution profile) it is feasible to set up incubations in an analogous fashion to steps 1−3, but incubate for a fixed time period before proceeding as steps 4 and 6. The single time point is sufficient to define the elution profile of the proteolytic activity. Again, include appropriate buffer blanks.

of a proteolytically-vulnerable protein or peptide that is labelled either with chromogenic/fluorogenic dyes or with a radionuclide. After exposure of the substrate to the protease, the substrate is separated from the products (usually by acid precipitation) and the amount of product is determined by virtue of the label. I describe here two protease assays that I have used extensively. The first employs azocasein, which is casein dyed red with a sulphanilamide dye (78). The second uses the B-chain of insulin, radiolabelled with ^{125}I (71,79). The former is relatively insensitive, but is easier to perform than the latter, which is perhaps up to three orders of magnitude more sensitive, albeit requiring the preparation of a radiolabelled peptide. *Method Table 10* describes the azocasein assay, *Method Table 11* the preparation of iodinated insulin B-chain and *Method Table 12* the use of the radiolabelled insulin B-chain in protease assays. Suitable ranging assays will be needed to determine the intrinsic level of proteolytic activity in the sample of interest.

Although such assays are useful for indicating the extent of proteolytic activity in particular preparations, they may detect proteases that have no effect on the protein of interest or indeed, may fail to detect proteases (particularly exopeptidases) that are responsible for proteolytic artefacts. Ultimately the parameter of interest is not so much the level of proteases in the preparations, but the activity of those that cause the artefacts. Some method is therefore needed to establish the involvement of proteases in the generation of proteolytically-modified forms of the protein. Realistically, the only valid approach is to monitor the appearance of proteolytic artefacts under conditions in which protease inhibitors have been added to the preparation. This has the added bonus of indicating the catalytic mechanism of the offending protease—information of value in defining the strategy to be used in full-scale purification. Again, no specific advice can be given as to the nature of the monitoring procedure; it will be closely allied to the method by which the problem was first indicated. These exploratory studies can be performed relatively quickly on a small scale.

To make the most effective selection of protease inhibitor, it is necessary to understand a little of the catalytic mechanisms of proteases. Mechanisms are better understood for endopeptidases than exopeptidases, although there are close similarities between some members of the two classes (76,80−82). Serine endopeptidases bring about hydrolysis of a peptide bond by attack upon the carbonyl carbon by a nucleophilic serine residue in the active site of the enzyme, the reactivity of which involves a histidine residue. The active site serine and histidine residues are both effective targets for inhibitors. Those that are commonly used are summarized in *Table 8*. Additionally, several microbial inhibitors of proteases are readily available with sufficiently tight binding to give effective inhibition; these are discussed later. Note that the most effective classical inhibitor of serine proteases, diisopropylphosphofluoridate (DipF) is a potent inhibitor of acetylcholinesterase and therefore is extremely toxic. A method of dispensing a bulk sample of DipF into a safer solvent is given in *Method Table 13*.

Cysteine proteases depend upon a nucleophilic cysteine residue that is again a target for suitable inhibitors (*Table 9*). One of these inhibitors (E-64) is an epoxide that seems to be highly specific for the nucleophilic cysteine residue. Others such as iodoacetic acid or iodoacetamide are less specific and can inactivate many enzymes that involve a cysteine residue in the active site.

Aspartic proteases employ a pair of aspartic residues to labilize and hydrolyse the scissile peptide bond. The most effective inhibitor of aspartic endopeptidases is the tight binding transition state analogue pepstatin (*Table 10A*).

Metallo-proteases are dependent upon the electron-withdrawing ability of a metal ion (to date, exclusively zinc) to labilize the peptide bond in the substrate. Metallo-protease inhibitors direct an electronegative moiety to the vicinity of the metal ion and are tight binding but reversible inhibitors (*Table 10B*). Simple chelators such as EDTA or 1,10-phenanthroline are also effective inhibitors, but may be undesirable as components of the purification buffer. In particular, 1,10-phenanthroline has a strong UV absorbance and can interfere with spectrophotometric determinations.

Exopeptidases often utilize mechanisms that are closely analogous to the endopeptidases. Although the general approach to inhibition of these enzymes is similar to that for endopeptidases, there are a few additional inhibitors that are specific for exopeptidases and which are worthy of consideration (*Table 11*).

Finally, it should be mentioned that a number of newly discovered and incompletely characterized proteases do not appear to fall cleanly into one of the established classes. These proteases may not respond archetypically to classic inhibitors and may require new approaches.

Table 8. Inhibitors of serine proteases.

A. *Irreversible inhibitors*

Diisopropylphosphofluoridate

Synonyms	Diisopropylfluorophosphate, DipF, DFP
Mol. wt	184.2
Effective concentration	0.1 mM
Stock/solvent	200 − 500 mM in dry propan-2-ol
Stability of stock solution	Several months at −70°C
Notes	Highly toxic; see *Method Table 13* and *Figure 8* for preparation of stock solutions. Active towards all serine proteases

Phenylmethanesulphonyl fluoride

Synonyms	PMSF
Mol. wt	174
Effective concentration	0.1 − 1 mM
Stock/solvent	20 − 50 mM in dry solvents (propan-2-ol, MeOH, EtOH)
Stability of stock solution	At least 9 months at 4°C
Notes	Not as effective or toxic as DipF, inhibits cysteine proteases (reversible by reduced thiols). Active towards all serine proteases

4-Amidinophenylmethanesulphonyl fluoride

Synonyms	APMSF, *p*-APMSF
Mol. wt	270.7 (APMSF.HCl.H_2O)
Effective concentration	5 − 50 μM
Stock/solvent	20 − 50 mM in water
Stability of stock solution	Stable when aliquoted at −20°C
Notes	Not as effective as DipF, but more effective than PMSF. Specific for trypsin-like serine proteases. No effect on acetylcholinesterase

3,4-Dichloroisocoumarin

Synonyms	3,4-DCI
Mol. wt	215.0
Effective concentration	5 − 100 μM
Stock/solvent	10 mM in dimethylformamide or dimethylsulphoxide
Stability of stock solution	Stable at −20°C
Notes	Active towards wide range of serine proteases. Slowly reversible. Not active towards β-lactamases

L-1-Chloro-3-[4-tosylamido]-7-amino-2-heptanone − HCl

Synonyms	TLCK, Tosyl lysyl chloromethyl ketone, Tos-Lys-CH_2Cl
Mol. wt	369.4 (hydrochloride)
Effective concentration	10 − 100 μM
Stock/solvent	10 mM in aqueous solution (pH 3.0 − 6.0)
Stability of stock solution	Prepare fresh as needed
Notes	Active towards some trypsin-like serine proteases

L-1-Chloro-3-[4-tosylamido]-4-phenyl-2-butanone

Synonyms	TPCK, Tosyl phenylalanyl chloromethyl ketone, Tos-Phe-CH_2Cl
Mol. wt	351.9
Effective concentration	10 − 100 μM
Stock/solvent	10 mM in methanol or ethanol
Stability of stock solution	Several months at 4°C
Notes	Active towards some chymotrypsin-like serine proteases

B. *Reversible inhibitors*

Leupeptin

Synonyms	*N*-Acetyl-Leu-Leu-Arg-al
Mol. wt	542.7 (hemisulphate, monohydrate)
Effective concentration	$1-100~\mu$M
Stock/solvent	10 mM in water or buffer
Stability of stock solution	1 week at 4°C, 1 month at -20°C
Notes	Leupeptin is an amino acid aldehyde, the aldehyde being contributed by an arginine residue. It inhibits trypsin-like serine proteases and most cysteine proteases

Antipain

Synonyms	[(S)-1-carboxy-2-phenylethyl]-carbamoyl-Arg-Val-Arg-al
Mol. wt	604.7
Effective concentration	$1-100~\mu$M
Stock/solvent	10 mM in water or buffer
Stability of stock solution	1 week at 4°C, 1 month at -20°C
Notes	Antipain is an amino acid aldehyde, the aldehyde being contributed by an arginine residue. It has a similar specificity to leupeptin

Chymostatin

Synonyms	Phe-(Cap)-Leu-Phe-al
Mol. wt	582.7
Effective concentration	$10-100~\mu$M
Stock/solvent	10 mM in DMSO
Stability of stock solution	Stable for months at -20°C
Notes	Chymostatin is an amino acid aldehyde, the aldehyde being contributed by a phenylalanine residue. It inhibits chymotrypsin-like serine proteases and most cysteine proteases

Elastatinal

Synonyms	None
Mol. wt	512.6
Effective concentration	$10-100~\mu$M
Stock/solvent	10 mM in water
Stability of stock solution	Stable for 1 week at 4°C, months at -20°C
Notes	Elastatinal is an amino acid aldehyde, the aldehyde being contributed by an alanine residue. It inhibits elastase-like serine proteases

5.3 **Inhibition of proteases**

By the time that proteolysis is correctly identified by inhibitor studies as a source of artefacts or losses, the solution is available. Knowledge of the catalytic class of protease permits inhibition of its activity *in situ* or sometimes, its removal from the solution (see Section 5.4). Inhibition *in situ* has in turn, two approaches. The first is to maintain in solution at all times an effective concentration of a reversible inhibitor. Opportunities for loss of inhibitor are many-fold and include dialysis, gel permeation and ion-exchange chromatography, or chemical or biological inactivation of the inhibitor itself. Many protease inhibitors are peptides or possess labile functional groups such as aldehydes, and may themselves be vulnerable to enzymic attack (83). It may be preferable to select a cocktail of inhibitors such as, for example, leupeptin, chymostatin and EDTA or

Method Table 13. Preparation of stock solutions of diisopropylphosphofluoridate.

---

1. DipF is normally supplied in 1 g quantities in a rubber crimp-sealed glass vial of 3 ml volume. The vial can develop a positive pressure that can cause the DipF to be ejected when puncturing the vial with a hypodermic needle. This procedure was developed as a safe method for dispensing DipF and diluting a whole vial into dry propan-2-ol as a safer, less volatile stock solution. Alternative methods may be preferable if only small amounts of DipF must be diluted at any one time. Wear gloves and a gas mask at all times, and perform all manipulations in a high efficiency fume hood. Keep a bucket of 0.1 M NaOH in the fume hood; DipF is rapidly inactivated in alkaline conditions.
2. Dry some propan-2-ol by adding anhydrous molecular sieve (any other method will do).
3. Chill the vial of DipF on ice.
4. Puncture the vial with a hypodermic needle, but do not let the needle enter the liquid (*Figure 8*).
5. Insert the second needle through the seal, connected by a length of tubing to the receiving vessel.
6. Using a syringe connected to the first needle, flush solvent through the DipF vial so that it flows into the receiving vessel. Use the syringe to force air through the system to drain the vial.
7. If necessary, add extra solvent to give the desired DipF concentration. The concentration will not be known exactly and some error is inevitable.
8. Store the aliquoted DipF solution at $-20°C$ or $-70°C$ in small vials, in a container containing solid sodium carbonate. This ensures that any leaks of DipF are neutralized rapidly.
9. Treat all tubing and materials, empty vials, etc with 0.1 M NaOH to destroy residual DipF.

1,10-phenanthroline. The value of such cocktails can only be determined by preliminary tests.

Reversible inhibitors may also be used as the first stage of a two-stage approach, in which the initial inhibitory mixture is reinforced by the addition of irreversible inhibitors. Note that the reversible inhibitors may compete with the irreversible inhibitors and it may be preferable to use irreversible inhibitors at the earliest possible stage. A combination of irreversible serine protease inhibitors, such as DipF, phenylmethyl-sulphonyl fluoride (PMSF) or DCI, and the irreversible cysteine protease inhibitor E-64, may be most effective, although there are no readily accessible irreversible inhibitors of metallo-proteases or aspartic proteases (the latter are much less of a problem). Chelators and pepstatin may be needed at all times if there is evidence that either of these classes of enzymes are a source of problems.

5.4 **Removal of proteases**

An alternative to inhibition of proteases *in situ* relies on affinity techniques to remove the proteases from a relatively crude preparation. In particular, α-2 macroglobulin,

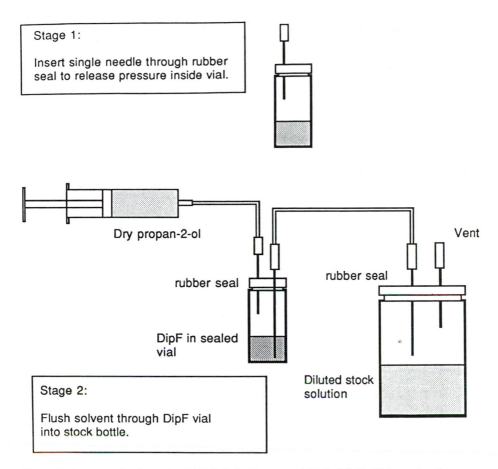

Figure 8. Apparatus for dispensing DipF in bulk. See text and *Method Table 13* for explanation.

a plasma protease inhibitor of very broad specificity, when coupled to an immobilized matrix is able to bind a large proportion of endopeptidases, but no exopeptidases. Other proteinaceous inhibitors of proteases are readily available and easily coupled to matrices such as CNBr-activated agarose or oxirane acrylic beads (details of coupling procedures will be found in Chapter 5). *Table 12* lists some inhibitors that have found application in the removal of proteases.

6. PURIFICATION STRATEGY—E.L.V.Harris

Before starting a purification the end use for the final product should be defined. With this in mind, minimum criteria of success for the purification can be set for quality, quantity and economy. In addition as much knowledge about the protein, its sources, its properties and its stability should be gathered. With this information to hand a strategy can be planned with a judicious combination of separation techniques most likely to lead to success. These initial questions have been covered in more detail in Section 1. The following key points are worth repeating.

Table 9. Inhibitors of cysteine proteases.

Iodoacetate	
Synonyms	IAA
Mol. wt	208.0 (Sodium salt)
Effective concentration	$10-50$ μM
Stock/solvent	$10-100$ mM in water
Stability of stock solution	Decomposes slowly; prepare fresh
Notes	IAA is not specific for the active site cysteine residue of serine proteases and can inhibit many other proteins and enzymes
E-64	
Synonyms	L-*trans*-Epoxysuccinyl leucylamido (4-guanidino)-butane
Mol. wt	357.4
Effective concentration	10 μM
Stock/solvent	1 mM in aqueous solution
Stability of stock solution	Stable for months at $-20\,^{\circ}$C
Notes	E-64 is an effective irreversible inhibitor of cysteine proteases that does not affect cysteine residues in other enzymes
Chloroacetyl-HO-Leu-Ala-Gly-NH$_2$	
Synonyms	ClCH$_2$CO-hydroxyleucyl-alanyl-glycinamide
Mol. wt	350.8
Effective concentration	1 mM
Stock/solvent	10 mM in water
Stability of stock solution	Unknown—prepare fresh
Notes	An irreversible inhibitor of cysteine proteases but also able to alkylate other classes of protease. Most effective at high pH values (>8.0)
Leupeptin, antipain, chymostatin.	See *Table 8*

(i) *Quality*. The desired end use for the protein will set limits on the acceptable levels of contamination. One hundred per cent purity is usually not practically achievable, removal of the final few percent contamination can take several steps with resultant loss in yield and increase in cost. For functional studies precautions will be necessary to maintain activity of the protein by minimizing proteolysis and denaturation. Information on the susceptibility of the protein to various pH's, ionic strengths, metal ions, etc. will be useful to avoid conditions that may cause loss of activity.

(ii) *Quantity*. The end use of the purified protein will define the quantity required. In turn this will define the scale of the purification and the acceptable yield. In cases where there is a choice of sources the quantity required may limit the choice.

(iii) *Economy*. Economy is of key importance in industrial applications, but should also be considered in the general laboratory. In addition to material costs, labour costs and reliability of the process should be considered.

Table 10. Inhibitors of aspartic and metallo-proteases.

A. *Aspartic proteinases*

Pepstatin

Synonyms	Pepstatin A
Mol. wt	685.9
Effective concentration	1 μM
Stock/solvent	1 mM in methanol
Stability of stock solution	Stable for months at $-20\,°$C
Notes	Pepstatin is a transition state analogue that is a potent inhibitor of cathepsin D, pepsin, renin and many microbial aspartic proteases

H-Val-D-Leu-Pro-Phe-Phe-Val-D-Leu-OH

Synonyms	None
Mol. wt	834.1
Effective concentration	1 μM
Stock/solvent	1 mM in methanol
Stability of stock solution	Unknown
Notes	Effective when coupled to immobilized matrix for affinity chromatography of aspartic proteases

B. *Metallo-proteases*

EDTA

Synonyms	None
Mol. wt	372.24 (disodium salt, dihydrate)
Effective concentration	1 mM
Stock/solvent	0.5 M in water, pH 8.5
Stability of stock solution	Stable for months at 4°C
Notes	EDTA acts as a chelator of the active site zinc ion in metallo-proteases but can also inhibit other metal ion-dependent proteases such as the calcium-dependent cysteine proteases. EDTA may interfere with other metal-dependent biological processes

1,10-Phenanthroline

Synonyms	Ortho-phenanthroline
Mol. wt	198.2
Effective concentration	1 − 10 mM
Stock/solvent	200 mM in methanol
Stability of stock solution	Stable for months at $-20\,°$C
Notes	1,10-Phenanthroline has a strong UV absorbance

Phosphoramidon

Synonyms	None
Mol. wt	543.6
Effective concentration	1 μM
Stock/solvent	1 mM in water
Stability of stock solution	Stable for 1 month at $-20\,°$C
Notes	Phosphoramidon is an inhibitor of few mammalian and many bacterial metallo-endopeptidases

Table 11. Inhibitors of exopeptidases.

Amastatin

Synonyms	[(2S,3R)-3-Amino-2-hydroxy-5-methyl-hexanoyl]-Val-Val-Asp-OH
Mol. wt	474.6
Effective concentration	$1-10 \ \mu M$
Stock/solvent	1 mM in methanol
Stability of stock solution	Unknown
Notes	Amastatin is an inhibitor of aminopeptidases, notable amino-peptidase N

Bestatin

Synonyms	[(2S,3R)-3-Amino-2-hydroxy-4-phenyl-butanoyl]-Leu-OH
Mol. wt	308.4
Effective concentration	$1-10 \ \mu M$
Stock/solvent	1 mM in methanol
Stability of stock solution	Unknown
Notes	Similar to specificity of amastatin

Diprotin A

Synonyms	H-Ile-Pro-Ile-OH
Mol. wt	359.3 (monohydrate)
Effective concentration	$10-50 \ \mu M$
Stock/solvent	1 mM in water, methanol or ethanol
Stability of stock solution	Unknown
Notes	Inhibitor of dipeptidyl aminopeptidase IV

Diprotin B

Synonyms	H-Val-Pro-Leu-OH
Mol. wt	327.4
Effective concentration	$50-100 \ \mu M$
Stock/solvent	$1-10$ mM in water, methanol or ethanol
Stability of stock solution	Unknown
Notes	Similar to diprotin A. Inhibitor of dipeptidyl aminopeptidase IV

Table 12. Protein inhibitors of proteases suitable for coupling to immobilized matrices.

Inhibitor	*Target proteases*
α-2 Macroglobulin	Most endopeptidases
Soybean trypsin inhibitor	Trypsin and chymotrypsin-like serine proteases
Lima bean trypsin inhibitor	Trypsin and chymotrypsin-like serine proteases
Egg-white ovomucoid	Trypsin and chymotrypsin-like serine proteases (specificity depends on source)
α-1 Proteinase inhibitor	Trypsin-like serine proteases
Pancreatic proteinase inhibitor (aprotinin)	Mammalian serine proteases
Egg-white ovostatin	Most metallo-proteinases, including gelatinase and collagenase
Egg-white cystatin	Cysteine proteases but not calpains

6.1 **Preserving activity**

Once released from its native environment an intracellular protein will be subject to many inactivating conditions. Within the cell the pH is maintained at $\sim 6.5-7.5$, the protein concentration is high (~ 100 mg ml^{-1}) and the reducing potential is high. On disruption into buffer the protein concentration is reduced and the protein experiences an oxidizing environment. In addition tissue disruption will lead to the breakdown of compartmentalization. Of particular importance is the disruption of lysosomes, which releases proteases into the environment and causes an acidification of the solution. Judicious choice of buffers must therefore be made to minimize denaturation, inactivation and/or proteolysis of intracellular proteins. Extracellular proteins are not usually as sensitive since they have evolved to maintain activity in the less controlled environment outside the cell.

6.1.1 *Minimizing denaturation*

Extremes of pH, temperature and organic solvents are the principal causes of denaturation. Within the cell most proteins would experience approximately neutral pH, thus buffers of pH $6-8$ should prevent pH denaturation. Most proteins would be stable over a broader pH range, about pH $5-9$ or greater. However, this should be tested for the protein of interest. It is usual to carry out the initial steps of a purification at 4°C to minimize proteolysis. At later stages room temperature (~ 20°C) is acceptable since few proteins are denatured at this temperature. Indeed some proteins are denatured by lower temperatures due to weakening of intramolecular hydrophobic bonds. Clearly high temperatures (usually above 40°C) should be avoided unless the protein is known to be heat-stable.

Other denaturants to avoid are organic solvents and chaotropic agents such as urea and guanidine hydrochloride. Although in some cases, if the protein is resistant to these denaturants, they can be exploited during the purification to denature contaminant proteins. Organic solvents can be used at low temperatures, where denaturation is minimal, to precipitate proteins.

6.1.2 *Minimizing inactivation*

The active site by its very nature is extremely reactive and therefore very susceptible to inactivation. This is particularly evident with enzymes containing a free sulphydryl group in the active site, which may be rapidly oxidized after disruption of the cell. Reducing reagents, such as 2-mercaptoethanol or dithiothreitol should therefore be added to all buffers used for purifying such proteins. 2-Mercaptoethanol should be used at $5-20$ mM and added to the buffer immediately prior to use. Since it oxidizes fairly rapidly its protective action is lost within 24 h. The protective action of dithiothreitol lasts longer and it is therefore more suitable for use in storage buffers (it is also less odorous than 2-mercaptoethanol). $1-5$ mM is usually an adequate concentration, and like 2-mercaptoethanol, dithiothreitol should be added to the buffers immediately prior to use. A stock solution at 1 M can be stored aliquoted at -20°C for 1 month, repeated freeze–thawing should be avoided.

The presence of metal ions can also inactivate sulphydryl groups, therefore a complexing agent, such as EDTA (0.1 — 1.0 mM), should be added to all buffers. However, some enzymes are metal-ion dependent, therefore EDTA should not be included in the buffers used for purifying these enzymes.

Many enzymes are stabilized by including their cofactors or substrates in buffers. With cofactor-requiring enzymes, the enzyme may be removed from its cofactor during purification (e.g. by dialysis or gel permeation chromatography) with a resultant total loss of activity. In most cases activity can be restored by adding the cofactor back; alternatively cofactor should be added to the buffers used.

6.1.3 *Minimizing proteolysis*

Speed and low temperatures (~ 4°C) during the initial stages are particularly important to minimize proteolysis. In addition a cocktail of protease inhibitors (Section 5) should be added to the extraction buffer and other buffers used in the initial steps. Lysosomes are a major source of proteases. Disruption of these organelles will cause acidification of the solution, and release of the proteases. Inclusion of sucrose, or maltose in the extraction buffer will stabilize the lysosomal membrane and minimize disruption. The buffer chosen for extraction should have sufficient buffering capacity to minimize acidification.

During the initial purification steps the majority of protease contaminants are usually removed. However, trace amounts of protease can still cause considerable proteolysis and some proteases can co-purify with the protein of interest; caution should therefore still be exercised even during the later stages. Since many proteases have molecular weights of 20 000 — 30 000, gel permeation chromatography is often used as an initial step in the purification of proteins with molecular weights significantly different from this (this method does however have disadvantages for use as an initial step, see Section 6.2).

6.1.4 *Other precautions*

Dilute protein solutions are often unstable, due to adsorption to surfaces and dissociation of subunits. Some proteins are more susceptible than others to adsorption to surfaces, particularly glass; in these cases it may be necessary to use siliconized containers, or polypropylene containers. Polystyrene containers should not be used for protein solutions, since most proteins will bind irreversibly to polystyrene (this phenomenon is exploited in ELISA assays). Protein concentrations should therefore be maximized, and contact with surfaces minimized. Addition of BSA ($\leq 0.1\%$) to the buffers will minimize loss, but this may well not be a desirable additive during protein purification. Inclusion of low levels ($\leq 0.1\%$) of non-ionic detergents, such as Tween 20 or Triton X-100, may also minimize adsorption losses (care should be taken that these additives do not interfere with purification steps to be used).

Glycerol is often included in buffers to minimize losses of activity; usually 10 — 20% for in-process buffers, and up to 50% for storage. Glycerol acts by reducing the water activity of the solution, thus reducing denaturation. In addition glycerol solutions (>20%) do not freeze at −20°C thus they can be stored at low temperatures with minimal damage to even proteins which cannot normally be frozen. Glycerol may interfere with subsequent purification steps due to its chemical nature and viscosity; caution must therefore be exercised. Alternatives to glycerol are sugars, such as glucose, or sucrose.

Protein samples must often be stored whilst waiting for assay results or overnight prior to the next purification step. Low storage temperatures, 4°C or below, are usually desirable, preferably with addition of bacteriostatic agents and protease inhibitors. If storage for longer than overnight is necessary, or the protein is particularly sensitive to proteolysis, freezing is preferable. However, some enzymes are inactivated by freezing. Addition of glycerol may minimize losses of activity. Rapid freezing in liquid nitrogen or dry-ice and methanol is preferable. During freezing pure water will separate and freeze first, thus concentrating salts and protein. Phosphate buffers should be avoided since the pH drops on freezing due to one salt freezing prior to the other. Storage at −25°C or below is preferable to storage in a domestic freezer which is usually only at −15°C. High concentrations of ammonium sulphate stabilize proteins, therefore many enzymes are best stored as a suspension in ammonium sulphate. For this reason an ammonium sulphate precipitation step is often carried out immediately prior to storage overnight as the centrifuged pellet.

6.2 Planning a purification strategy

The goal of a purification strategy is to obtain a maximum yield of protein with maximal purity, cost-effectively. In order to achieve this a judicious choice of purification steps must be made and these should be ordered to minimize the number of steps involved.

6.2.1 Choice of techniques

Each protein has a unique combination of properties which can be exploited for purification. Thus, by combining a series of steps which exploit several of these properties the protein can be purified from a mixture. *Table 13* lists the most commonly used purification techniques and the properties they exploit. Each technique should be evaluated for its capacity, resolving power, the probable protein yield, and cost (*Table 13*).

(i) *Capacity*. The capacity of the technique is defined as the amount of sample (in terms of volume and protein concentration) which can be handled. A key requirement early in the purification is often to reduce the volume, thus high capacity techniques such as precipitation are often used first. Of the chromatography steps those involving absorption have highest capacity. Gel permeation has a low capacity and is therefore often not appropriate for early stages.

(ii) *Resolution*. The resolution of a technique determines how efficiently it separates proteins from one another. The precipitation steps have a low resolution, whilst the chromatography steps are more highly resolving. Affinity chromatography often shows extremely high resolution and purification factors of > 1000 are frequently achieved (84). Use of high resolution techniques minimizes the number of steps required and becomes increasingly important as the protein becomes purer.

Table 13. Properties of the commonly used purification techniques.

Technique	Property exploited	Capacity	Resolution	Average yield	Cost	Sample composition	Product composition
pH precipitation	Charge	High	Very low	Medium	Low	>1 mg ml^{-1}	Small volume, High concentration
Ammonium sulphate precipitation	Hydrophobicity	High	Very low	High	Low	>1 mg ml^{-1}	High I, Small volume, High concentration
Aqueous two-phase extraction	Mixture, Bioaffinity	High, High	Very low, High	High, Variable	Low, High	Can contain solids	High polymer concentration (e.g. PEG)
Ion-exchange chromatography	Charge	Medium	Medium	Medium	Medium	Low I, Correct pH	High I, Different pH
Hydrophobic interaction chromatography	Hydrophobicity	Medium	Medium	Medium	Medium	High I	Low I, Different pH
Chromatofocusing	Charge/pI	Low	High	Medium	High	Low I	Presence of ampholytes
Dye affinity chromatography	Mixture	Medium	High	Medium	Medium	Low I, Neutral pH	High I, Different pH
Ligand affinity chromatography	Bioactivity	Medium–low	Very high	Low	High	Dependent on ligand	Potentially denaturing conditions
Gel permeation chromatography	Size	Very low	Low	High	Medium	Low volume	Diluted

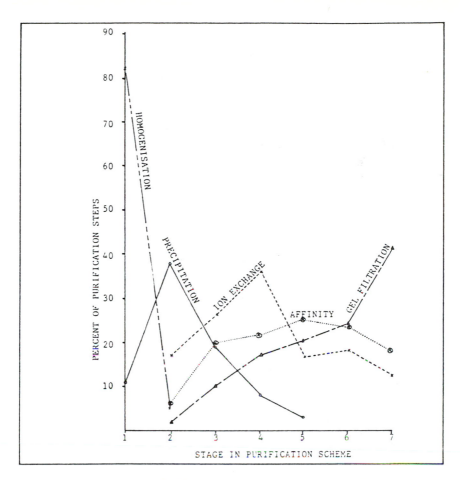

Figure 9. Analysis of the methods of purification used at successive steps in the purification schemes reported in 100 papers published during 1984. (Reproduced with permission.)

(iii) *Protein yield.* Due to the nature of the interactions and the conditions used the techniques will show a range of average yields. Precipitation with ammonium sulphate and aqueous two-phase extraction usually gives yields of more than 80%. Affinity methods often result in lower yields (~ 60%) due to the harsh conditions usually used for elution.

(iv) *Cost.* Affinity techniques are usually expensive, for this reason they are not often used as an initial purification step. A cheaper technique is usually used first to remove many of the contaminants such as particulate matter, lipids and DNA, to avoid damage of expensive affinity columns. If the affinity ligand is a protein, removal of proteases from the mixture prior to application will preserve the life of the column.

These factors must be balanced against one another and what is required at each stage of the purification. Thus in the early stages capacity and low cost are important, whilst at the later stages high resolution is important. The individual techniques are covered in detail in the following chapters.

6.2.2 *Ordering the techniques*

Analysis of several published purification strategies has indicated that homogenization is the first technique used in most strategies (84). The preferred sequence of use is precipitation, followed by ion-exchange, then affinity chromatography with a final gel permeation step (*Figure 9*). This strategy is a logical one. Each chromatography technique exploits a different property of the protein. Precipitation which can handle the large initial volumes and protein concentrations is used first. Ion-exchange is used to remove many contaminants prior to use of the more expensive affinity column. Gel permeation is then used as a final clean-up when its low capacity is not a problem.

Other logical strategies can be devised. In addition to exploiting different properties of the protein, the number of steps can be limited by ensuring that the product from one technique can be applied in the next without further manipulation. Thus, for a sample to bind to ion-exchange the ionic strength must be low and the pH correct. A redissolved precipitate from an ammonium sulphate precipitation may therefore have too high an ionic strength to bind to an ion-exchange matrix. Hydrophobic interaction chromatography would be more suitable since high ionic strength is required for binding. Important requirements for samples and the composition of the product for the techniques are given in *Table 13*.

A classic example of how not to design a purification strategy is cited in ref. 85. The process included ammonium sulphate precipitation, metal chelate chromatography, hydrophobic interaction chromatography, gel permeation, ion-exchange and chromatofocusing. In addition several dialysis steps were required to make the sample composition suitable for the next step! A more efficient strategy would be to apply the redissolved ammonium sulphate precipitate onto the hydrophobic interaction column. The gel permeation column should then follow to separate high and low molecular weight components and also exchange into the buffer prior to chromatofocusing. Ion-exchange chromatography is redundant since it exploits the same property as chromatofocusing (i.e. charge). If necessary to achieve the required level of purity the metal chelate column could be inserted after the hydrophobic interaction column.

A universal strategy for all proteins cannot be given since the materials available and the requirements for each application will differ. Some guidelines and suggestions are given in *Figure 10*. Typical purification strategies for a variety of proteins are described in the companion to this volume *Protein Purification Applications: A Practical Approach*. Computer software is available for teaching the design of purification strategies without the risk of losing one's protein (86,87). However, there is no substitute for trial-and-error in learning the necessary understanding. Providing sufficient supply is available each appropriate technique can be tested for its suitability. Assays for yield and fold purification will indicate whether the yield is acceptable and whether one technique is more effective than another. Analysis by gel electrophoresis will also indicate how pure the protein is and how many more contaminants are still present. In addition the molecular weight of the contaminants can be determined if SDS−PAGE is used, therefore indicating whether gel permeation chromatography may remove the contaminants. In this way a choice can be made between alternative techniques and the next step can then be evaluated. From analysis of 100 published purification processes the following statistics were calculated (85). On average four purification steps are

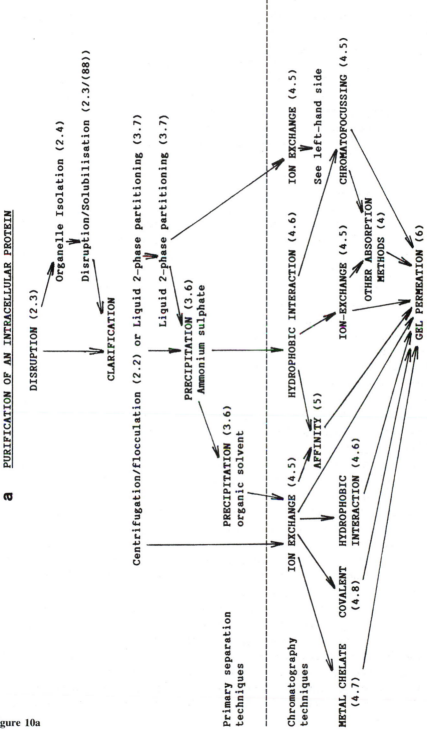

Figure 10a

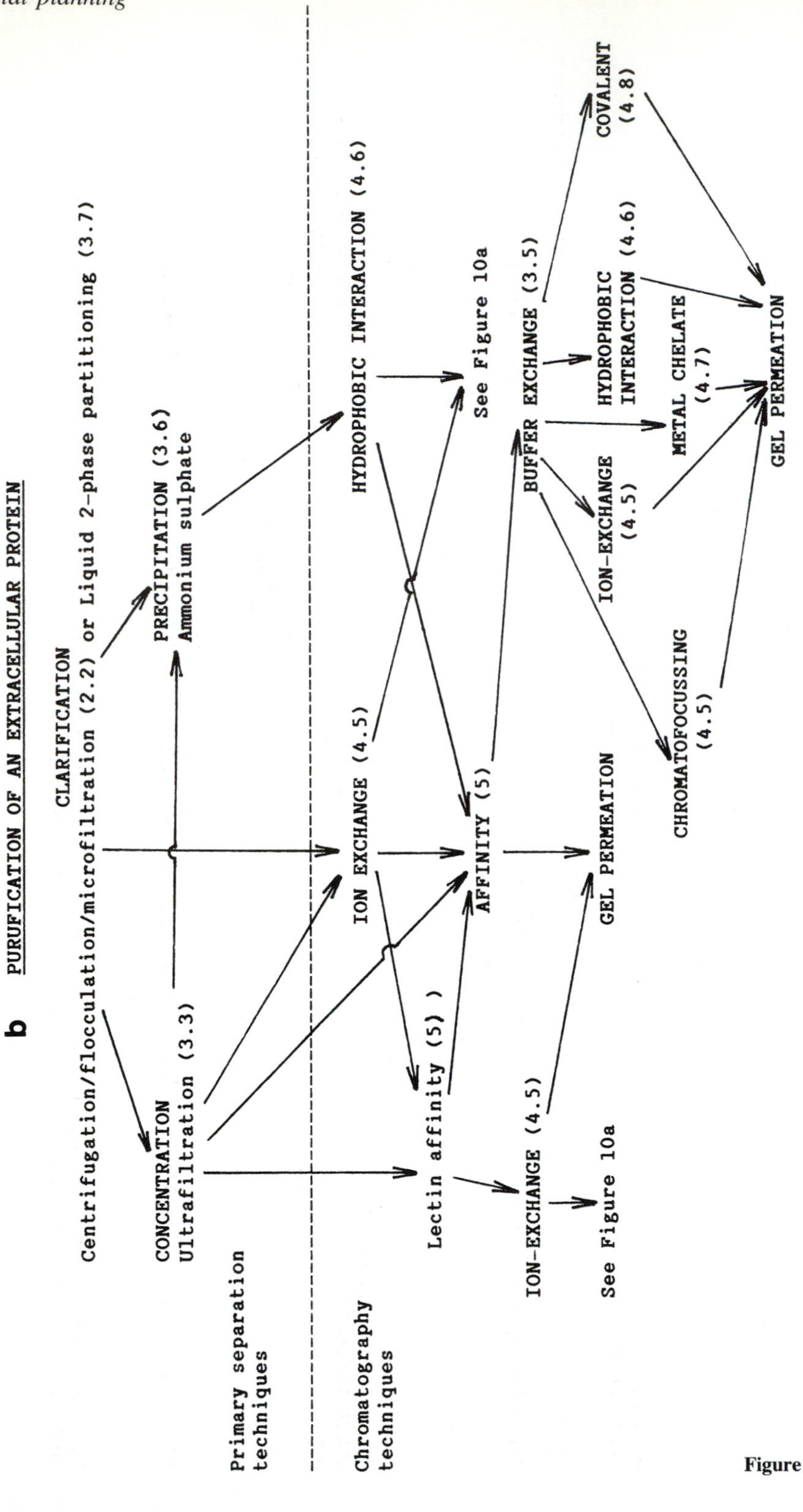
b
PURUFICATION OF AN EXTRACELLULAR PROTEIN
CLARIFICATION
Centrifugation/flocculation/microfiltration (2.2) or Liquid 2-phase partitioning (3.7)
CONCENTRATION
Ultrafiltration (3.3)
PRECIPITATION (3.6)
Ammonium sulphate
Primary separation techniques
Chromatography techniques
Lectin affinity (5))
ION EXCHANGE (4.5)
HYDROPHOBIC INTERACTION (4.6)
ION-EXCHANGE (4.5)
See Figure 10a
AFFINITY (5)
See Figure 10a
GEL PERMEATION
BUFFER EXCHANGE (3.5)
CHROMATOFOCUSSING (4.5)
ION-EXCHANGE (4.5)
HYDROPHOBIC INTERACTION (4.6)
METAL CHELATE (4.7)
COVALENT (4.8)
GEL PERMEATION

Figure 10b

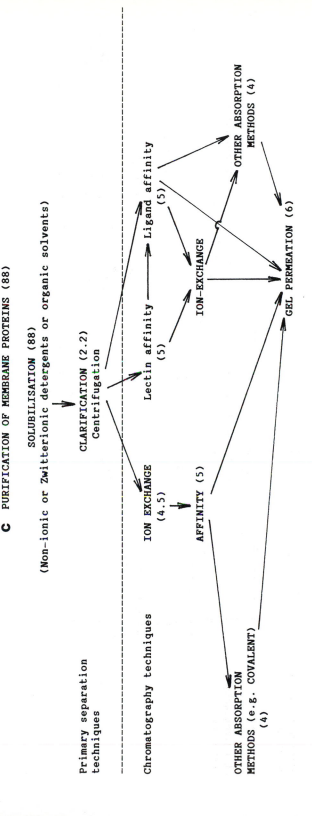

C PURIFICATION OF MEMBRANE PROTEINS (88)

SOLUBILISATION (88)

(Non-ionic or Zwitterionic detergents or organic solvents)

Primary separation techniques

CLARIFICATION (2.2)
Centrifugation

Chromatography techniques

ION EXCHANGE (4.5)

AFFINITY (5)

Lectin affinity (5)

Ligand affinity (5)

ION-EXCHANGE

OTHER ABSORPTION METHODS (4)

GEL PERMEATION (6)

OTHER ABSORPTION METHODS (e.g. COVALENT) (4)

(*N.B.* HYDROPHOBIC INTERACTION has limited use due to presence of detergents or organic solvents.)

Figure 10. Suggested purification strategies: figures in brackets refer to chapter and section numbers where information on the technique can be found. (**a**) for an intracellular protein; (**b**) for an extracellular protein; (**c**) for membrane proteins.

required to purify to homogeneity from an initial average purity of 0.016%. The overall yield was 28%, corresponding to an average yield per step of 74%. The overall fold purification achieved was 6380, corresponding to an average fold purification per step of 8.

6.3 Troubleshooting

The usual problem encountered is a sudden loss in activity. This may be due to:

(i)   failure of the assay;

(ii)   incorrect buffer composition, causing denaturation and/or inactivation;

(iii)   impure reagents, causing denaturation and/or inactivation;

(iv)   proteolysis—this can be identified as the problem either by assaying for proteases (Section 5) or by monitoring the purification by SDS−PAGE and checking for the appearance of small molecular weight bands. Proteases may co-purify even to the later stages of the purification of a protein. If this is a problem then a cocktail of inibitors can be added or the type of protease can be identified and a specific inhibitor added (Section 5);

(v)   precipitation—a precipitate may be evident in a solution, but if precipitation occurred within the column this will not be so evident. Precipitation may be caused by use of the wrong buffer, for example a low pH buffer, or a low ionic strength buffer. The protein within these precipitates may well be denatured and therefore will not be easily recovered;

(vi)   presence of an inhibitor—if an inhibitor of the enzyme or the assay is present in the buffers this will cause an apparent sudden loss in activity;

(vii)   loss of activator or cofactor—if the loss of activity cannot be explained by any of the above then this may be the cause. Cofactors and activators can become separated during the purification, often following dialysis or gel filtration. Activity is usually regained by recombining the fractions or adding a prepared solution containing the cofactor or activator.

In some instances an increase in yield is observed. This is due to removal of an inhibitor of the enzyme or the assay.

7. ACKNOWLEDGEMENTS

Financial support of the Muscular Dystrophy Group of Great Britain and Northern Ireland is acknowledged by M.J.Dunn.

R.J.Beynon is grateful for the many conversations with colleagues which have led to the ideas expressed in Section 5. Thanks are also due to Maggie Cusack for supplying details of the iodination of insulin B-chain. Supported by grants from the Medical Research Council and the Science and Engineering Research Council.

8. REFERENCES

1. Brewer,S.J. and Sassenfeld,H.M. (1989) In *Protein Purification Applications: A Practical Approach.* Harris,E.L.V. and Angal,S. (eds), IRL Press, Oxford.

2. Uhlen,M., Moks,T. and Abrahamsen,L. (1988) *Biochem. Soc. Trans.,* **16**, 111.

3. Naito,H.K. (1982) In *CRC Handbook of Clinical Chemistry.* Werner,M. (ed.), CRC Press, Florida, Vol. 1, p. 109.

4. Gomori,G. (1955) In *Methods in Enzymology.* Colowick,S.P. and Kaplan,N.O. (eds), Academic Press, New York and London, Vol. 1, p. 138.

5. McKenzie,H.A. (1986) In *Data for Biochemical Research*, 3rd edition. Dawson,R.M.C. *et al.* (eds), Clarendon Press, Oxford, p. 426.
6. Good,N.E., Winget,G.D., Winter,W., Connolly,T.N., Izawa,S. and Singh,R.M.M. (1966) *Biochemistry*, **5**, 467.
7. Good,N.E. and Izawa,S. (1972) In *Methods in Enzymology*. San Pietro,A. (ed.), Academic Press, New York and London, Vol. 24, p. 53.
8. Gaspar,L. (1980) In *Methods of Protein Analysis*. Kerese,I. (ed.), Ellis Horwood, Chichester, p. 30.
9. Warburg,O. and Christian,W. (1941) *Biochem. Z.*, **310**, 384.
10. Scopes,R.K. (1974) *Anal. Biochem.*, **59**, 277.
11. Gornall,A.J. and Bardawill,C.S. and David,M.M. (1949) *J. Biol. Chem.*, **177**, 751.
12. Goldberg,M.L. (1973) *Anal. Biochem.*, **51**, 240.
13. Itzhaki,R.F. and Gill,D.M. (1964) *Anal. Biochem.*, **9**, 401.
14. Lowry,O.H., Rosenbrough,N.J., Farr,A.L. and Randall,R.J. (1951) *J. Biol. Chem.*, **193**, 265.
15. Peterson,G.L. (1979) *Anal. Biochem.*, **100**, 201.
16. Smith,P.K., Krohn,R.I., Hermanson,G.T., Mallia,A.K., Gartner,F.H., Provenzano,M.D., Fujimoto,E.K., Goeke,B.J., Olson,B.J. and Klenk,D.C. (1985) *Anal. Biochem.*, **150**, 76.
17. Bradford,M.B. (1976) *Anal. Biochem.*, **72**, 248.
18. Spector,T. (1978) *Anal. Biochem.*, **86**, 142.
19. Sedmak,J.J. and Grossberg,S.E. (1977) *Anal. Biochem.*, **79**, 544.
20. Compton,S.J. and Jones,C.G. (1985) *Anal. Biochem.*, **151**, 369.
21. Krystal,G., Macdonald,C., Munt,B. and Ashwell,S. (1985) *Anal. Biochem.*, **148**, 451.
22. Hames,B.D. and Rickwood,D. (eds) (1989) *Gel Electrophoresis of Proteins: A Practical Approach.* 2nd Edn. IRL Press, Oxford, in press.
23. Righetti,P.G. (1983) *Isoelectric Focusing: Theory, Methodology and Applications*. Elsevier, Amsterdam.
24. Allen,R.C., Saravis,C.A. and Maurer,H.R. (1984) *Gel Electrophoresis and Isoelectric Focusing of Proteins: Selected Techniques*. de Gruyter, Berlin.
25. Dunn,M.J. (ed.) (1986) *Gel Electrophoresis of Proteins*. Wright, Bristol.
26. Andrews,A.T. (1986) *Electrophoresis: Theory, Techniques, and Biochemical and Clinical Applications.* Clarendon, Oxford.
27. Margolis,J. and Kenrick,K.G. (1969) *Nature*, **221**, 1056.
28. Rothe,G.M. and Maurer,W.D. (1986) In *Gel Electrophoresis of Proteins*. Dunn,M.J. (ed.), Wright, Bristol, p. 37.
29. Anderson,N.L. and Anderson,N.G. (1978) *Anal. Biochem.*, **85**, 341.
30. Altland,K. and Altland,A. (1984) *Electrophoresis*, **5**, 143.
31. Reynolds,J.A. and Tanford,C. (1970) *J. Biol. Chem.*, **243**, 5161.
32. Swank,R.T. and Munkres,K.D. (1971) *Anal. Biochem.*, **39**, 462.
33. Anderson,B.L., Berry,R.W. and Teber,A. (1983) *Anal. Biochem.*, **132**, 362.
34. Laemmli,U.K. (1970) *Nature*, **227**, 680.
35. Ornstein,L. (1964) *Ann. N.Y. Acad. Sci.*, **121**, 321.
36. Davis,B.J. (1964) *Ann. N.Y. Acad. Sci.*, **121**, 404.
37. Wyckoff,M., Rodbard,D. and Chrambach,A. (1977) *Anal. Biochem.*, **78**, 459.
38. Booth,A.G. (1977) *Biochem. J.*, **163**, 165.
39. Poduslo,J.F. and Rodbard,D. (1980) *Anal. Biochem.*, **101**, 394.
40. Panyim,S. and Chalkley,R. (1969) *Arch. Biochem. Biophys.*, **130**, 337.
41. Franklin,S.G. and Zweidler,A. (1977) *Nature*, **266**, 273.
42. Bonner,W.M., West,M.P. and Stedman,J.D. (1980) *Eur. J. Biochem.*, **109**, 17.
43. Zweidler,A. (1978) *Methods Cell Biol.*, **17**, 223.
44. Jovin,T.M. (1973) *Biochemistry*, **12**, 871.
45. Chrambach,A. and Jovin,T.M. (1983) *Electrophoresis*, **4**, 190.
46. Righetti,P.G., Gelfi,C. and Gianazza,E. (1986) In *Gel Electrophoresis of Proteins*. Dunn,M.J. (ed.), Wright, Bristol, p. 141.
47. Burghes,A.H.M., Dunn,M.J. and Dubowitz,V. (1982) *Electrophoresis*, **3**, 354.
48. Alltand,K. and Rossman,U. (1985) *Electrophoresis*, **6**, 314.
49. Righetti,P.G., Gelfi,C., Bossi,M.L. and Boschetti,E. (1987) *Electrophoresis*, **8**, 62.
50. Altland,K., von Eckardstein,A., Banzhoff,A., Wagner,M., Rossman,U., Hackler,R. and Beder,P. (1987) *Electrophoresis*, **8**, 52.
51. O'Farrell,P.H. (1975) *J. Biol. Chem.*, **250**, 4007.
52. Dunn,M.J. (1989) In *Advances in Electrophoresis*. Radola,B.J., Dunn,M.J. and Chrambach,A. (eds), VCH Verlagsgesellschaft, Weinheim, Vol. 1, in press.
53. Neuhoff,V., Stamm,R. and Eibl,H. (1985) *Electrophoresis*, **6**, 427.
54. Merril,C.R., Harasewych,M.G. and Harrington,M.G. (1986) In *Gel Electropheresis of Proteins.* Dunn,M.J. (ed.), Wright, Bristol, p. 323.

55. Harris,H. and Hopkins,D.A. (1976) *Handbook of Enzyme Electrophoresis in Human Genetics*. North Holland, Amsterdam.
56. Towbin,H., Staehelin,T. and Gordon,J. (1979) *Proc. Natl. Acad. Sci. USA*, **76**, 4350.
57. Gershoni,J.M. (1989) In *Advances in Electrophoresis*. Radola,B.J., Dunn,M.J. and Chrambach,A. (eds), VCH Verlagsgesellschaft, Weinheim, Vol. 1, in press.
58. Beisiegel,U. (1986) *Electrophoresis*, **7**, 1.
59. Khyse-Andersen,J. (1984) *J. Biochem. Biophys. Methods*, **10**, 203.
60. Bjerrum,O.J. and Schafer-Nielsen,C. (1986) In *Electrophoresis '86*. Dunn,M.J. (ed.), VCH, Verlagsgesellschaft, Weinheim, p. 315.
61. Hancock,K. and Tsang,V.C.W. (1983) *Anal. Biochem.*, **133**, 157.
62. Moermans,M., Daneels,G. and De Mey,J. (1985) *Anal. Biochem.*, **145**, 315.
63. Moermans,M., De Raeymaeker,B., Daneels,G. and De May,J. (1986) *Anal. Biochem.*, **153**, 18.
64. Batteiger,B., Newhall,W.J. and Jones,R.B. (1982) *J. Immunol. Methods*, **55**, 297.
65. Burghes,A.H.M., Dunn,M.J., Statham,H.E. and Dubowitz,V. (1982) *Electrophoresis*, **3**, 177.
66. Bjellqvist,B., Ek,K., Righetti,P.G., Gianazza,E., Gorg,A., Postel,W. and Westermeier,R. (1982) *J. Biochem. Biophys. Methods*, **6**, 317.
67. Bond,J.S. and Beynon,R.J. (1987) *Mol. Asp. Med.*, **9**, 173.
68. Cohen,G.N. and Holzer,H. (1979) In *Limited Proteolysis in Microorganisms*. Cohen,G.N. and Holzer,H. (eds), US Dept Health, Education and Welfare, p. 79.
69. Pringle,J.R. (1979) In *Limited Proteolysis in Microorganisms*. Cohen,G.N. and Holzer,H. (eds), US Dept Health, Education and Welfare, p. 191.
70. Pringle,J.R. (1975) In *Methods in Cell Biology*. Prescott,D.M. (ed.), Academic Press, New York, 12, p. 149.
71. Beynon,R.J. (1987) In *Methods in Molecular Biology*. Walker,J. (ed.), Humana Press, Clifton, New Jersey, 3, p. 1.
72. Khairallah,E.A., Bird,J.W.C. and Bond,J.S. (1984) In *Intracellular Protein Catabolism*. Alan R.Liss, New York.
73. Beynon,R.J., Place,G.A. and Butler,P.E. (1985) *Biochem. Soc. Trans.*, **13**, 306.
74. Citri,N. (1973) *Advances in Enzymology*, **37**, 397.
75. Barrett,A.J. and McDonald,J.K. (1980) In *Mammalian Proteases: A Glossary and Bibliography. Endopeptidases*. Academic Press, London, 1.
76. Barrett,A.J. and Salvesen,G.S. (1986) In *Proteinase Inhibitors*. Elsevier/North Holland Biomedical Press, Amsterdam.
77. Wagner,F.W. (1986) In *Plant Proteolytic Enzymes*. Dallng,M.J. (ed.), CRC Press, Boca Raton, Florida, p. 17.
78. Beynon,R.J., Shannon,J.D. and Bond,J.S. (1981) *Biochem. J.*, **199**, 591.
79. Fulcher,I.S. and Kenney,J. (1983) *Biochem.J.*, **211**, 743.
80. Barrett,A.J. (1986) In *Plant Proteolytic Enzymes*. Dallng,M.J. (ed.), CRC Press, Boca Raton, Florida, p. 1.
81. McDonald,J.K. (1985) *Histochem. J.*, **17**, 773.
82. McDonald,J.K. and Barrett,A.J. (1986) In *Mammalian Proteases: A Glossary and Bibliography. Exopeptidases*. Academic Press, London, 2.
83. Brown,C.P. and Beynon,R.J. (1983) *Biosci. Rep.*, **3**, 179.
84. Bonnerjea,J., Oh,S., Hoare,M. and Dunnill,P. (1986) *Bio/Technology*, **4**, 954.
85. Sofer,G. and Mason,C. (1987) *Bio/Technology*, **5**, 239.
86. Booth,A.G. (1987) *Protein Purification: A Strategic Approach*. Hames,B.D. (ed.), IRL Press, Oxford.
87. Pharmacia-LKB (1988) *The Protein Purifier*.
88. Findlay,J. (1989) In *Protein Purification Applications: A Practical Approach*. Harris,E.L.V. and Angal,S. (eds), IRL Press, Oxford.

CHAPTER 2

Clarification and extraction

1. INTRODUCTION—E.L.V.Harris

The first step in the purification of a protein is the preparation of an extract containing the protein in a soluble form. How this extract is prepared will depend on the cellular location of the protein, that is, is it extra- or intracellular, or is it contained within a cellular organelle? In addition, if the protein is insoluble, for example if it is membrane-bound or present in an inclusion body, a method to solubilize the protein will be required.

For an extracellular protein, such as one secreted into fermentation medium or present in serum or urine, cells and any other particles must be removed to prevent interference in subsequent purification steps. To achieve this a clarification step, such as centrifugation or microfiltration, is required (see Section 2). If the extract is to be loaded onto a column it is important that all particulate matter is first removed to prevent fouling and blocking of the column. However, if the next step is a precipitation step it may be possible to ignore a low level of particulate matter, which will often aggregate during the precipitation step and can then be discarded. Care should be taken during the clarification step to minimize cell lysis which will cause unnecessary contamination.

For an intracellular protein the cells must be disrupted by one of the methods described in Section 3. Pre-treatment may be necessary. For example, tissues may require prior removal of fat, or mincing into smaller pieces. Cells grown in fermenters will need concentrating by clarification to enable efficient cell disruption. Several potential problems may be consequent on disruption, due to the destruction of intracellular compartmentalization. Proteases present in organelles, such as lysosomes, may be released upon disruption. It is advisable therefore to include a cocktail of protease inhibitors in the extraction buffer (see Chapter 1) and work as quickly as possible, ideally with buffers and equipment pre-chilled to 4°C. Fat particles may be formed, which will float to the surface during centrifugation; these can be removed by coarse filtration through glass wool or a fine mesh cloth. DNA released by disruption will increase the viscosity of the extract; this can be precipitated with a polycationic macromolecule, such as protamine sulphate, or degraded by addition of nucleases. Mechanical cell disruption may cause local overheating with consequent denaturation of protein. The extract and equipment should therefore be pre-chilled and several short bursts of disruption used instead of one long burst. Short bursts will also minimize foaming and shearing, thereby minimizing denaturation. After disruption the protein will encounter an oxidizing environment which may cause inactivation, denaturation or aggregation. Addition of 2-mercaptoethanol (5−20 mM) and/or EDTA (1−5 mM) to chelate metal ions will minimize oxidation.

The choice of buffer in which the cells are disrupted is influenced by the stability and solubility of the protein of interest and by the requirements of the subsequent purification steps. Water may be used but the resultant ionic strength and pH may be low. Many proteins are insoluble at low ionic strength and/or low pH, or may be inactivated by these conditions. A buffer approximating to physiological conditions (i.e. ~ pH 7, 0.15 M sodium chloride) will allow efficient extraction of most proteins. Typical examples are 20−50 mM sodium phosphate, pH 7−7.5 or 0.1 M Tris−HCl, pH 7.5 with 0−0.15 M NaCl. Inclusion of cofactors (e.g. pyridoxal phosphate or metal ions) or substrates may help to stabilize enzymes. For isolating organelles an iso-osmotic buffer, containing for example sucrose or sorbitol, is required.

After disruption, the extract is clarified by one of the methods described in Section 2. The majority of the protein of interest should be in the liquid phase, but some will remain in the residue due to the trapped liquid. To minimize the percentage lost in the residue a larger volume of buffer should be used during disruption, or the residue may be re-extracted by resuspension in buffer and reclarification. For most animal tissues the volume of buffer used for disruption should be 2−2.5 times the volume of the tissue. A larger volume would result in a more dilute extract which may give rise to problems during subsequent purification steps. In contrast plant cells release much liquid from vacuoles and intercellular spaces and therefore a smaller volume of buffer can be used for extraction.

If the protein of interest is located within a cellular compartment a considerable degree of purification can be achieved by isolating the compartment before releasing the protein. For example, some proteins are located in the periplasmic space of *Escherichia coli* [e.g. recombinant human growth hormone (1), IgG light chain (2)] and enzymic or osmotic shock treatment can be used to release these proteins, leaving the inner cell wall intact (3,4). For those proteins located in organelles, the organelle may be isolated and the proteins then released by treatment with detergents (see below) or by disruption (e.g. osmotic shock or ultrasonication). However, yields of organelles are often low and therefore the advantage of a purer extract should be balanced against a lower final yield. An additional advantage may be minimization of problems associated with the disruption of intracellular compartmentalization (e.g. release of proteases from lysozymes). Isolation methods for some of the types of organelles most frequently isolated (nuclei, microsomes, mitochondria and chloroplasts) are given in Section 4 (see also refs 5−8).

Some proteins will remain insoluble or associated with the particulate matter under normal extraction conditions. These fall into three main categories: membrane-bound proteins, extracellular matrix proteins and proteins in inclusion bodies.

Membrane-bound proteins are associated with the lipid bilayer of the cell membranes (e.g. plasma membrane, endoplasmic reticulum and nuclear membrane). They may be integrated into the hydrophobic phase (i.e. intrinsic membrane proteins), or associated with the membrane through interactions with other proteins or exposed regions of phospholipid (i.e. extrinsic membrane proteins). To extract intrinsic membrane proteins the phospholipid bilayer must be disrupted with detergent or organic solvent; alternatively the extracellular portion of the protein may be released by proteolytic cleavage from its membrane anchor. Intrinsic membrane proteins will often require their hydrophobic domains 'shielding' with detergent, to maintain solubility throughout their purification. Thus, purification of membrane proteins is often not as straightforward as that of soluble proteins. For example, ammonium sulphate precipitation can result in the detergent, Triton X-100, together with the proteins, floating to the surface. The purification of membrane proteins is therefore dealt with in a separate chapter in the companion to this volume, *Protein Purification Applications: A Practical Approach* (9). Extrinsic membrane-bound proteins can often be released in a soluble form by altering the ionic strength of a buffer or in more resilient cases by partial denaturation (9).

Many extracellular matrix proteins (e.g. elastin, collagen and keratin) remain insoluble due to numerous inter-chain cross-links. These inter-chain bonds may be covalent, for example, those between lysine residues in elastin and collagen, or non-covalent as found in keratins. In order to solubilize these proteins, the protein chains must be broken either by proteolysis or chemical hydrolysis. For example, collagen may be boiled in water to give gelatin, a mixture of soluble polypeptides; alternatively, treatment with pepsin will yield a soluble collagen preparation (10).

The third type of insoluble proteins is those found in inclusion bodies. These are frequently formed by proteins which are expressed in *E.coli* by genetic engineering, particularly if the protein is not normally expressed in *E.coli* or a closely related bacteria. Thus, this problem is frequently encountered when a protein from a eukaryote is expressed in *E.coli*, [e.g. recombinant bovine growth hormone (11), and prochymosin (12)]. To solubilize these proteins they must first be denatured and then allowed to slowly refold. A frequently used regime is denaturation by heating in 6 M guanidine hydrochloride, pH 8.0, in the presence of a reducing agent. The denatured protein is then rapidly diluted or dialysed against an appropriate buffer (e.g. 50 mM Tris−HCl, pH 8.0) at an optimum redox potential (obtained by use of glutathione in both oxidized and reduced forms). Frequently these inclusion bodies consist of relatively pure protein, thus by isolating them from the cytoplasm prior to denaturation a considerable degree of purification is achieved. Purification of proteins in inclusion bodies is covered briefly in *Protein Purification Applications: A Practical Approach* (13), and in more detail in references 3 and 4.

2. CLARIFICATION—P.N.Whittington

Clarification is used at several stages during purification of a protein to remove particulate matter (e.g. cells, organelles, debris or precipitated macromolecules) from the surrounding fluid (e.g. fermentation medium or buffer). The protein of interest may be present either in the particulate matter (e.g. an intracellular protein will be located within the cells) or in the surrounding fluid (e.g. a secreted protein).

The two most commonly used techniques for clarification are centrifugation, possibly following a flocculation step, and microfiltration. They may be used in conjunction to ensure all particles are removed from suspension. For instance, even if a centrifuge step is 99% efficient 10^7 cells ml^{-1} will remain in suspension from an original concentration of 10^9 ml^{-1}; microfiltration would then recover the remaining cells. Alternative techniques and the types of equipment used in clarification are shown in *Table 1*. Several of these are suitable for large-scale processes (e.g. disc-stack centrifuge, rotary vacuum drum). Some of those listed are still at early stages of their development.

Table 1. Separation techniques.

Physical property exploited	Density			Particle size	Surface activity hydrophobicity
Process	Sedimentation		Filtration	Flotation	2-phase aqueous extraction
	Gravity	Centrifugation			
Equipment types	Settling tank	Disc-stack Multi-chamber Bucket/tube Decanter Cyclones	Microfilter Ultrafilters Rotary vacuum Drum Filter Plate and frame	Dissolved gas Dispersed gas Chemical Electrolytic Biological	Mixer Settlers Counter-current

Table 2. Features of biological cells.

Cell type	Shape	Size (μm)	Density (kg m^3)	Relative shear resistance	Typical separation process
Bacteria	rods cocci	0.5−3	1050−1080 1050−1090	high fairly high	Centrifugation; cross-flow filtration.
Yeasts	spheres	5−10	1050−1090		Dead end filtration; centrifugation.
Filamentous fungi Actinomycetes	long filaments	1 × 100's	1050−1090	medium	Rotary vacuum filtration.
Plant	irregular	20−100	1050−1090	low	Microfiltration; low speed centrifugation; flotation.
Animal	non-defined	10−40		very low	
Cell flocs.	assorted	10−100's	1010−1080	variable	Centrifugation; gravity settling.
Cell debris	assorted	0.4	1010−1200	low	Centrifugation; Microfiltration; 2-phase aqueous partitioning
Protein precipitates	assorted	0.1−100's	1010−1200	medium	Centrifugation; Microfiltration; Ultrafiltration.

The size and density of the particles determines the ease of recovering/removing them and which technique is most suitable. In addition the resistance to shear must also be taken into account. These properties are shown in *Table 2* for a variety of particles. Bacterial cells and cell debris are the most difficult to recover by either centrifugation or microfiltration due to their small size, low density, compressibility and cohesivity. Flocculation can be used to improve the ease of recovery of these particles. Conversely, mammalian cells are large, but fragile, and therefore to recover them intact gentle methods are required.

2.1 Centrifugation

Centrifugal sedimentation is the application of radial acceleration to a particle suspension by rotational motion. Particles denser than their suspending medium will move rapidly to the perimeter of the centrifuge. Conversely, particles or liquid lighter than the bulk fluid will move rapidly inwards. The Brownian diffusion forces, which hinder or prevent the settling of very fine particles under gravity, are overcome in a centrifugal field. The theory and practice of centrifugation is discussed below and also in ref. 5.

The sedimentation velocity, V_g, of a particle in a less dense fluid under the influence of gravity is given by Stokes' law:

$$V_g = D^2 \frac{\varrho_1 - \varrho_2}{18 \, \mu} g \qquad (1)$$

where D = particle diameter, ϱ_1 = density of particle, ϱ_2 = density of liquid, μ = fluid viscosity, and g = gravitational acceleration.

In centrifugation, the acceleration term g is replaced by $\omega^2 r$, where r is the distance of the particle from the centre of rotation and ω is the angular velocity of the centrifuge (radians $\sec^{-1}$):

$$\omega = \frac{\pi n}{30}$$

where n is the rotational velocity of the rotor in revolutions per min.

For yeast cells in water where:

$$D = 8 \, \mu m = 8 \times 10^{-6} \text{ m};$$
$$\varrho_1 - \varrho_2 = 50 \text{ kg m}^{-3}$$
$$\mu = 10^{-3} \text{ kg m}^{-1} \sec^{-1},$$

the sedimentaton rate can be calculated using equation 1. Thus, $V_g = 1.75 \times 10^{-6}$ m $\sec^{-1}$. In a centrifuge with radius 0.2 m rotating at 5000 r.p.m., the centrifugal force increases the sedimentation rate to V_s where:

$$V_s = V_g \frac{r\omega^2}{g}$$

thus $V_s = 9.8 \times 10^{-3}$ m $\sec^{-1}$.

From the above equation it can be seen that the sedimentation velocity of a particle can be increased by:

(i)   increasing particle diameter (D);

(ii) increasing the difference between the densities of the particle and the liquid, $(\varrho_1 - \varrho_2)$;

(iii) increasing the speed of the centrifuge, (ω);

(iv) decreasing the viscosity of the suspending fluid (μ).

In addition, increasing the time the particle is exposed to the centrifugal force will cause the particle to move further from the centre of rotation, since:

$$V_s \, t = \text{distance settled.}$$

With tube centrifuges this is increased by increasing the centrifugation time, whilst for continuous flow centrifuges (e.g. disc-stack) the residence time of the particle must be increased.

For ease of comparison between centrifuge types and as a standardized convention, centrifugal force is normally quoted in terms of 'g' units (i.e. relative to gravity). Usually g_{av} is quoted, which is the centrifugal force taken at the mid-point radius of the tube or bowl. Occasionally g_{max} will be quoted, which is the centrifugal force at the outer radius of the tube, or the periphery of the bowl.

2.1.1 *Equipment*

Various types of centrifuge are currently available. At the laboratory scale (up to a few litres) the tube centrifuges are used. For the larger volumes handled at pilot or plant scale continuous flow centrifuges (e.g. disc-stack) are used, where the suspension to be separated is continuously pumped in and the clarified solution is continuously removed. The highest centrifugal forces are achieved in tube centrifuges, up to 400 000 g (c.f. $8000 - 12\ 000$ g for a disc-stack centrifuge).

Centrifuges are best located in a separate room, since they do produce a certain amount of noise. Exceptions are the small bench top tube centrifuges which can be quiet and are usually only used for short periods of time. For some applications contained operation is advantageous (e.g. handling pathogenic bacteria) and many of the tube centrifuges and continuous flow centrifuges are designed for such applications. The most serious thing that can go wrong with a centrifuge is an imbalance of the rotor. In a tube centrifuge, this may be caused by a tube cracking during a run or by incorrect loading in the first place. However, most modern centrifuges are equipped with sensors to detect imbalance and will cut out the power to the rotor. In addition all rotor chambers are heavily armoured, thus preventing the escape of the rotor if it leaves the spindle.

Centrifuges and their rotors should be frequently (ideally after each use) cleaned with water and wiped dry. This is particularly important if there has been a spillage, or if a solution with a high salt concentration (e.g. ammonium sulphate precipitations) has been used, since salt is extremely corrosive to metals, particularly aluminium. Proprietary protective solutions are available from rotor manufacturers which help minimize corrosion. A corroded rotor is dangerous and must be discarded.

(i) *Tube centrifuges.* These have a relatively simple design consisting of a motor driven rotating spindle to which the tube holding rotor is coupled. Manufacturers include Beckman, Sorvall-DuPont, MSE-Fisons and Leybold-Heraeus. Centrifuges range from small bench top centrifuges with low maximum speeds to large floor standing ultracentrifuges with high maximum speeds. Most centrifuges are fitted with a timer and a 'hold' facility for longer centrifuge runs. Many centrifuges are refrigerated, often a desirable property for protein and cell separations, and some are programmable allowing highly controlled manipulations of acceleration and deceleration rates.

Several rotor types are available, differing in their mode of holding the tubes: 'swing-out', which hold the tube horizontally whilst rotating, and vertically when stationary; fixed angle; and, less commonly, vertically held. The material that the rotor is made of will depend on the maximum g force at which it is used. For low-speed use (≤ 6000 g) rotors are usually made of steel, aluminium or bronze; at higher speeds aluminium or titanium is used, whilst the fastest rotors ($< 600\ 000$ g) are made of titanium. Centrifuge tubes are usually made from glass, polypropylene, polycarbonate and other plastics. Only glass tubes are resistant to all solvents, sterilants and disinfectants that might be used. Tubes of ordinary glass (Pyrex) will only withstand $3-4000$ g, for forces up to 15 000 g special toughened glass (Corex) is used (available from DuPont). The manufacturers' recommendations for maximum g forces and resistance to chemicals and heat sterilization should be checked for all plastic tubes. Polypropylene tubes are more chemically resistant and withstand higher g forces than polycarbonate. With plastic tubes, the following precautions should be taken: the shape of the tube should be the same as the hole in the rotor, thus for conical bottomed tubes a conical shaped adaptor should be used; at the very high g forces used in ultracentrifuges it is important that the tube is completely filled with liquid. Plastic tubes have the advantage of being cheaper, and thus disposable, compared to Corex glass tubes.

Rotors must always be balanced carefully to avoid expensive damage to the centrifuge and/or the rotor itself. Thus, tubes of equal weight must be placed opposite one another, although in rotors with six (or multiples of six) tube holes, three tubes can be equally spaced apart to balance the rotor. For small volume tubes, the volume in each tube can be adjusted to equality by eye; for volumes of > 200 ml balancing is best done by weighing. It is important to remember that densities of liquids can vary, thus a given volume of water will not weigh the same as an equal volume of 80% saturated ammonium sulphate (density ~ 1.2). A larger volume of water can be used to balance the ammonium sulphate solution. However, it is best to split the ammonium sulphate solution into two tubes, since it is actually the inertia, not the weight, that should be equal, and inertia will increase as the particles sediment.

Typically bacteria in low viscosity media (such as saline buffer, or minimal salts media) are usually sedimented by centrifuging at $2-3000$ g for $10-15$ min. In higher viscosity solutions higher centrifugal forces and/or longer times would be required. For cell debris, or bacterial or fungal spores higher speeds (up to 12 000 g) and longer times ($30-45$ min) are required, due to their smaller size and lower density, respectively. For protein precipitates 15 000 g for 10 min or 5000 g for 30 min is usually sufficient. Theoretical calculations can be used to give an indication of the time and speed required, though in practice there is no substitute for preliminary testwork.

(ii) *Disc-stack centrifuge.* The most appropriate type of pilot scale centrifuge for biotechnological applications, given the nature of the solids and liquids involved, is the disc-stack centrifuge. Basket, multi-chamber and scroll-type centrifuges are slower, and in addition small solids removal and contained operation is less efficient. Slow

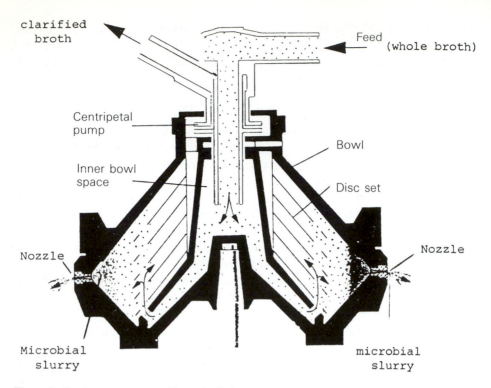

Figure 1. Continuous separator with nozzle discharge.

speed centrifugation is the only method suitable for shear-sensitive mammalian cell recovery, and the imperforate bowl and multi-chamber type machines have been used for this application (14). They are also usually designed for much larger scales of operation; the interested reader is referred to (15) for a comprehensive review of these centrifuge types.

Disc-stack centrifuges contain a stack of conical discs in the bowl, which increase the settling area and rotate with the bowl (*Figure 1*). The feed suspension is dispersed and accelerated radially from the centre of the bowl to the area between the disc-stack and the bowl wall. The suspension flows radially inwards through the channels between the discs towards the bowl outlet at the top. The denser particles in this suspension are forced radially outwards by the centrifugal force against the inner surface of the discs. Having 'settled' against the disc the particle is forced to slide down along the disc surface and out to the bowl perimeter (the solids collecting space). After a period of time the solids collecting space fills with cells, which must then be removed before they interfere with the separation on the discs and are re-entrained in the feed. There are several methods for performing this operation including: manual removal from a solid bowl, continuous discharge via peripheral nozzles and intermittent solids ejection via peripheral ports or an opening bowl. Where intermittent solids removal is used, the interval between solids removal operations must be carefully calculated if good separation and efficient use of the centrifuge is to be achieved.

Centrifuge speeds up to around 12 000 r.p.m. are attained by the newer machines, enabling high flow rates [several hundred litres per hour, e.g. 200 litres h^{-1} *E.coli*

(16)] and good separation. The latest machines (e.g. Alfa-Laval BTPX) are completely contained to eliminate aerosols, have reduced shear inlet zones (essential for some cell and floc types) and can be cleaned, steam sterilized and cooled. Feed concentrations up to 15% solids can be handled.

2.1.2 *Optimizing centrifuge performance*

Several parameters can be changed to improve the efficiency of centrifugation. These were briefly outlined in the introductory section but will be discussed in more detail here.

(i) *Particle size.* Sedimentation velocity increases with the square of particle size. To increase particle size flocculation can be induced (see Section 2.2.1). Particle size is usually measured by one of the following methods: microscopic analysis using a graticule eyepiece, use of Coulter counter in which size is measured electrically or by laser light scattering.

(ii) *Density difference.* Increasing the difference between the densities of the particle and the solution increases the sedimentation velocity. Cells and particularly organelles are often only slightly denser than their surrounding fluid due to the presence of lipid. The density difference may be increased by dilution of the suspension or changing the solute composition of the fluid.

(iii) *Liquid viscosity.* Higher viscosities lead to reduced sedimentation velocity and are often encountered after cell disruption when macromolecules (e.g. DNA, protein) are released into the medium. Increasing the temperature will reduce the viscosity, but may lead to protein denaturation. Dilution of the suspension will also decrease the viscosity in addition to increasing the density difference. Alternatively, the viscous macromolecules may be removed by precipitation with pH, salt or alcohol treatment, or by treatment with DNase, RNase or specific proteases (clearly these treatments must not affect the protein of interest).

(iv) *Grade efficiency.* This is an important parameter for continuous or semicontinuous centrifuges. Due to the distribution of particle sizes in a suspension, some particles will require longer residence times than others for efficient separation. Thus, in a given distribution of particle sizes there will be a distribution of recovery efficiency, termed the grade efficiency. The grade efficiency of a disc-stack centrifuge is determined by the speed of the centrifuge, the number of discs, the difference between the radii of the inner and outer discs, the angle of inclination of the discs and the relevant properties of the particle suspension (see Stokes' law).

(v) *Differential centrifugation.* Selective, sequential centrifugation procedures are used for the clarification and subfractionation of eukaryotic cell homogenates to isolate organelles such as nuclei and mitochondria. A typical sequence of centrifuge steps is shown in *Figure 2*; more detailed procedures for isolation of organelles are given in Section 4 of this chapter.

2.2 **Flocculation**

Flocculation is the process whereby destabilized particles, or particles formed by destabilization, are induced to come together, make contact and subsequently form large(r) agglomerates (17).

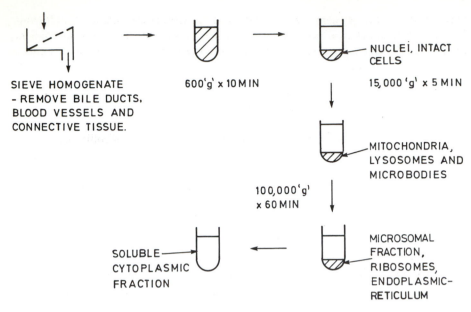

SIEVE HOMOGENATE
- REMOVE BILE DUCTS,
BLOOD VESSELS AND
CONNECTIVE TISSUE.

600'g' x 10 MIN

NUCLEI, INTACT
CELLS

15,000 'g' x 5 MIN

MITOCHONDRIA,
LYSOSOMES AND
MICROBODIES

100,000'g'
x 60 MIN

SOLUBLE
CYTOPLASMIC
FRACTION

MICROSOMAL
FRACTION,
RIBOSOMES,
ENDOPLASMIC-
RETICULUM

Figure 2. Typical sequence of centrifuge steps for isolation of cellular organelles.

Colloidal particles (~ 1 μm diameter) such as bacteria, eukaryotic cell organelles and cell debris, are able to maintain a dispersed state in suspension due to the overriding influence of interfacial forces and the diminishing influence of gravity. Therefore flocculation is often required to facilitate clarification of such suspensions. Before flocculation of these particles can occur their colloidal dispersion must be destabilized (termed coagulation) so that the flocculated particles cannot redisperse prior to separation from the fluid. The coagulation process exploits the interfacial or surface phenomena which also stabilize the colloidal dispersion. Principal among these are the electrostatic surface charge carried by the particles, Brownian motion and the degree of hydration of the particle's surface. Typically, coagulation is induced by adjusting the chemical environment of the particle suspension: for example, addition of metal ions (as salts) to reduce/reverse charge repulsion; addition of small particles ('seeding') to increase the frequency of collisions due to Brownian motion (termed perikinetic flocculation); or the use of surfactants to alter the hydrophobic/hydrophilic nature of the particle surface and so reduce hydration stabilization. These techniques may be employed individually or in combination to achieve flocculation. However, for appreciable floc formation some form of induced turbulence or fluid velocity gradient is required to promote particle collisions and subsequent attachment (termed orthokinetic flocculation).

The increased particle size of the floc will enhance or enable centrifugal or gravity settling operations and improve filtration rates. However, a flocculation step may be disadvantageous due to denaturation of protein, toxicity of some polymeric flocculants or adverse effects on subsequent protein purification due to the presence of unbound polymer.

Significant flocculation of particles in suspension is brought about by inter-particle collisions followed by particle attachment. To optimize flocculation, therefore, the hydrodynamics of the system must be optimized to increase the number of collisions. The principle parameters affecting these hydrodynamics are the fluid velocity gradient, or shear rate (G) and the conditioning time (t). The rate of reduction in particle numbers is given by the Smoluchowski equation:

$$\frac{-\mathrm{d}n}{\mathrm{d}t} = \frac{16}{3} \alpha \ G \ R^3 \ N^2$$

where R = particle radius, N = particle number, and α = fraction of collisions resulting in permanent attachment.

A series of tests should be carried out to optimize a coagulation and flocculation process and minimize denaturation.

2.2.1 Optimizing coagulation

Coagulation can be induced by the following.

(i)   Chilling to below 20°C (particularly effective with yeast cells).
(ii)   Adjusting the pH, usually the optimum is pH 3−6. Rapid uneven adjustment can be particularly effective.
(iii)   Increasing the ionic strength, often in conjunction with adjusting the pH.
(iv)   Increasing the number of particles.
(v)   Addition of multivalent metal ions (e.g. Al^{3+} or Fe^{3+}).
(vi)   Addition of polyelectrolytes.

Coagulation efficiency is usually tested in a beaker with some means of stirring the solution at a controlled speed. Ideally beakers should be larger than 600 ml volume to allow accurate addition of coagulants and minimize adsorption of polyelectrolytes to the glass wall. The magnitude and control of stirrer speed is important, since rates of particle collisions will be proportional to speed; usually coagulant is added at high stirrer speed, followed by slower stirrer speed to promote flocculation. Suitable methods for measuring coagulation/flocculation efficiency are as follows.

(i)   Measurement of settling rate (e.g. by measurement of the particle settling velocity or by monitoring the decrease in turbidity of the suspension).
(ii)   Measurement of rate of filtration through an appropriate filter.
(iii)   Measurement of clarity of a filtrate.
(iv)   Measurement of particle size before and after coagulation (see Section 2.1.2).
(v)   Monitoring the number of primary particles (e.g. cells) remaining in suspension (a measure of the completeness of flocculation).

Methods (i), (iv) and (v) are more suitable if centrifugation is to be used for removing the flocculate, whilst methods (ii), (iii), (iv) and (v) are best if filtration is to be used. Other process-specific tests can be devised by the researcher concerned, which reflect the important criteria involved in the separation.

After testing a variety of coagulants the results are plotted graphically, for example, settling rate versus coagulant dose or pH (*Figures 3* and *4*). Alternatively coagulant dose and pH form the two axes and settling rate is plotted as contour lines, thus depicting areas of optimal coagulation (*Figure 5*).

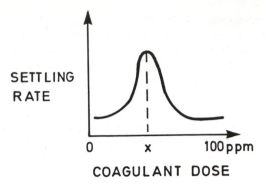

Figure 3. Typical profile obtained for settling rate versus coagulant dose. x, optimal coagulant dose.

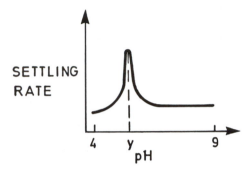

Figure 4. Typical profile obtained for settling rate versus pH of the solution. y, optimal pH for coagulation.

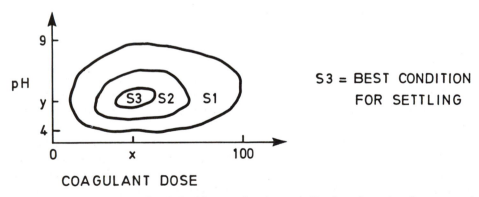

Figure 5. Typical contour profile obtained for coagulant dose and pH, where the contour lines represent settling rate. x, optimal coagulant dose; y, optimal pH; and S3, optimal conditions for coagulation and settling.

Stock solutions of coagulant should be prepared regularly, preferably weekly, particularly for iron salts and polyelectrolytes (manufacturers' literature often indicates shelf-life). For a metal salt coagulant, the concentration is usually expressed as that of the metal ion, thus 500 mg l^{-1} corresponds to 5.83 g l^{-1} alum [$Al_2(SO_4)_3 \cdot 16H_2O$]. A typical final dosage of Al is 5 mg l^{-1} (5 p.p.m.). See *Method Table 1* for the procedure for optimizing coagulation.

Method Table 1. Optimizing coagulation.

1.	Prepare and assemble required coagulants and chemical solutions.
2.	Place 500 – 1000 ml of a representative suspension sample in each stirrer beaker.
3.	Agitate samples at maximum impeller speed (e.g. 300 r.p.m.).
4.	Adjust pH of each sample by adding acid or alkali (e.g. for six samples, try pH 4, 5, 6, 7, 8 and 9).
5.	Add coagulant solutions simultaneously to each sample and start stop-watch.
6.	Maintain rapid stirring for a set time period (e.g. 1 min).
7.	Reduce stirring speed to the minimum required to maintain flocs in suspension (e.g. 50 r.p.m.). Alternatively try a range of speeds and determine the optimum.
8.	Maintain slow, conditioning stirring for a set time period (e.g. 15 min).
9.	Stop stirring.
10.	Carry out floc analysis as appropriate (e.g. allow to settle, take sample for sizing, flow through turbidometer).

Table 3. Types of flocculants.

Naturals	*Synthetics*
Starch	Polyacrylamides
Gums	Polyamines/imines
Tannin	Cellulose derivatives (e.g. CMC)
Alginic acids	Ion-exchange resin beads
Sugar/sugar acid polymers	Polydiallyldimethyl ammonium chloride
Chitosan (polyglucosamine)	

2.2.2 *Optimizing flocculation*

Flocculants are usually added to a particle suspension which has already been destabilized by addition of coagulant, thus increasing both the rate of aggregation and the size and strength of the flocs. Some flocculants, for example polyelectrolytes, can be used without prior addition of a coagulant.

A variety of flocculants are currently used (*Table 3*). A preliminary choice can often be made on the basis of cost, availability and toxicity. A particular class of flocculants can often be subdivided on the basis of charge and molecular size, thus low and high molecular weight ($10^5 - 15 \times 10^6$) forms of polyacrylamide are available which are cationic, anionic or non-ionic. Some polymers are amphoteric, thus their charge is pH-dependent and can be anionic or cationic. Initial trials with polymers of the same molecular weight and charge density, but with different charge types, will identify whether an anionic, cationic or non-ionic polymer is the most suitable. Further tests with polymers of varying molecular weight and charge density will identify the optimal flocculant.

Maximum concentrations of stock polymer solutions are limited by high viscosities, and are usually between 0.5 and 1% w/v. Stock solutions are usually diluted prior to

Method Table 2. Optimization of flocculation.

Follow the method for optimization of coagulants (*Method Table 1*) steps 1−5.
6.   Maintain rapid stirring for 5 min.
7.   Add flocculant solution, and mix rapidly for 1 min.
8.   Reduce stirring speed (e.g. 50 r.p.m.) and maintain for 15 min.
9.   Analyse flocculation by an appropriate method (e.g. allow to settle or filter).

---

use to ease accurate addition of flocculant. Manufacturers' recommendations on preparation and storage of stock solutions and maximum final concentrations (usually 0.01%) should be noted. Polymers obtained as powders should be carefully dissolved, usually by addition of a small amount of methanol or ethanol prior to rapid addition of water with simultaneous vigorous stirring. Final concentrations of coagulant or flocculant should be expressed as mg g^{-1} weight of cells, or mg m^{-2} cell surface area to enable direct comparison of test results and consistent dosing. Flocculants are tested by the same methods used for coagulants. The procedure for optimization of floccu-lation is given in *Method Table 2*.

In addition to flocculant type and concentration the shear rate and conditioning time (i.e. time after flocculant addition prior to separation) are important parameters to optimize (see Section 2.2). Increased conditioning time results in increased particle size and decreased number, whereas increased shear rate will have the opposite effect. Other properties of the flocs, for example, strength (i.e. resistance to shear or compression), and rheological properties of the sediment or filter cake, are also influenced by conditioning.

Flocculation is usually carried out in an agitated vessel [e.g. stirred beaker, jar flocculator (Voss Instruments Ltd)], a Couette flocculator (which consists of two concentric cylinders with a small gap between them and the inner cylinder is rotated) or a laminar flow tube flocculator (in which the flocculating suspension is passed through a tube). These latter types of equipment are usually custom built (see ref. 17).

2.3 Microfiltration

Microfiltration operates by the application of hydrostatic pressure across a semi-permeable membrane filter. The pressure difference across the filter forces the smaller solutes and water through the filter pores as permeate, whilst retaining and concentrating the cell/particle suspension (the retentate). To prevent fouling of the pores, in tangential flow microfiltration the feedstream is passed across the membrane surface, thus 'sweeping' particles off the membrane.

Microfiltration membranes are usually impermeable to particles between 0.02−10 μm but permeable to smaller particles (cf. for ultrafiltration membranes the pore sizes range from 1−20 nm, see Chapter 3). However, under certain circumstances (for example if fouling of the pores occurs) smaller molecules may also be retained. Microfiltration membranes are usually isotropic (i.e. the pore sizes are the same throughout the thickness of the membrane), although recently anisotropic membranes have become available (in these membranes the microfiltration membrane of defined pore size is ~0.5 μm thick and is supported by a deep, large pore layer ~150 μm thick).

Microfiltration is principally used for:

(i)   cell and cell debris removal;
(ii) concentration of cell suspension prior to disruption;
(iii) recycling biomass and removing product from continuous fermentations;
(iv) sterile filtration.

The main advantages are:

(i)   high separation efficiency;
(ii) high flux rates and therefore rapid processing;
(iii) cell size and density have little effect on performance, and therefore flocculation is not required (cf. centrifugation);
(iv) high level of containment;
(v)   compact, simple and reliable equipment.

The theoretical principles operating during microfiltration are not well understood at present. Two theoretical models have been proposed which allow predictions to be made for factors affecting microfiltration (see refs 18−20 for descriptions of the models). Both models predict that the permeate flux will increase in direct proportion with transmembrane pressure and inversely with viscosity. Thus, by operating at higher temperatures the fluid viscosity will be decreased and the flux rate increased (however, care should be taken not to cause heat denaturation of the proteins). Similarly the flux rate may be increased by prior use of nucleases to break down the viscous DNA released during cell disruption. Increasing the pore size will also cause the flux rate to increase. In contrast to ultrafiltration, the flux rate in microfiltration processes is not as dependent on the concentration of the solution, due to the larger particle size (the permeability

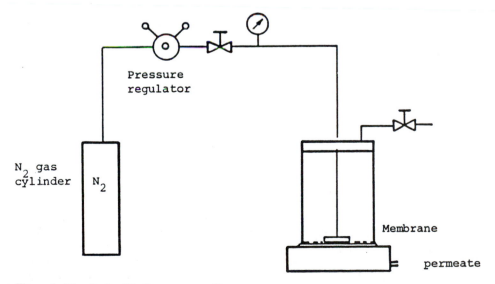

Figure 6. Stirred microfiltration pressure cell.

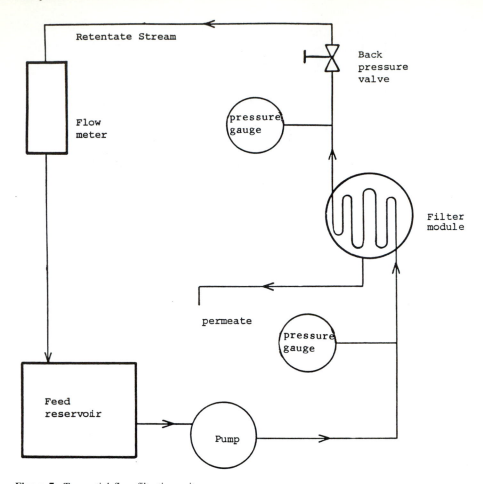

Figure 7. Tangential flow filtration unit.

of the concentrated layer above the filter is proportional to the square of the particle size). Increasing the cross-flow velocity will also increase flux rates by minimizing fouling (however, higher velocities lead to higher shear rates and care should be taken not to cause shear-induced denaturation of the proteins).

2.3.1 *Equipment*

Typical filter systems are shown in schematic form in *Figures 6* and *7*. *Figure 6* depicts a stirred cell filtration system, *Figure 7* a cross-flow filtration system.

(i) *Stirred cell modules.* These units have reservoir volumes within the cell from 2 ml to 300 ml. Larger volume reservoirs (5 – 10 litres) can be used to supply these kinds of module, though the small filter area limits the scale of operation. Thermostatic jackets on some units allow the temperature to be controlled. Polarization or fouling is minimized by the stirrer located above the filter surface. The driving force for the filtration system shown in *Figure 6* is nitrogen supplied from a pressurized cylinder. This allows precise

Table 4. Membrane filter materials.

Material	Pore sizes (μm)	Comments
Plastics		
Cellulose−acetate, −nitrate	0.2 5.0	Analytical uses: hydrophilic; can be autoclaved though pore sizes may change.
PTFE (Teflon)	0.2, 1.0	Low fouling; autoclavable; solvent-resistant.
PVDF	0.1, 0.2, 0.45, 0.65	Most common; hard wearing (1−2 years life); chemically resistant; usually hydrophobic; but can be made hydrophilic; 'sterilizable'.
Others		
PVC, nylon, cellulose, polypropylene, polyester, polycarbonate	0.1−10	
Inorganics		
Carbon/zirconia, sintered stainless steel, alumina	0.1−30	Robust; heat stable; costly; high flows and pressures possible; usually tubular units.

Table 5. Membrane module types (ref. 20).

		Comments	*Suppliers*
Laboratory scale			
Stirred cell	Disc filters	10−500 ml batch continuous filtration possible; limited filter area	Domnick Hunter Sartorius Millipore Amicon Gelman
Cross-flow unit— spiral, radial or zig-zag flow channels	Disc filters	Continuous filtration, usually batches of a few litres	As above
Process scale			
Flat sheet		Wide range of module types and sizes. Multi-unit systems for large-scale usage. Filters in cassette form, 0.5 m^2 area is typical	Millipore Dorr-Oliver Sartorius Rhone-Poulenc
Tubular		Wide bore tubes: 6−25 mm i.d. Turbulent flow; usually plastic supported membranes, some inorganic filters	PCI SFEC Abcor
Hollow-fibre		0.2−1.5 mm i.d. fibres. Many fibres (up to 3000) in a single module—up to 5 m^2 area	Amicon Microgon
Pleated		Membrane cast onto support and covered in turbulence-promoting mesh. Pleated around central permeate tube.	Gelman

control of the pressure gradient and avoids pump effects, such as pressure pulses and shear damage, on the feed suspension. Typical pressure gradients are between 15 and 60 p.s.i. (1−4 bar).

(ii) *Cross-flow filtration modules.* A wide variety of module and filter types are currently available (see below) which use tangential fluid flow as the mechanism for controlling the extent of fouling. The range of conditions under which a system can be operated is determined by its design and the filter material. Thus, membranes made from inorganic materials can be operated at higher transmembrane pressures and cross-flow velocities than the majority of 'plastic' membranes. The pressure and tangential flows are normally produced by a pump (e.g. peristaltic or gear type), working against the filter and a back pressure valve (see *Figure 7*). Temperature control is effected by controlling the temperature of the feed reservoir and lagging the equipment (or by an in-line heat exchanger). Typical operating pressures vary from 5 to 60 p.s.i. (0.3−4 bar), with cross-flow velocities up to 4 m sec^{-1} (calculated from flow rate/channel area). Flux rates for microfilters vary between 20 and 300 litres m^{-2} h^{-1}. Higher flux rates can be obtained at higher temperatures (plastic filters made of polysulphone or polypropylene are stable to 60°C, inorganic filters to over 100°C). However, protein denaturation and precipitate formation, leading to protein loss and filter fouling, places an upper limit on the operating temperature (perhaps as low as 20°C).

(iii) *Filter types.* Microfiltration equipment has been primarily developed from reverse osmosis and ultrafiltration equipment (see Chapter 3); in many cases the equipment is identical, only the pore size is different. *Tables 4* and *5* present most of the equipment types currently available; detailed information is available from manufacturers. Micro- or ultrafiltration membranes can be used for small particle filtration applications, often ultrafilters are used to provide a molecular sieving operation in addition to particle removal. Those most commonly used for microfiltration are the 0.2 and 0.45 μm pore size plastic membranes, either in flat sheet, or hollow fibre form. Although 0.45 μm filters give higher initial flux rates, often 0.2 μm filters are preferable, since they tend to show reduced fouling for 1 μm particles, such as bacterial cells, and so can sustain higher flux rates and concentration factors.

The recently developed inorganic filters tend to have larger pore sizes (8−15 μm) and show high flux rates with high cross-flow velocities. A major advantage is their resistance to steam sterilization. The nature and quantity of protein adsorption to inorganic filters is an important factor in filter choice, which merits careful consideration.

Correct equipment choice can only be made following a certain amount of testwork to check its suitability for the specific process application. Never rely entirely on manufacturers' claimed performance data, usually produced with 'ideal' or at least 'clean' feed streams. For example:

Manufacturer's test data: 100 litres *E. coli* suspension in defined media concentrated to 15 litres using 0.5 m^2 of 0.45 μm pore membranes in 20 min;

Author's test: 100 litres *B. subtilis* in complex media and 1% anti-foam concentrated to 15 litres using 1.0 m^2 of 0.45 μm pore membranes in 20 h.

Both tests used the same module and filter type (flat sheet cassettes) and demonstrate the difference between an ideal and a real process situation.

2.3.2 *Optimization of microfiltration*

To assess the efficiency of a microfiltration process the flux rate must be measured. This can be done either by use of an in-line flowmeter on the permeate line or by timed measurement of the volume or weight of the permeate. It is important that the initial flux rate and the steady state flux rate are both measured, to assess the formation of the concentrated layer on the membrane surface, as well as fouling. The effect of the individual parameters discussed below on the steady state flux rates can then be assessed. With a well optimized process a concentration factor of at least 10-fold should be attained before flux rates become diminishingly low.

If high flux rates are to be maintained over long process times or there is repeated use of the membrane, some form of membrane cleaning must be carried out. Methods

Table 6. Chemical cleaning agents for microfilters.

NaOH	0.1 M
NaCl	1−2 M
NaOCl	<200 p.p.m.
Pepsin	1%
Trypsin	1%
Detergents (Alconex, Sparkleen	0.1%
Urea	6 M
HCl	0.1 M
HNO$_3$	0.25−0.3%

of membrane cleaning can involve physical, chemical or biological treatments, or a combination of them all. Physical methods include water washing, backflushing and recycling, sponge ball treatment and ultrasonic treatment. Inorganic membranes can be cleaned by heat treatment. Many chemical cleaning reagents can be used; some are listed in *Table 6*. Manufacturers' guidelines on membrane compatibility with cleaning agents must be noted; some manufacturers provide proprietary cleaning agents. Cleaning efficiency can be assessed by the subsequent increase in flux rate, ideally back to the initial flux rate of the new membrane.

(i) *Transmembrane pressure.* It is preferable to keep the pressure difference across the filter to the minimum consistent with providing sufficient driving force to obtain reasonable solvent flux rates. As the transmembrane pressure is increased so the solvent flux rate increases; however, above a certain pressure difference, determined by the filter and the feed suspension, there is no further increase in flux. Higher pressures cause compaction of any fouling layer of particles and thus lead to losses in flux rates.

(ii) *Cross-flow velocity.* Increasing the cross-flow or tangential velocity of the feed suspension over the membrane surface will reduce the thickness of the fouling layer and thus increase the permeability of the membrane. This increases the rate of solvent flux through the filter. High cross-flow velocities allow the transmembrane pressure to be increased, which in turn also increases solvent flux. Very high cross-flow velocities, however, can cause shear denaturation of proteins and thus precipitation. Further, the high pressure drop produced in the flow channels of the membrane increases pumping costs (an important factor to consider for larger process scales).

(iii) *Filter characteristics.* Informed filter selection can prevent many subsequent processing problems which cannot be solved by fluid management. Important filter properties to consider include the following.

(1)   Pore size—the pore size relative to the particle size determines the probability of pore blockage.
(2)   Overall permeability, which directly affects flux rate.
(3)   Thickness, which contributes to the filter's permeability—for a given pore size a thick filter will have a lower flux than a thin filter.
(4)   Surface character—the surface chemistry of the membrane, determined by its chemical composition, determines the type and extent of interaction with

components of the feed suspension (particulate or dissolved). Fouling is an example of an adverse interaction. Principal factors are the hydrophobic/hydrophilic character and the electrostatic charge on the surface. Hydrophobic filters tend to adsorb hydrophobic solutes, such as anti-foams and proteins, leading to formation of a fouling layer and loss in flux. Positively-charged filters will tend to attract oppositely-charged cells and solutes (thus, alteration of the pH and ionic conditions should be tested to minimize failing).

(iv) *Particle concentration*. Flux decreases with increasing concentration of solids in the feed suspension, due to the higher concentration of particles at the filter surface. However, the flux for particulate suspensions is not as concentration-dependent as is ultrafiltration of protein solutions, until a critical concentration is reached, where there is a sudden loss in flux rate due to inhibition of the back diffusion of particles away from the filter surface (18). Trial runs will determine the point at which this occurs.

(v) *Temperature*. Increasing temperature reduces the viscosity of the feed solvent (for water, ~3% per °C rise), hence increasing the flux rate (by 3% per °C) (19). Therefore, microfiltration should be carried out at the highest temperature at which the protein is stable.

(vi) *Hydrodynamic conditions*. These are determined by the cross-flow velocity of the suspension produced by the pumping conditions, and the geometry of the flow channels in the filter system (or stirring in a stirred cell system). Turbulent conditions are favourable in a filtration system since they reduce the thickness of the fouling layer on the membrane surface, and enhance mass transfer. High shear rates are also favourable, within defined limits [see above (ii)].

3. DISRUPTION—T.Salusbury

Disruption may be achieved by chemical, physical or mechanical steps in order to release intracellular proteins prior to purification. Alternatively, components from the cell membrane may be removed to allow protein leakage (e.g. by use of detergents, or organic solvents—see *Protein Purification Applications: A Practical Approach*, edited by E.L.V.Harris and S.Angal). This section discusses many of the disruption techniques available and gives guidelines for their use. The performance of each technique will depend upon cell type, culture conditions and the pre-treatment used.

3.1 **Pre-treatment**

Most animal proteins of interest are located in specific organs or muscles. The surrounding fatty tissue can often interfere with protein isolation from the disruptate and so must be scrupulously removed. Both organs and tissue should first be homogenized to produce a cell paste. A Colworth Stomacher is suitable for soft animal tissue. A Waring blender is suitable for harder material, although grinders, mincers or mills need to be used for really hard material. The homogenate can then be disrupted to release protein.

Hard plant tissues (such as seed) are best homogenized by milling or grinding. Again, removal of plant constituents which would cause processing problems is advisable. Green leaf and root tissue can usually be homogenized in a hammer mill, although clumps

Table 7. The relative susceptibility of various cell types to disruption.

	Sonication	*Agitation*	*Liquid extrusion*
Animal cells	7	7	7
Plant cells	7	7	7
Gram-negative bacteria	6	5	6
Gram-positive bacilli	5	4	5
Yeast	3	3	4
Gram-positive cocci	3	2	3
Spores	2	1	2
Mycelia	1	6	1

The numbers indicate relative susceptibilities (7, very susceptible; 1, very resistant).
Redrawn from Edebo (21).

of fibrous tissue are a common cause of blockage. With cultured cells, either prokaryote or eukaryote, the process starts with a cell harvesting step (i.e. clarification, see Section 2). For an extracellular protein, this produces a cell-free liquor, which can be used directly for subsequent purification steps. With an intracellular protein, cell disruption and debris separation must precede the purification steps.

3.2 Factors affecting ease of disruption

The relative ease with which cells can be disrupted by mechanical laboratory-scale methods is outlined in *Table 7*. These susceptibilities should be considered when choosing a method of cell disruption. Cells exhibit considerable morphological, structural and size variation, depending on age and culture conditions. Bacteria occur as spheres, rods, helices and filaments. Fungi may be mycelial, pseudomycelial (often cultured in pellet form) or unicellular (yeasts). Cultures can be highly heterogeneous, being composed of single, paired, grouped or chained cells. Agglomeration of these cells will form filaments or clumps. The typical sizes of cells are given in *Table 8*.

The cell wall is responsible for strength, rigidity and shape and is the major barrier to release of any intracellular proteins. It is thus pertinent to consider the structure of the cell wall, as well as cellular contents that may either denature the desired protein on disruption, or hamper its isolation and purification.

3.2.1 *Animal cells*

Animal cells are enclosed solely by a plasma membrane which is weakly supported by a cytoskeleton. They thus lack the rigidity of bacterial and yeast cells. This fragility and their relatively large size make them vulnerable to shear and easy to disrupt.

3.2.2 *Plant cells*

Plants have a cell wall composed of insoluble carbohydrate complexes, lignin or wax, which surrounds the plasma membrane. Cellulosic microfibrils cover the plasma membrane in a random manner, imparting some protection against shear, although the

Table 8. Representative diameters of cells.

Bacteria (discrete cells)	$0.7-4.0\ \mu m$
(flocs)	$10\ \mu m$ to several mm
Fungi (filaments)	$2-7\ \mu m$
(spheres)	$5-10\ \mu m$
Filamentous bacteria	Several mm in length
Animal cells (culture)	$\sim 10\ \mu m$
(tissue cells)	Variable
Plant cells (culture)	$\sim 100\ \mu m$
(tissue cells)	Variable

large size of the cells makes them more vulnerable. Phenolic compounds, which can bind and inactivate proteins, are present in vacuoles and will be released by disruption. Prompt extraction of the desired protein is thus essential.

3.2.3 *Bacteria*

The plasma membrane of both Gram-positive and Gram-negative bacteria is surrounded and protected by an outer cell wall, which owes its rigidity and strength to peptidoglycan. In Gram-positive bacteria the major component of the cell wall is peptidoglycan (40−90%), together with teichoic acids, teichuronic acids and other carbohydrates. This layer can be 20−50 nm thick, depending on the species and growth conditions. In contrast, the peptidoglycan layer in Gram-negative bacteria is only 2−3 nm thick and is surrounded by a lipopolysaccharide layer containing proteins and lipids. Lysozyme is frequently used to break down peptidoglycan and if used in an isotonic medium will remove the outer cell wall and release the contents of the periplasmic space (located between the cell wall and the plasma membrane). Gram-negative bacteria are less susceptible to lysozyme, due to the lipopolysaccharide layer; their susceptibility can be increased in the presence of calcium-chelating agents (e.g. EDTA), which destabilize the lipopolysaccharide structure.

3.2.4 *Filamentous fungi and yeast*

The cell walls of filamentous fungi and yeasts are composed of about 80−90% polysaccharide. In the filamentous fungi this is chitin cross-linked with glucans, and in lower fungi and yeast it is mannan and glucan. The remainder of the cell wall is composed of protein and lipid. Mature fungi have no discrete cell wall structures, instead the layers blend into one another. The inner fungal wall has a microfibrillar skeleton, embedded in an amorphous polysaccharide matrix. The wall strength is determined by the microfibril length, number, diameter and orientation. This arrangement is also found in modern ultra-strong composite materials, such as glass-fibre reinforced plastics.

3.3 **Consequences of disruption**

Mechanical methods of disruption generate heat. Thus, it is essential to pre-chill the slurry, ideally to 4°C, since most proteins are denatured by heat.

Once the cell is disrupted, control of metabolic regulatory mechanisms is lost. Thus the desired protein may be degraded by intrinsic catabolic enzymes (e.g. proteases), necessitating prompt extraction, addition of inhibitors and/or cooling to reduce their effects. Also, upon disruption, long strands of nucleic acids and structural protein are freed, causing an increase in the viscosity of the disruptate. The addition of suitable nucleases and proteases (e.g. ficin, papain or trypsin) breaks them down and 'thins' the homogenate (however, use of proteases is limited by the susceptibility of the protein of interest). Altering the pH of the disruptate can also reduce viscosity.

Since disruption is only one of many steps in the isolation of a desired protein, methods should be chosen which facilitate subsequent purification steps. Although cell disruption is often termed 'homogenization', or is carried out in 'homogenizers', it is important to distinguish between the two terms. For example, if the cell wall is broken up into a homogenate of tiny fragments, these will form a stable emulsion which is difficult to separate from the desired protein. A cruder disruptate, from which the desired protein can easily be separated from larger 'chunks' of cell debris, 'split' or 'holed' cells is preferable.

An assay for detecting the protein released is required, so that the course and efficiency of the disruptive process can be monitored. This can be

(i)   a crude assay of the supernatant for total protein (see Chapter 1). Monitoring the adsorption at 280nm is possible, as long as other intracellular constituents do not interfere;

(ii)   an assay of the activity of the desired protein; or

(iii)   an assay of cell viability (such as a rapid bioluminescent assay of intracellular ATP). A decline in cell viability indicates disruption.

3.4 Methods of tissue and cell disruption

3.4.1 *Disruption by physical methods*

In general, the tissues of higher organisms, mammals, higher plants and multicellular algae are tough. Small quantities, however, can be disrupted satisfactorily without using harsh treatment. Once disrupted, animal and plant cells release enzymes readily, although in some cases the addition of detergent (e.g. SDS or Triton) to release membrane-bound enzymes is advantageous.

(i) *Mixers and blenders*. Laboratory mixers and liquidizers resemble domestic models, but are more robust. The Waring blender (Christison Scientific Equipment) is a popular design of laboratory device. Since the vessel is made of steel, it will retain low temperatures if pre-chilled, counteracting the heating effects of disruption. Organs and tissues may be disrupted and simultaneously homogenized in blenders. These techniques are not usually satisfactory for the disruption of more robust microbial cells; a rotor-stator blender is more suitable. These latter devices (e.g. the 'Polytron' range, Christison Scientific Equipment) have a high kinematic frequency, generated by unit acceleration of material, which causes highly turbulent power densities to form in the slits and teeth of the rotor-stator. Scissor and impact effects are also generated by a toothed ring. Rotor-stators can disrupt lumps of tissue without blockage, and then homogenize this disruptate.

The usual method of operation is as follows.

(1)   Chill the blender head and a tall, thin vessel.

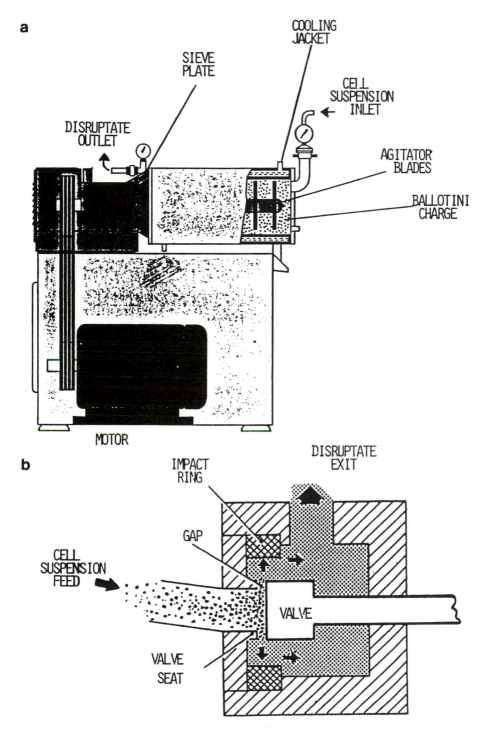

a

SIEVE
PLATE

COOLING
JACKET

CELL
SUSPENSION
INLET

DISRUPTATE
OUTLET

AGITATOR
BLADES

BALLOTINI
CHARGE

MOTOR

b

IMPACT
RING

DISRUPTATE
EXIT

GAP

CELL
SUSPENSION
FEED

VALVE

VALVE
SEAT

Figure 8. (a) Bead mill, and (b) liquid extrusion cell disrupter.

(2)   Resuspend the cells in isotonic, neutral buffer at 4°C, to about 30% cell volume. Ensure there is sufficient liquid to cover the blender rings.

(3)   Pour cell paste into the vessel and place the vessel in ice.

(4)   Start the blender gradually, allowing for 'climb' of the paste and foam formation.

(5)   Monitor temperature rise and stop the blender periodically to prevent excessive foaming (since air/liquid interfaces denature protein).

(ii) *Agitation with abrasives.* At laboratory scale, most small volumes (< 1.0 ml) of cell pastes can be ground in a pestle and mortar with alumina or sand to produce some release of intracellular components. Cells can also be disrupted by shaking pastes in flasks containing abrasives.

Ball-milling provides an excellent means of disintegrating larger volumes (< 300 ml of cell suspension). Ball mills are rapidly vibrating vessels which contain glass beads ('ballotini') that cause grinding, and collision, leading to cell disruption. The device has a double-walled cylindrical grinding tank with a sieve plate at the base to retain the ballotini charge (*Figure 8a*). The extent of disruption depends on the rate of stirring or vibration, the ratio of beads to cells, the size of the beads and the contact time (22).

Contained small-scale devices can be used for disrupting pathogenic organisms. For small batch operations of a few millilitres the Retsch Mixer Mill Type MM2 (supplied by Glen Creston) is ideal. Disposable reaction vials are filled with cell suspension and ballotini, and placed in the mill and shaken. Up to 6.0 ml of a 40% cell slurry can be disrupted in this device in more than 5 min, with only a 5°C increase in suspension temperature.

For larger scale work devices such as the Dyno-Mill (Glen Creston) are preferred. This is a much larger device in which beads are agitated by rapidly rotating discs. The KDL model Dyno-Mill can disrupt 300 ml of slurry in batches, or 600 ml continuously. Heat generation is a problem with these machines, so they must be cooled by being operated in a cold room, or by the circulation of coolant through a cooling jacket. A cell paste concentrated to between 30 and 60% cell wet volume should be used. The cell paste should be chilled to 4°C, since operating at ambient temperatures reduces the rate of protein release. Agitation with beads is capable of disrupting the toughest of cells. The limitations of these devices are that high agitator speeds, very concentrated cell pastes and low or multiple throughputs are required. Cooling is required and colloidal glass from the beads is often present in the process stream which can lead to protein separation problems.

The technique has limited application at very large scale, and is best used for those microorganisms which show a high degree of resistance to disruption by other methods, particularly yeast *Staphyloccus* sp., *Micrococcus* sp. and *Streptococcus* sp. (23), and suspensions where viscosity increases sharply on disruption. As indicated in *Table 7*, agitation by, for example, ball milling is particularly appropriate for disruption of mycelial cell suspensions.

(iii) *Liquid extrusion.* This method relies on the principle that forcing a cell suspension at high pressure through a narrow orifice will provide a rapid pressure drop. This is a very powerful means of disrupting cells. By varying the pressure applied, cells may be completely or only partly disrupted (the latter usually being sufficient for the release of periplasmic enzymes from bacteria).

Table 9. Disruption of bacterial cell paste by liquid extrusion.

Assay	Feed	Pass 1	Pass 2	Pass 3	Pass 4
Free protein: (% of total)	5.7	6.9	78.9	89.0	100
Temperature (°C):	21.1	29.1	34.7	41.0	46.1
Viscosity (mPa)	1.31	2.22	3.34	3.56	3.51

E. coli (NCIB 9132) cells were separated from their growth medium by membrane filtration, and resuspended in phosphate buffer at an approximate cell concentration of 60% w/v. The suspension was then passed through an APV Manton-Gaulin 15M-8BA homogenizer, and recirculated without cooling, at a pressure of 710 kgf cm^{-2}. Samples were withdrawn at each pass, and the supernatant protein concentration compared with the total protein concentration of the broth. Temperature and viscosity increases were also measured. No further protein was released after the fourth pass. (Author's unpublished data, 1985).

The most common types of equipment are Gaulin (APV) and Rannie (Silkeborg). These homogenizers consist of a reciprocating positive-displacement, piston-type pump which incorporates an adjustable, restricted orifice discharge valve (*Figure 8b*). The cell suspension is forced through the adjustable valve at high pressure and low velocity. The suspension collides with the valve face, and is directed away at right angles, with a resultant rapid increase in flow velocity and shear. The stream then emerges from an annulus, undergoing rapid decompression.

The author's experience of disrupting *E. coli* cell pastes (*Table 9*) demonstrates the requirement for multiple passes to achieve complete release of intracellular components. When this is necessary, the device is operated with batch recycle, or continuous recycle with bleed. With multiple passes, heating becomes a problem. This can be minimized by pre-chilling the cell slurry and cooling the equipment. Protein release falls as the feedstock is cooled, however, and frozen lumps or foreign bodies must be eliminated from the homogenizer feed to prevent a blockage. Viscosity also increases as DNA is released into solution, although the addition of nucleases can reduce this. Alternatively, in larger scale operations, a second valve is used to exert further shear forces on the disruptate and this substantially reduces the disruptate viscosity.

For laboratory or small pilot scale disruptions of 7 ml or more the small volume Microfluidizer (Microfluidic Corp, Christison Scientific Equipment) fits a useful niche, midway between conventional laboratory-scale batch methods and continuous, large-scale devices. Powered by compressed air, the feed stream is pumped at 250 ml min^{-1} into a chamber in which fluid sheets interact at ultra-high velocities and pressures. The fixed microchannels within the chamber provide an extremely focused interaction zone of intense turbulence causing the release of energy amid cavitation and shear forces. By controlling the intensity of the interaction, a uniform sized disruptate is obtained.

(iv) *Solid extrusion*. This method involves the extrusion at high pressure of a frozen cell paste through a narrow orifice. The shear forces and the abrasive properties of ice crystals disrupt the cells. A device known as the X-press, which consists of two identical cylindrical chambers, separated by a perforated disc, is usually used for solid extrusion disruption. It is an excellent method for obtaining intracellular proteins from bacteria, but its performance is dictated by the temperature, type and concentration of cell paste. It is unsuitable for preparing proteins that are damaged by multiple passes.

The usual technique is as follows.

(1) Pre-cool the press to 0°C.
(2) Fill the press with microbial paste, with the valves open to exclude air (which can squirt out explosively with the last of the suspension).
(3) Close the valves and apply pressure.
(4) At the desired pressure, cautiously crack open the release valve and maintain the desired pressure by continued pumping during extrusion of disruptate.
(5) Collect the disruptate on ice.

(v) *Ultrasonication*. When frequencies of 20 kHz and above are applied to solutions, they cause 'gaseous cavitation', i.e. areas of rarefaction and compression which rapidly interchange. As the gas bubbles collapse, shock waves are formed. Sonication in batch or continuous processing has been employed successfully for the disruption of many types of microbial cells but others, such as *Staphylococcus* sp., are relatively resistant. A number of ultrasonic disintegrators are available (e.g. Fritsh ultrasonic disintegrator, Christison Scientific Equipment and numerous devices supplied by Ultrasonics).

The protein release constant has been reported to be almost proportional to the acoustic power input and independent of cell concentration. At very high cell concentrations, however, there is insufficient mixing and disruption decreases. Successful disruption is dependent on the correct choice of pH, temperature and ionic strength. The selection of these parameters is empirical, varying with both the organism and the protein of interest. Cell pastes of greater than 20% w/v should not be used. About 170 Watts acoustic power at 20 kHz is usually required for near-complete disruption. Surfactants such as SDS can be used, since they aid protein release, but they can also reduce cavitation.

Cell breakage occurs at an exponential rate, dependent on exposure time. As the contact time increases, however, heat is produced and free radicals and ions are produced, which may also cause protein denaturation. The cell paste should therefore be kept on ice and sonication should be carried out in bursts of 30 sec or less (programmable timers are available for this). Although this is a versatile method for laboratory-scale work, ultrasonic treatment is not widely used at large scale.

3.4.2 *Non-physical methods of cell disruption*

(i) *Heat treatment*. Heat treatment is being used extensively in algal, fungal and bacterial single cell protein (SCP) production, where bulk foodstock protein is required, rather than a biologically active molecule. Direct heating in aqueous suspension, spray drying, drum drying and direct steam injection are all common techniques. Increased temperatures cause autolysis, since control of intracellular metabolism is lost.

On a smaller scale, heat treatment has been used successfully for the extraction of heat-stable enzymes from bacteria. Nucleic acid strands are heat-labile, and rapidly denatured, reducing the viscosity of the disruptate. The rate of autolysis will decrease with temperature, depending on the heat sensitivity of the organism.

Autolysed yeast can be produced by the following technique.

(1) Prepare 200 ml volumes of a 15% cell paste in isotonic, neutral phosphate buffer.
(2) Place in a shaking waterbath at 50°C.

(3)   After 10 min, stimulate autolysis by the addition of NaCl (to a final concentration of 5%).

(4)   Reduce the temperature to 45°C and allow the flasks to shake for 60 min.

(5)   Clarify the resultant disruptate.

(ii) *Freeze—thaw*. With certain susceptible microbes and eukaryote cells, repeated freezing and thawing results in extensive membrane lesions. This is accompanied by the release of periplasmic and intracellular proteins. Intracellular ice formation appears to be the major cause of membrane leakage. Cell type, age, final freeze temperature and the rates of heating and cooling are important. This is a simple method, and since it can take place in a closed system, it is suitable for protein extraction from pathogens. It is not used much in practice, because many organisms are resistant to rupture, especially when accumulated intracellular solutes prevent freezing. In addition, it is time-consuming and costly to scale-up. Enzymes which are cold-labile can also be inactivated. Storing cells at −27°C for 7 days in a freezer is an ideal method of disrupting susceptible cells.

Freeze-drying cells can also cause disruption. This is usually achieved as follows.

(1)   Prepare a 20% cell paste and rapidly freeze it in liquid nitrogen.

(2)   Gently vacuum-dry the frozen paste to a constant weight.

(3)   Resuspend the dried cells in isotonic, neutral phosphate buffer at 20°C.

(iii) *Dessication*. Dessication damages the cytoplasmic membrane, increasing its permeability. This is usually achieved by freeze-drying, spray-drying, air or vacuum-drying. Chemical dehydration with ethanol, methanol or acetone is also possible; however, this method is limited by the stability of the desired protein.

(iv) *Explosive decompression*. In this method, the cell suspension is saturated at high pressure with a soluble gas, usually nitrogen. The pressure is then rapidly relieved, often through a bursting disc, and the resultant explosive decompression ruptures cell walls and membranes. Foaming should be kept to a minimum to prevent protein denaturation. This method has never really developed beyond laboratory scale, since it is relatively inefficient. However, it is suitable for the extraction of proteins from pathogens, since it is contained. It is also particularly suitable for animal cell disruption. Vessels are usually 'home-made'. A typical operating procedure is given below.

(1)   Prepare a 20% cell paste, containing 0.05% anti-foam.

(2)   Pour the paste into the chilled vessel, close and seal all outlets.

(3)   Pump in nitrogen to the vessel's maximum pressure limit (usually ~60 bar).

(4)   Place the vessel in ice and allow 20 min to pass, whilst shaking the vessel to equilibrate the pressure between the gas and liquid phases.

(5)   Attach a tube to the outlet nozzle, and rapidly expel the vessel contents into a chilled vessel.

(v) *Osmotic shock*. Proteins may be released from plant and animal cells, and some Gram-negative bacteria by osmotic shock. The technique is as follows.

(1)   Harvest the cells from their growth media.

(2)   Resuspend the cells in a chilled neutral buffered solution of high osmotic pressure (usually containing 20% sucrose).

(3)   Allow the cells to equilibrate (30 min) and then harvest the cells by centrifugation.

(4)   Resuspend the pellet of cells in pure water at 4°C.

The resultant shock is often insufficient to burst the cell, but it allows some proteins to be released from the cells. It is thus possible to avoid, or greatly reduce, further purification steps, since under the right conditions only the desired protein, and a few contaminants, may be released.

This method is inappropriate for Gram-positive bacteria, since they are more resistant to osmotic shock. In practice, this method is only used at laboratory scale, since large volumes of chilled liquids would make large-scale operations uneconomic.

(vi) *Lytic enzymes.* Enzymes provide a mild and selective method of disrupting cell walls; the cell membrane is then easily disrupted by osmotic shock or a gentle mechanical treatment. Lysis can be achieved in a crude manner by standing a cell suspension at elevated temperature and waiting for autolysis to liberate the contents. Controlled results, however, can only be obtained by the use of specific lytic enzymes. These are usually available only in small amounts at high cost and thus are ruled out for large-scale use. The enzymes commonly in use include trypsin, or other proteases, neuroaminidase and lysozyme. Biocatalysts Ltd supply enzymes and enzymic preparations for both large- and small-scale work.

For bacteria, lysozyme is the only enzyme which can be used at large scale, since it can be produced in bulk from egg-white. It catalyses the hydrolysis of the β-1,4-glycosidic bonds in peptidoglycan and is therefore more effective with Gram-positive than Gram-negative bacteria. As mentioned previously, Gram-negative species should be treated with EDTA prior to suspension in a buffered lysozyme solution to increase their susceptibility. Lytic enzyme treatment is particularly effective when followed by mild osmotic shock.

As an example, catalase can be prepared from the lysozyme-susceptible bacterium *Micrococcus lysodeikticus* by the following method.

(1)   Harvest and resuspend *M. lysodeikticus* cells in 0.5% NaCl solution, to a final cell concentration of 5% dry weight to volume.

(2)   For each 1.0 g dry weight of cells, add 0.0125 g dry egg-white preparation.

(3)   Stir gently at room temperature for 20 min; lower temperatures reduce the rate of reaction of the lysozyme.

(4)   Add an equal volume of ethanol (95%), and remove the cell debris and precipitated proteins by centrifugation.

(5)   Increase the ethanol concentration to 75% v/v to precipitate catalase.

Yeast cell walls contain glucans as their major structural component. This is arranged in a net structure, with other complex cell wall polymers. Effective enzyme lysis is best achieved with either crude enzymes or defined enzyme mixtures, since each individual enzyme has restricted access to its appropriate substrate. Highly effective sources of lytic enzymes are *Cytophaga* sp., *Streptomyces* sp., *Actinomyces* sp. and *Helix pomatia* (the edible snail). Commercial preparations of polysaccharide hydrolases are widely available (e.g. Biocatalysts Ltd, Cellulase CP from Sturge Ltd and Novozym from Novo Ltd).

Lytic enzyme treatment is a viable alternative or complement to mechanical or chemical methods of cell disruption where the isolation of an intracellular component is best initiated by a gentle, highly specific procedure. Lytic enzyme treatment is particularly applicable where organisms are resistant to mechanical disruption, or contain sensitive products. The disadvantages are that enzyme preparations are expensive, and lost at each disruption. They also have to be removed from the disruptate in subsequent processing steps.

(vii) *Alkali treatment*. Where the desired protein is stable above pH 11.0, extraction by alkali addition is an inexpensive method of disruption. This technique is possible at large scale, since cheap, bulk chemicals (aqueous ammonia or a concentrated NaOH solution) can be used. Organisms and proteases are rapidly inactivated, so pathogens or genetically engineered cells can be utilized with reduced loss of the desired protein.

L-Asparaginase can be extracted from *Erwinia carotovora* cell paste by the following method.

(1)   Adjust the pH with NaOH to 11.5.
(2)   Allow the cells to equilibrate for 30 min.
(3)   Reduce the pH to 6.5 with acetic acid and clarify the resultant suspension by centrifugation.

(viii) *Detergents and solvents*. Detergents dissociate proteins and lipoproteins from the cell membrane, thus membrane-bound and intracellular components are released. Bile salts, sodium laurylsulphate, sodium dodecylsulphate and Triton are frequently used. Their action is very dependent on pH and temperature. Foaming, protein denaturation and precipitation are distinct disadvantages which may limit their use.

Cholesterol oxidase has been extracted from *Nocardia* sp. with a Triton/buffer solution (24). Detergent treatment is often improved by pre-treatment with solvents, such as acetone, which initiate and stimulate autolysis. It is widely used in the production of enzymes from yeast autolysate (e.g. β-D-galactosidase).

Invertase, a relatively hardy enzyme, can be released from yeast by the following protocol.

(1)   Resuspend a 60% cell paste in warm (40°C) toluene (0.1 M).
(2)   Allow 30−60 min for the yeast to 'liquify', then add papain to complete the digestion.
(3)   Adjust the pH to 4.5 and precipitate invertase by adding ethanol (95%).

Further discussion on the use of detergents and solvents in the purification of membrane-bound proteins can be found in ref. 9.

4. ISOLATION OF ORGANELLES

4.1 **Isolation of Nuclei**—G.H.Goodwin

Numerous procedures have been described for the preparation of nuclei from animal tissue culture cells, most of which involve lysing the cells using detergents and/or mechanical procedures followed by differential centrifugation to pellet the nuclei (see ref. 25 for a review of methods). In the past, highly purified nuclei were required in order to identify chromosomal structural entities such as histones, HMG proteins and scaffold proteins. However, it has been recognized that many important but loosely bound nuclear factors may be washed out of the nuclei during extensive purification procedures. Now that most of the structural components of the nuclei have been characterized, attention has turned to identifying and purifying nuclear factors involved in transcription and replication. In these cases, functional assays are used to monitor the activities of interest and thus highly purified nuclei are not essential and indeed may be detrimental in that proteins may be lost or degraded during extensive nuclear purification schemes. Thus, the emphasis is to produce reasonably pure nuclei rapidly for extraction of labile factors.

4.1.1 *Equipment*

Glass Dounce homogenizers with teflon plungers (Jencons) are required for homogenizing tissues, lysing cells and resuspending pellets. A range of sizes, holding $5-50$ ml, and clearances between 0.15 and 0.95 mm are necessary. Some tissues are more easily homogenized using blenders with rotating blades. Domestic kitchen blenders are perfectly suited to this task when fairly large amounts of tissue (> 10 g) are to be homogenized. Otherwise, there are laboratory blenders with varying blade and container sizes (e.g. Virtis blenders). All stages of the purification should be monitored by light microscopy, using a stain such as 0.2% aqueous methylene blue to stain nuclei and cytoplasmic debris.

4.1.2 *General considerations*

To minimize protein degradation, all operations are carried out at 4°C and protease inhibitors must be added. Most commonly, a stock solution of 50 mM phenylmethyl-sulphonyl fluoride (PMSF) is made up in iso-propanol and this is added to the buffers just prior to use, such that the final concentration of PMSF is 0.5 mM. (Add to the buffer with vigorous stirring to avoid precipitation of the PMSF). The ionic composition of the buffers is important. Extraction of proteins from nuclei increases with ionic strength (most non-histone proteins are extracted with 0.35 M NaCl) and the tendency of nuclei to lyse and aggregate increases with ionic strength. It is therefore advisable to keep the ionic strength of buffers below 0.1. Divalent metal ions are required to stabilize nuclei. Since calcium activates endogenous nucleases, magnesium is usually used in the $2-10$ mM range. If it is important that the nuclear DNA be uncleaved by endogenous nucleases, the method of Hewish and Burgoyne (26) should be used (*Method Table 4*), in which divalent ions are completely excluded by chelating agents and the nuclei are stabilized with spermine and spermidine. We have also found that using $MgCl_2$ and 0.5 mM ethylenebis (oxyethylenenitrilo)-tetra-acetic acid (EGTA) (to complex Ca^{2+} ions) is usually adequate to inhibit nucleases in many tissues and cell types.

Method Tables $3-7$ detail five protocols for isolating nuclei from different cell types. The first two use ultracentrifugation through 2.2 M sucrose to obtain nuclei with minimal cytoplasmic contamination. These two methods are recommended for isolating nuclei from tissues such as liver, spleen, kidney and thymus. The other protocols (*Method Tables $5-7$*) use low speed centrifugations to prepare and wash nuclei; the final product is generally not as pure as that obtained by ultracentrifugation through dense sucrose.

The first step in the preparation of nuclei involves homogenizing the tissue or washed tissue culture cells to lyse the cells. Soft tissues (such as liver) and tissue culture cells are best homogenized with a Dounce homogenizer. A blender is used for more fibrous

Method Table 3. Preparation of thymus nuclei by centrifugation through 2.2 M sucrose[a].

---

1. Rinse the thymus tissue with PBS[b] and mince the tissue with a mincer or, for small amounts of tissue, cut with scissors.
2. Homogenize with ~20 vol of 0.25 M sucrose, 10 mM Tris−HCl, pH 7.5, 10 mM $MgCl_2$, in a blender for 1−3 min.
3. Filter through two layers of surgical gauze to remove connective tissue and centrifuge at 2000 g for 10 min.
4. Resuspend the pellet in the same volume as used in step 2 of 2.2 M sucrose, 10 mM Tris−HCl, pH 7.5, 10 mM $MgCl_2$, by blending for 30 sec.
5. Centrifuge at 30 000 g[c] for 1 h.
6. Resuspend the pellet of nuclei[d] in 0.25 M sucrose, 10 mM Tris−HCl, pH 7.5, 10 mM $MgCl_2$, using a loose-fitting Dounce homogenizer. Approximately 10 mg of DNA is obtained from 1 g of tissue.

---

[a]This procedure can also be used to isolate nuclei from tissues such as liver, kidney and spleen.

[b]The PBS used in Method Tables 3−7 is that described in the Appendix but does not contain the $MgCl_2$ and $CaCl_2$.

[c]For some tissues (e.g. liver) the nuclei should be pelleted at 100 000 g.

[d]The final preparation of nuclei should be checked by electron microscopy, since thymocytes have a very small cytoplasm which is difficult to observe by light microscopy and it is therefore not easy to judge that during the preparation all the cells have been lysed and the cytoplasmic material removed by the centrifugations.

Method Table 4. Preparation of liver nuclei based on the method of Hewish and Burgoyne (26).

---

1. Mince livers from animals starved overnight, using scissors.
2. Homogenize with ~10 vol of 0.34 M sucrose in buffer A (60 mM KCl, 15 mM NaCl, 0.15 mM spermine, 0.5 mM spermidine, 15 mM Tris−HCl, pH 7.4, 0.2 mM EDTA, 0.2 mM EGTA) using a Dounce homogenizer.
3. Filter through two layers of gauze and centrifuge for 15 min at 2000 g.
4. Resuspend the pellet in half the above volume of buffer A containing 2.2 M sucrose and layer the mixture over four 30 ml pads of buffer A containing 2.2 M sucrose in centrifuge tubes.
5. Mix the interfaces and centrifuge at 100 000 g for 1 h.
6. Resuspend the pellet in 0.34 M sucrose in buffer A and centrifuge at 2000 g for 5 min. Approximately 50 mg of DNA is obtained from 50 g of tissue.

---

tissues such as thymus. Homogenization is continued until all the cells are broken open (as checked by light microscopy). Liver cells are readily lysed by homogenization but with refractory cells lysis is assisted if a non-ionic detergent (0.1−0.5% v/v Triton X-100 or Nonidet NP-40) is included in the buffer at this step. However, if nuclei are

Method Table 5. Preparation of nuclei from embryonic chick red blood cells.

1.	Peel off the eggshell and outer membrane of 9 – 15 day old fertilized eggs, leaving the embryo plus yolk within the intact inner membrane. Cut a prominent artery leading from the embryo to the membrane and allow the blood to drain out. Transfer the blood rapidly into PBS containing 5 mM EDTA.
2.	Filter through two layers of surgical gauze.
3.	Centrifuge at 2000 *g* for 10 min and resuspend the cells in 10 – 20 vol PBS – EDTA using a loose-fitting Dounce homogenizer.
4.	Centrifuge and wash the cells twice with PBS.
5.	Centrifuge and resuspend in RSB buffer (10 mM Tris – HCl, pH 7.6, 10 mM NaCl, 3 mM $MgCl_2$) containing 0.5% v/v Nonidet NP-40 using a loose-fitting Dounce homogenizer.
6.	Gently stir for 30 min and centrifuge at 4000 *g* for 10 min or longer if necessary.
7.	Wash the pellet four times with RSB buffer by resuspending with a Dounce homogenizer and centrifuging at 2000 *g* for 10 min. Approximately 10 mg of DNA is obtained from a dozen 15 day old chick embryos.

Method Table 6. Preparation of nuclei from tissue culture cells.

1.	Collect cells grown in suspension by centrifugation and wash with PBS by gentle resuspension and recentrifugation. If cells are grown attached to dishes or roller bottles, decant the medium and rinse the cells with PBS. Scrape the cells off the PBS and collect by centrifugation.
2.	Wash the cells once with RSB buffer (10 mM Tris – HCl, pH 7.6, 10 mM NaCl, 3 mM $MgCl_2$).
3.	Resuspend the cells in 10 – 20 vol of RSB buffer containing 0.1 – 0.5% Triton X-100 and homogenize with a loose-fitting Dounce homogenizer until all the cells have lysed.
4.	Pellet the nuclei by centrifugation at 2000 *g* for 10 min and wash the nuclei twice with RSB buffer by resuspending with Dounce homogenizer and centrifuging at 2000 *g* for 10 min. Approximately 5 mg of DNA is obtained from 10^9 cells.

being isolated for the preparation of a transcription extract, detergents should be avoided and lysis accomplished by using a tight-fitting Dounce homogenizer (27 – 29).

Tissue homogenates are then filtered through gauze to remove connective tissue. In the following steps cytoplasmic contaminants are removed by centrifuging the nuclei through dense sucrose (2.0 – 2.4 M sucrose) or by repeated washes with low ionic strength buffers with or without 0.25 M sucrose. In the case of avian red blood cells (*Method Table 5*), fairly extensive washes are required to remove haemoglobin, but with most tissue culture cells (*Method Table 6*) one or two washes after lysis are adequate.

Thymus tissue is the most convenient source for the preparation of nuclei and nuclear proteins since it has very little cytoplasm and, compared with other cells and tissues, it gives by far the highest yield of nuclei per weight of starting tissue. Having little

Method Table 7. Preparation of thymus nuclei.

---

1. Rinse the thymus tissue with PBS and mince the tissue.
2. Homogenize in a blender running at low speed for ~3 min with 10−20 vol of 0.25 M sucrose, 10 mM Tris−HCl, pH 7.6, 5 mM $MgCl_2$, 0.2−0.5% Triton X-100, and filter through two layers of gauze.
3. Centrifuge at 2000 g for 10 min and repeat step 2.
4. Centrifuge and wash the pellet twice with 0.25 M sucrose, 10 mM Tris−HCl, pH 7.5, 5 mM $MgCl_2$. Approximately 10 mg of DNA is obtained from 1 g of tissue.

---

cytoplasm, centrifugation through 2.2 M sucrose is not essential and a rapid thymus nuclear isolation procedure is given in *Method Table 7*. Proteolytic and nucleolytic activities are also low in thymus cells from most species, except calf. Thymus tissue is thus the tissue of choice when preparing histones, HMG proteins, RNA, and DNA polymerases.

Nuclei from liver and tissue culture cells are a good source of transcription factors (e.g. ref. 29), though the preparation of the pure proteins in sufficient quantities for microsequencing may prove to be very expensive in that large numbers of cells will be required. Since avian erythrocytes (from embryos and adult chickens) are nucleated, they have proved to be very useful in the study of globin gene expression and are a convenient source of nuclear proteins which bind to specific sequences flanking the genes (e.g. ref. 30). Nuclei from adult chicken erythrocytes can be prepared using the procedure described in *Method Table 5* and since large quantities of chicken blood are readily obtainable it will be feasible to purify many of these factors in the quantities required for microsequencing.

4.2 **Isolation of microsomes**—N.Lambert

Due to limitations on space this section covers the isolation of microsomes from only one source (rat liver, the most commonly used source). Although these methods may not be directly applicable to microsome isolation from all other sources, the discussion will give the reader a feel for the important parameters affecting the status of the final microsome preparation. Details of methods for isolation from other sources are given in the following reviews (31−36).

Microsomes principally derive from the endoplasmic reticulum (ER) which comprises an extensive and complex membrane network. The ER accounts for approximately 15% of the total cell volume, about 40% of the total membrane area, 20% of the total protein, 50% of the total phospholipid and about 60% of the RNA in hepatocytes. The proteins/enzymes located within the ER are involved in such diverse functions as xenobiotic metabolism, synthesis and transport of many proteins, and the synthesis of cholesterol, phospholipids and triglycerides. Upon cell breakage, even by mild homogenization, the ER is extensively disrupted, undergoing fragmentation and vesiculation. It is the resulting closed membranous vesicles, isolated from the homogenate, that are termed 'microsomes'. The microsomal fraction will not contain solely ER membranes, nor will the fraction contain the total ER membranes (see Section 4.2.2). It should be noted that unlike many other subcellular fractions such as nuclei, mitochondria and lysosomes, microsomes *per se* do not correspond to an *in vivo* physiological structure. To put it bluntly, microsomes are an artefact of isolation.

Microsomes are heterogenous in size (~ 100 − 300 nm diameter), density and surface charge, but all the vesicles possess the same orientation, i.e. the outside and inside membrane surfaces correspond to the cytoplasmic and luminal surfaces of the ER, respectively. In normal livers about 60% of the ER surface is covered with ribosomes and is termed 'rough' ER, (the remainder being termed 'smooth' ER); microsomes derived from the rough ER are, thus, studded on the outside with ribosomes. Except for the presence of ribosomes, the only differences between types of microsomes are quantitative. Biochemical analysis shows liver microsomal membranes to be composed of approximately 70% protein and 30% lipid.

There are several distinct methods available for the isolation of microsomes from a liver homogenate, with almost as many minor (but not necessarily insignificant) modifications as there are scientists working with these vesicles. It is evident from the literature that 'microsomes' prepared in one laboratory can have significantly different properties to those isolated in others. The choice of isolation procedure depends largely on available resources and the kind of studies to be subsequently undertaken (e.g. large- or small-scale preparations, or the need for undamaged vesicles). In this section three different methods are described, and the merits and applications of each discussed.

4.2.1 *Preparation of the liver homogenate*

The following initial procedures are common to all microsomal preparations and probably have a larger bearing on the properties of the final product than the specific steps discussed further on.

(i) *Choice of animals and husbandry.* The following methods use rat liver as the source; however, if one is contemplating using more than 500 g of liver, larger livers (e.g. bovine or porcine) should be considered. For quantitative studies on microsomal proteins/enzymes, it is important to establish the strain, age and sex of the animals to be used, and once chosen these factors should be kept constant for all experiments. Similarly the type of feed and bedding used can affect certain microsomal components; even residual detergent from cage cleaning has been known to increase the activities of certain cytochrome P450 related enzymes. Usually the animals are starved for 16 − 20 h before they are killed to deplete the glycogen store, which can have adverse effects on the subsequent isolation steps. It should be remembered that starvation itself will affect the animal's metabolism, for example microsomal glucose-6-phosphatase activity increases upon starvation.

(ii) *Liver isolation.* Animals should be killed quickly, causing them minimum anxiety. Not only is this humane; subjecting the rat to stress could significantly affect its metabolism. Cervical dislocation is a common practice for killing rats, however, livers obtained from decapitated rats will contain less blood, and therefore microsomes prepared from these animals are less likely to be contaminated by blood. The administration of lethal doses of anaesthetics for killing could have marked effects on the animal's biochemistry. For killing large numbers of animals, gassing with carbon monoxide or dioxide should be considered. Storage of isolated livers at subzero temperatures will result in some organelle damage and should be avoided if possible.

Freshly excised livers are washed in ice-cold homogenization buffer (see below) to remove as much blood as possible. Perfusion of the livers also reduces the problem of blood contamination. After paper-blotting and weighing, the liver is usually finely minced with scissors prior to homogenization.

(iii) *Homogenization.* Since microsomes are formed as a direct consequence of homogenization this step is critical to the status of the final microsomal preparation. Once a successful protocol has been selected, it should be adhered to rigorously. Use of the same homogenizing apparatus each time is strongly recommended. Most methods described employ an homogenizing medium containing sucrose (0.25 M), to which microsomal vesicles are freely permeable, thus avoiding osmotic damage. Sucrose at this concentration also reduces microsomal aggregation, although higher concentrations (>0.5 M) decrease still further the possibility of aggregation (32). Obviously one should ascertain whether sucrose has a detrimental effect on the protein of interest (e.g. sucrose inhibits microsomal ATPase). The addition of ions (especially divalent cations) to the medium can promote microsomal aggregation (32), so unless specifically required, ions should be omitted. If degassed distilled water is used, the natural buffering capacity of a liver homogenate will result in a pH around 7.0, so unless a different pH is required the addition of buffers is not absolutely necessary.

The scissor-minced liver is homogenized at 4°C, usually with a Potter–Elvehjem homogenizer (or similar apparatus) with a motor driven teflon pestle (Scientific Supplies). This results in good hepatocyte breakage, with minimal organelle damage. The gap between the vessel and pestle is $0.15-0.25$ mm; the exact size is not critical, but should not be too tight (wear and tear will alter the size). The important factors are the speed at which the pestle turns, the speed of vertical motion of the pestle relative to the vessel and the number of passes performed. (One pass equals one up and one down movement). The more extensive the homogenization, the greater is the disruption of cells and recovery of the microsomes, but organelle damage and consequent contamination of the microsomal fraction also increases. As a guideline, four passes at 400 r.p.m. give a low yield of microsomes with high purity, whilst 10 passes at $1000-2000$ r.p.m. give a high yield with low purity (32).

From a $200-250$ g starved rat, a liver weighing about 10 g is obtained. Typically the maximum volume for a Potter homogenizer is 50 ml, thus if the liver is homogenized in 2 vol of medium, only about 15 g liver can be homogenized at one time. Hence, for large-scale preparations one should coordinate killing, liver removal and homogenization, so that dead animals and excised livers are not 'lying around' for excessive periods of time. Alternatively the use of larger, more powerful homogenizers [e.g. Waring blenders (Gallenkamp) or Ultraturrex (Scientific Supplies)] can be employed, although mechanical damage and contamination of the resulting microsomes will be significantly increased.

(iv) *Preparation of post-lysosomal supernatant.* The homogenate is usually centrifuged gently at $600-800$ g for 10 min. Rehomogenizing the resulting pellet, recentrifuging and pooling the supernatants significantly increases the final yield of microsomes. The final 600 g pellet (sometimes termed 'crude nuclear fraction'), containing unbroken

Method Table 8. Isolation of rat liver microsomes by ultracentrifugation.

---

1. Kill two male Sprague—Dawley rats (starved overnight, 150—200 g body wt) by cervical dislocation, and dissect out the livers.
2. Wash livers 3—4 times in ice-cold STKM buffer (250 mM sucrose, 50 mM Tris—HCl, pH 7.5, 25 mM KCl, 5 mM $MgCl_2$) paper-blot and weigh ($\sim 8-10$ g liver/rat). Perform all subsequent procedures on ice or at 4°C.
3. Finely scissor-mince the livers, and add 2 volumes STKM.
4. Homogenize the tissue in a 50 ml Potter—Elvehjem glass homogenizer with a motor driven teflon pestle by making three passes[a] at ~ 800 r.p.m.
5. Centrifuge the homogenate at 680 *g* for 10 min. Retain the supernatant.
6. Rehomogenize the pellet in an equal volume of STKM as described in step 4.
7. Recentrifuge at 680 *g* for 10 min and pool the supernatant with that obtained in step 5. Discard the 'crude nuclear' pellet.
8. Centrifuge the combined supernatants at 10 000 *g* for 10 min. Carefully pipette off the post-lysosomal supernatant and discard the 'lysosomal' pellet.
9. Centrifuge the post-lysosomal supernatant (~ 55 ml) at 100 000 *g* for 60 min.
10. Carefully pipette away the reddish supernatant (cytosol) and any lipid floating on the surface. The sticky, gelatinous reddish-brown pellet is the crude microsomal fraction.
11. Gently resuspend the microsomal pellet in ~ 20 ml 0.15 M Tris—HCl buffer, pH 8.0, using an all-glass hand homogenizer (Gallenkamp).
12. Centrifuge at 100 000 *g* for 60 min. Resuspend the pellet (washed microsomes) in STKM as described in step 11 to a concentration of $10-20$ mg protein ml^{-1}. Discard the supernatant (Tris wash).

---

[a]See text for definition of a 'pass'.

cells, nuclei, the majority of the mitochondria, heavy lysosomes, and fragments of ER, is discarded, and the supernatant recentrifuged at $10-12\ 000$ *g* for 10 min. The pellet (mainly lysosomes and small mitochondria) is discarded and the supernatant (the post-lysosomal supernatant), containing predominantly microsomes and cytosol, is used as the starting material for isolating microsomes.

4.2.2 *Isolation of microsomes from the post-lysosomal supernatant*

(i) *Preparation of microsomes by ultracentrifugation.* Ultracentrifugation is the most commonly employed method for isolating microsomes. *Method Table 8* gives full experimental details of one method for the preparation of rat liver microsomes using ultracentrifugation (37). From the post-lysosomal supernatant microsomes are collected by centrifugation at 100 000 *g* for 60 min. Significantly higher '*g*' forces, or longer centrifugation times, greatly increase the risk of vesicle damage, and therefore should be avoided. The higher viscosity of sucrose solutions over 0.25 M requires harsher centrifugation conditions to pellet the microsomes; hence this is not recommended. The largest rotors commonly available for use at such *g* forces are 8 × 50 ml, and therefore the maximum volume of supernatant processed in one run is approximately 350 ml,

Table 10. Distribution of protein and marker enzymes upon subcellular fractionation of rat liver.

		Mitochondrial		Lysosomal	Microsomal			Cytosolic
Fraction	Protein	GDH	SDH	AP	NDPase	NCCR	PDI	LDH
Homogenate	2175	185[a]	34.3[b]	57.7[c]	640[d]	13.3[e]	14[f]	3550[g]
Crude nuclear								
pellet	44	75	69	21	31	24	37	13
Lysosomal pellet	8.5	7	7	34	10	11	8	4
Cytosol	25	5	4	15	9	3.5	10	65
Tris wash	2	0.5	0	2	1	0.4	1.5	6
Washed microsomes	9.5	1	0	10	31	45	28	0.7
% Recovery	89	88.5	80	81	82	83.9	84.5	88.7

Fractions were obtained as outlined in *Method Table 8*. Homogenate values represent total mg protein and total enzyme activities per 10 g liver, all other values are percentages of that for the homogenate. The marker enzymes used are: GDH, glutamate dehydrogenase; SDH, succinate dehydrogenase; AP, acid phosphatase; NDPase, nucleoside diphosphatase; PDI, protein disulphide-isomerase; NCCR, NADPH cytochrome *c* reductase; LDH, lactate dehydrogenase. Details of assays are given in ref. 37.
[a] μmol 2-oxoglutarate utilized min^{-1}.
[b] μmol 2,6 dichlorophenolindophenol reduced min^{-1}.
[c] μmol *p*-nitrophenol phosphate hydrolysed min^{-1}.
[d] μmol inorganic phosphate released min^{-1}.
[e] μmol cytochrome *c* reduced min^{-1}.
[f] PDI units are arbitrary.
[g] μmol NADH oxidized min^{-1}.

(equivalent for example, to 100 g of liver homogenized in three volumes of medium with no re-extraction of the initial pellet). For processing larger volumes of supernatant, either more centrifuges are required, or the centrifugation must be done in several batches, thereby significantly increasing the preparation time and involving the coordination of post-lysosomal supernatant production to avoid its storage for long periods. These problems can be circumvented by centrifuging at about 40 000 *g* for 1−2 h, thus enabling the use of larger capacity rotors. This procedure will not pellet the smaller microsomal vesicles, but when handling several 100 g of liver, this may not be an important factor.

The 100 000 *g* pellet contains substantial elements of the cytoplasm, and cytosolic proteins, many of which are basic, and thus strongly attracted to the negatively charged microsomal surface (especially in the presence of sucrose) (32). These contaminating proteins are washed off by resuspending the microsomes in 150 mM Tris−HCl buffer, pH 8.0, and repelleting at 100 000 *g* for 60 min (0.5 M KCl is often included). Up to a third of the protein in the initial pellet can be removed without any major loss of the typical microsomal proteins; however, the efficiency of washing is drastically reduced in the presence of sucrose. One wash is sufficient; further recentrifugation and resuspension unnecessarily prolongs the isolation, and greatly enhances the risk of vesicle damage. Whether this washing removes loosely associated membrane proteins that are part of the ER is not known.

Table 10 shows the distribution of protein and marker enzymes in the various fractions obtained using the protocol in *Method Table 8*. It can be seen that a substantial proportion

Method Table 9. Preparation of rat liver microsomes by gel-filtration.

---

1. Obtain two rat livers and treat them as described in *Method Table 8* steps 1−2.
2. Finely scissor-mince the livers and add 1.5 vol STKM.
3. Homogenize the tissue in a 50 ml Potter−Elvehjem glass homogenizer with a motor driven teflon pestle by making six passes at ∼800 r.p.m.
4. Centrifuge the homogenate at 680 g for 10 min and discard the pellet.
5. Centrifuge the supernatant at 12 000 g for 10 min, pipette off the post-lysosomal supernatant and discard the pellet.
6. Using a Pasteur pipette carefully load the post-lysosomal supernatant (∼25 ml) onto a Sepharose-2B (Pharmacia) column (18.5 × 4.5 cm) pre-equilibrated with STKM.
7. Elute the column with STKM at a flow rate of ∼1 ml min^{-1}. The microsomes elute after ∼100 min in the void volume as a thick milky-brown suspension. The reddy cytosol is retarded on the column and can be washed off with further buffer.

---

of the microsomal enzyme activities reside in the crude nuclear pellet. This 'microsomal' activity derives from two sources: cells unbroken by homogenization, and ER that is associated with mitochondria—the so called MITO-ER or rapidly sedimenting ER (38,39). The percentage of ER found in this fraction depends upon homogenization conditions and the physiological state of the cells, but values of around 50% have been reported. Whether the MITO-ER has a specific composition, or function, as compared to the ER of microsomes is still a matter of conjecture.

Only about 10% of the luminal content proteins, protein disulphide-isomerase (PDI) and nucleoside diphosphatase (NDPase), are found in the cytosol, indicating the limited extent of microsomal damage incurred using this protocol. The final microsomal preparation constitutes approximately 10% of the total homogenate protein with a yield of 45% for the integral microsomal membrane protein, NADPH cytochrome c reductase (NCCR), and a slightly lower recovery (∼30%) for the luminal microsomal enzymes, PDI and NDPase, because of losses due to membrane damage.

(ii) *Preparation of microsomes by gel-filtration.* This method, first described by Tangen *et al.* (40), involves fractionating the post-lysosomal supernatant on a gel-filtration column to separate the microsomes from the soluble material; a protocol based on this method (37) is given in *Method Table 9.* A maximum of roughly 50 ml supernatant can be loaded onto the column size described, which is equivalent to about 30 g initial weight of liver. For larger scale preparations either a larger column, or more than one column, will be required. The former will significantly increase the time of preparation and the latter will require attention to be given simultaneously to several columns. For smaller preparations, smaller columns can be employed. The use of fraction collector(s) and pump(s), although desirable, is not essential as one can visually determine when the opaque, fawn-coloured microsomes elute, and 1−2 ml min^{-1} flow rates can easily be achieved by gravitational methods using Sepharose-2B resin. When the columns are not in use they should be stored at about 4°C in the presence of a bacteriostat, such as 0.05% w/v sodium azide.

(iii) *Preparation of microsomes by Ca^{2+} ion precipitation.* The use of precipitation by Ca^{2+} for the preparation of liver microsomes was first described by Kamath *et al.* (41). Subsequently several variations have been developed (42). The method involves the addition to a post-lysosomal supernatant, of Ca^{2+} ions (typically 8 mM $CaCl_2$) which bind to, and aggregate, the microsomal vesicles enabling them to be harvested using relatively mild 'g' forces. Typically 10 000 g for 10 min is used (43), although much lower 'g' forces can be employed (42).

(iv) *Comparison of the different methods.* With the exception of predictably high Ca^{2+} levels in the microsomes prepared by Ca^{2+} precipitation, all three methods described can yield microsomes with similar biochemical compositions. Typically, microsomal preparations are comprised of 75−80% ER membranes, 7−8% plasma membranes, approximately 5% mitochondria, 1% lysosomes and 1% peroxisomes (32). Vesicle integrity of the microsomes prepared using the three different approaches is similar, as determined by the 'latency' of certain microsomal enzymes [thus, the activity of these enzymes cannot be detected (e.g. PDI) or is low (e.g. NDPase), without solubilizing the vesicle membranes (37,44)]. It is advisable to test the authenticity of prepared microsomes using simple marker enzyme assays, no matter which method is chosen for isolating microsomes.

The ultracentrifugation method has the advantage of being the most widely used procedure, and unlike the Ca^{2+} precipitation method the final microsomal preparation is not aggregated. The disadvantages of the method are

(1)   the need for an expensive ultracentrifuge and rotors;
(2)   the difficulties in centrifuging very large volumes at 100 000 g, and
(3)   the enhanced risk of vesicle damage of exposure to high g forces and resuspension.

The gel-filtration method is the gentlest of the three procedures, involving no aggregation, ultracentrifugation or resuspension, and has the benefit of not requiring an ultracentrifuge. The major drawback of the column method is that it does not lend itself readily to large-scale preparations ($>$100 g liver). The yield of microsomal protein per g liver and the concentration of the final microsomal suspension are both significantly reduced compared with the other methods. This stems from not re-extracting the initial 600 g pellet, in order to minimize the volume of post-lysosomal supernatant. Secondly the microsomes are diluted during elution, and the 'total' column eluate containing microsomes is usually not collected.

The advantages of the Ca^{2+} method are its speed, ease of application to large-scale preparations, and, like the gel-filtration method, no requirement for an ultracentrifuge. A drawback of the method is that the resulting microsomes contain substantial amounts of Ca^{2+}, and are aggregated and thus, to some extent, 'damaged'.

(v) *Storage of isolated microsomes.* If subsequent experiments solely involve preparing a soluble extract from which a protein is to be purified, then freezing at −20°C for a few days is adequate. For long-term storage (several weeks), temperatures of −70°C or lower and the inclusion of protease inhibitors (see Chapter 1) are recommended.

Method Table 10. Isolation of rat liver mitochondria.

1.	Kill two 120−150 g Sprague−Dawley rats by spinal dislocation (usually male rats, strain CD-1, are used and starved for 24 h prior to use).
2.	Remove the livers quickly and chop coarsely. Wash in 4−5 50 ml volumes of cold (4°C) buffer A (200 mM sucrose, 10 mM Hepes−NaOH, pH 7.4) to remove blood.
3.	Homogenize the livers in ~120 ml of buffer A at 4°C using a motor driven teflon 'Potter' homogenizer at a rotational rate ≤1200−1500 r.p.m. for 3−5 passes of 10 sec each.
4.	Dilute to 250 ml with cold buffer A and centrifuge at 2000 g for 10 min. Discard the pellet.
5.	Remove fat and lipid with a pasteur pipette from the supernatant and centrifuge at 9000 g for 10 min. Discard the supernatant and loosely packed material.
6.	Resuspend the crude mitochondrial pellet in 1−3 ml of cold buffer A, taking care not to disturb the blood cells at the bottom of the tube. These are easily identified by their red-black colour.
7.	Dilute to 80 ml. Repeat the 9000 g centrifuge step twice more. The final mitochondrial pellet will be a pale buff colour.
8.	Resuspend in buffer A to give a protein concentration of 80−100 mg ml^{-1}. Yields are typically 100−150 mg of protein.

If microsomes are to be stored and vesicle integrity maintained, they should be frozen as pellets at −20°C or lower (45). Even so, total intactness will only be retained for a few days. For further discussion of microsomal storage see Brown and Hallinan (45).

4.3 Isolation of mitochondria and mitochondrial enzymes—D.Griffiths

This section describes the isolation of mitochondria from rat liver and bovine heart muscle (the most commonly used sources) and the subsequent purification of proteins involved in ATP synthesis and related reactions. In particular points which often cause trouble are commented on. For additional information the reader is referred to refs 46 and 47.

Throughout this section centrifugation forces refer to that at the mid-point of the centrifuge tube (g_{av}). All operations are at 4°C unless otherwise stated. Brand names have been given for the equipment used; if these are not available the conditions required for other types should be optimized. The reader is referred to the following references for details of appropriate enzyme assays: ATP hydrolysis (48,49); ATP−^{32}P$_i$ exchange (50); and P$_i$−^{32}P$_i$ exchange (51). Protein concentrations are usually measured by the Lowry method (see Chapter 1) for rat liver mitochondria (52,53) and by a Biuret assay modified for membrane proteins (54).

4.3.1 *Isolation of rat liver mitochondria*

The method described in *Method Table 10* is a modification of that described by Pedersen *et al.* (55), except that sucrose is substituted for KCl as an osmotic agent; many other

Method Table 11. Isolation of rat liver mitoplasts.

---

1. Prepare rat liver mitochondria as described in *Method Table 10*. At the final step resuspend to a protein concentration of 100 mg ml^{-1}.

2. Mix with an equal volume of digitonin solution (1.2% w/v) in buffer B (700 mM sucrose, 200 mM mannitol, 2 mM Hepes−NaOH, pH 7.4, 0.05% w/v heated-treated Fraction V BSA. Stir slowly on ice for 15−18 min.

3. Dilute 3-fold with buffer B and centrifuge at 10 000 g for 10 min. Discard the supernatant (or use as a source of outer membrane enzymes).

4. Gently resuspend the pellet in buffer B to ~50% of the original mitochondrial suspension volume (step 1).

5. Centrifuge at 10 000 g for 10 min. Resuspend the pellet at a protein concentration of 100 mg ml^{-1}.

6. If required store at $\leq -60°C$ after rapid freezing in liquid nitrogen.

---

variations in buffer composition and general procedures have been used. Tris should be avoided if high levels of respiratory control are required. For many studies, especially if the mitochondria are to be used for protein purification, the method described gives suitably pure mitochondria. However, residual contamination may be removed by centrifuging the mitochondria through gradients of either sucrose or Percoll (56).

Removal of calcium from buffers is essential since it activates phospholipase A-2, and, more seriously, induces uncoupling of mitochondrial respiration. Thus, sucrose solutions should preferably be left to stir overnight with a small quantity of cation exchange resin. In addition, EGTA may be added to 1 mM in all buffers (EDTA should not be used, since this chelates Mg^{2+} as well as Ca^{2+}, and will cause leakiness of the membranes).

All equipment used for mitochondrial preparation should be thoroughly washed with glass-distilled and de-ionized water, and should be designated for mitochondrial preparation only. Traces of detergent may lead to uncoupling of respiration and therefore re-usable plasticware should not be treated with detergent and glassware should be thoroughly rinsed with distilled water.

The pestle of the 'Potter' homogenizer should have a clearance of 0.25 mm, and have curved grooves of approximately 0.1−0.2 mm depth cut into the teflon head.

4.3.2 *Isolation of rat liver mitoplasts*

Mitoplasts (mitochondria devoid of an outer membrane) are produced by treating mitochondria with low concentrations of detergent. These are particularly useful for the purification of enzymes involved in oxidative phosphorylation, since enzymes, such as adenylate kinase, which otherwise often co-purify, are removed. The method given in *Method Table 11* is adapted from that of Schnaitman and Greenawalt (57).

The purity of the mitoplast preparation can be assessed by measuring the activities of monoamine oxidase (MAO) (an outer mitochondrial membrane enzyme) and adenylate kinase (AK) (an enzyme present in the intramembrane space) (58). Activities of these enzymes should be 5% (MAO) and 2% (AK) of their original mitochondrial values;

Method Table 12. Purification of phosphate/proton symporter.

---

1. Prepare rat liver mitoplasts (300 mg protein) as described in *Method Table 11*. If these have been stored frozen prior to use thaw on ice.
2. Add an equal volume of buffer C [40 mM KCl, 20 mM KP_i^a, 2 mM EDTA, pH 7.2 (with KOH), 4°C]. Incubate on ice for 10 min.
3. Dilute 1.6-fold with buffer D [20 mM KCl, 10 mM KP_i, 1 mM EDTA, pH 7.2 (with KOH), 4°C] to a protein concentration of 30 mg ml^{-1}. Incubate on ice for 5 min.
4. Dilute with buffer D to a protein concentration of 3.6 mg ml^{-1}. Incubate on ice for 5 min.
5. Centrifuge at 20 000 g for 10 min. Discard the supernatant.
6. Resuspend the pellet to a protein concentration of 50 mg ml^{-1} in buffer D. Add an equal volume of 6% v/v Triton X-114, 6 mg ml^{-1} diphosphatidyl glycerol in buffer D, and incubate for 20 min on ice.
7. Centrifuge at 130 000 g for 35 min. Discard the pellet.
8. Apply the supernatant to columns of dry hydroxyapatite (0.5 g) prepared in Pasteur pipettes. Elute with 1% v/v Triton X-114, 2 mg ml^{-1} cardiolipin in buffer D. Collect 1 ml fractions.
9. Pool the initial fractions. Apply aliquots containing $1-2$ mg protein to a column of DEAE-Sepharose CL-6B (Pharmacia-LKB) equilibrated in 1% v/v Triton X-114, 1 mg ml^{-1} cardiolipin in buffer D. Wash with 12 ml of equilibration buffer and elute with 50 ml of a $0-0.3$ M NaCl gradient in equilibration buffer.
10. Pool the fractions containing the phosphate transport protein. Apply aliquots containing 0.5 mg protein to a 3 ml column of Affi-gel 501 or other similar organomercurial gel equilibrated in de-ionized water. Wash with 8 ml of 1% v/v Triton X-114, 1 mg ml^{-1} cardiolipin in buffer D. Elute with 50 ml of a $0-25$ mM gradient of 2-mercaptoethanol in wash buffer.
11. Pool fractions containing the active phosphate transport protein. Store in liposomes as described in *Method Table 13* for $4-6$ days.

$^a KP_i = K_2HPO_4$ plus KH_2PO_4.

if higher values are obtained an additional wash and centrifuge step sometimes helps.

Fraction V BSA often contains contaminating protease activities. These should be destroyed by heating a 10% w/v BSA solution in 0.1 M sodium phosphate buffer, pH $6.5-7.5$, at $55-60$°C for 3 h.

4.3.3 *Purification of rat liver phosphate/proton symporter*

The method given in *Method Table 12* is adapted from that of Kaplan *et al.* (58). The enzyme should be assayed by $P_i - ^{32}P_i$ exchange (51,59); this assay should be used to optimize conditions for sonication.

4.3.4 *Isolation of bovine heart mitochondria*

Various procedures have been described for the isolation of bovine heart mitochondria (61,62). Although liver tissue is often preferred for preparation of mitochondria for

Method Table 13. Preparation of phosphate transporter containing liposomes.

---

1. Dry a volume of stock solution containing 230 mg azolectin in a Corex glass tube with a stream of nitrogen (the stock solution in chloroform should be stored under nitrogen at $-20°C$ either in brown glassware or in the dark).
2. Redissolve in peroxide-free diethyl ether (see ref. 60). Dry under a stream of nitrogen and then under vacuum for 3 h.
3. Add 2.1 ml of buffer E [50 mM KCl, 20 mM Hepes, 10 mM KP_i, 1 mM EDTA, pH 6.6 (with KOH), 20°C]. Flush the tube with nitrogen or argon and seal it.
4. Place the tube in a water bath sonicator, for up to 30 min, to form liposomes (the dispersion will appear transparent when liposomes have formed).
5. Mix 20 μg of purified phosphate transporter with 550 μl of liposomes and vortex.
6. Rapidly freeze in liquid nitrogen and store at $-80°C$ for up to 6 days.
7. To use thaw the liposomes and sonicate for 3×6 sec with 10 sec intervals between bursts, using a probe sonicator. Dilute 1.5-fold with buffer E and stand for 10 min.

Method Table 14. Isolation of bovine heart mitochondria.

---

1. Obtain bovine hearts ($2-20$) from freshly slaughtered animals and immediately place on ice.
2. Wash free from blood with ice-cold 250 mM sucrose. Remove fat and connective tissue and dice the muscle tissue into ~ 2 cm cubes. Mince in an electric mincer.
3. Resuspend in buffer F (170 mM KCl, 1 mM EDTA, 10 mM Tris$-$HCl, pH 7.4, 4°C) at a ratio of $220-230$ g tissue per litre of buffer.
4. Homogenize using a high frequency disperser (e.g. Tekmar SD-45) for 2.5 min at 10 000 r.p.m.
5. Centrifuge at 14 000 g for 15 min. Discard the pellet.
6. Filter the supernatant through four layers of muslin. Centrifuge in a continuous flow centrifuge (e.g. Sharples) at ~ 100 ml min^{-1} and 50 000 r.p.m.
7. Resuspend the pellet in buffer G (250 mM sucrose, 5 mM Tris$-$HCl, pH 7.8). Adjust the pH to 7.8 with 1 M Tris base.
8. Centrifuge at 14 500 r.p.m. for 30 min. Discard supernatant.
9. Wash gently with buffer G and resuspend at a protein concentration of 100 mg ml^{-1} in buffer G. Adjust the pH to 7.5. Store mitochondria at $-20°C$. Yields of mitochondria are $0.4-0.5\%$ w/w.

---

physiological studies, heart muscle is the tissue of choice for preparation of submitochondrial particles, or for isolation of inner membrane enzymes and electron transfer complexes. This is due to (i) the relative ease of preparing large quantities (~ 100 g) of mitochondria, and (ii) the better quality of enzyme produced compared to liver tissue. The reason for the latter difference is not clear, although the higher levels of free fatty acids in liver mitochondria preparations have been implicated.

The procedure described in *Method Table 14* is adapted from that described by Joshi

Method Table 15. Isolation of electron transport particles.

1.	Thaw ~20 g of bovine mitochondria prepared as described in *Method Table 14*.
2.	Resuspend in buffer H (250 mM sucrose, 10 mM Tris$-H_2SO_4$, pH 7.5, 4°C) at a protein concentration of 20 mg ml^{-1}. Adjust the pH to 7.5 with 2 M H_2SO_4.
3.	Centrifuge at 8000 *g* for 10 min. Discard the supernatant. Remove the upper 'fluffy' layer of the pellet, which consists of broken mitochondria (usually ~20% of the total mitochondria). Use these mitochondria for preparing oligomycin-insensitive ATPase (see *Method Table 17*).
4.	Resuspend the mitochondria in buffer H at 20 mg ml^{-1}. Add 1.5 ml of 1 M $MgCl_2$ and 1.6 ml of 60 mM ATP per 100 ml of suspension. Adjust the pH to 7.5 with H_2SO_4 or KOH.
5.	Sonicate 150 ml aliquots for 3 × 20 sec at maximum power output, with cooling to 4°C between bursts.
6.	Centrifuge at 12 500 *g* for 10 min.
7.	Remove the supernatant and centrifuge at 114 000 *g* for 45 min. Discard the supernatant.
8.	Resuspend the ETP$_H$ particles in buffer H at a protein concentration of 20 mg ml^{-1}. Store at −60°C until required.

Method Table 16. Purification of oligomycin-sensitive ATP synthase.

1.	Thaw on ice ~2 g of ETP$_H$ prepared as describe in *Method Table 15*.
2.	Centrifuge at 114 000 *g* for 45 min. Resuspend in buffer H (250 mM sucrose, 10 mM Tris$-H_2SO_4$, pH 7.8, 4°C).
3.	Dissolve lysolecithin at 4 mg ml^{-1} in buffer H. Add an equal volume to the ETP$_H$ suspension. Stir on ice for 30 min (for consistency use the same size flea and same stirring rate for each preparation).
4.	Add an equal volume of buffer I (100 mM Mes$-$KOH, pH 6, 4°C) and mix by inversion. Leave to stand for 20 min.
5.	Centrifuge immediately at 105 000 *g* for 15 min.
6.	Recentrifuge the supernatant at 140 000 *g* for 150 min. Discard the supernatant.
7.	Resuspend the gelatinous pellet in buffer J (250 mM sucrose, 1 mM DTT, 10 mM Tris$-$acetate, pH 7.5, 4°C).
8.	Rapidly freeze aliquots in liquid nitrogen and store at −60°C or below.

and Sanadi (63), and has been routinely used for the purification of electron transport particles, vesicular oligomycin-sensitive ATP synthase, and F_1 oligomycin-insensitive ATPase. Similar procedures are also applicable to the preparation of porcine heart mitochondria (64).

Lower yields of mitochondria are obtained if domestic food processors (e.g. those produced by Braun) are used for homogenizing. It is also important to check the pH periodically, and adjust if necessary to pH 7.4−7.5 with either 1 M HCl or 1 M Tris.

4.3.5 *Isolation of electron transport particles (ETP$_H$)*

The method given in *Method Table 15* is based on that of Beyer (65). Sonicators used

Method Table 17. Purification of oligomycin-insensitive ATPase.

---

1. Resuspend ~25 g of disrupted mitochondria (prepared as described in *Method Table 15*) in buffer K (250 mM sucrose, 10 mM Tris−HCl, pH 7.5, 4°C) to a final protein concentration of 31 mg ml^{-1}.

2. Cool ~75 ml aliquots in ice-water. Sonicate for 4 min. Do not allow the temperature to rise above 30°C.

3. Centrifuge at 112 000 g for 15 min.

4. Recentrifuge the supernatant at 112 000 g for 90 min. Discard the supernatant.

5. Resuspend the pellet in buffer K at 20 mg ml^{-1} protein. Equilibrate to 25°C.

6. Sonicate 40 ml aliquots at 140 W for 50 sec. Do not allow the temperature to rise above 40°C.

7. Centrifuge at 110 000 g for 90 min at 25°C. Discard the supernatant.

8. Resuspend the pellet in 100 mM sucrose, 4 mM ATP, 2 mM EDTA at 30 mg ml^{-1} protein. Adjust the pH to 9.2 with ammonium hydroxide and incubate overnight at 25°C.

9. Sonicate ~75 ml aliquots at 150 W for 10 min using a 250 ml Rosette cell immersed in a water bath at 40°C.

10. Centrifuge at 110 000 g for 90 min at 25°C. Discard the pellet.

11. Adjust the pH of the yellow-coloured supernatant to pH 5.4 with 3 M acetic acid.

12. Centrifuge at 10 000 g for 5 min. Discard the pellet.

13. Adjust the pH of the supernatant to pH 8 with 2 M Tris base (complete this step within 10 min).

14. Equilibrate DEAE-Sephadex A-50 as follows: after swelling in water, wash with 400 ml of 200 mM Tris−HCl, pH 8. Decant any liquid. To the slurry (~180−210 ml) add 1 ml of 500 mM disodium-EDTA, 5 ml of 200 mM ATP, 10 ml of 500 mM Tris−HCl, pH 8 and distilled water to give a total volume of 250 ml. Pour the slurry into a 5 cm diameter column and wash with 100 ml of buffer L (4 mM ATP, 2 mM EDTA, 20 mM Tris−HCl, pH 8).

15. Apply the crude enzyme solution to the DEAE-Sephadex A-50. Wash with 100 ml of buffer L, then 400 ml of buffer L containing 0.1 M Na$_2$SO$_4$. The F$_1$-ATPase is eluted with buffer L containing 0.15 M Na$_2$SO$_4$, usually after ~60 ml of buffer.

16. Pool fractions containing the F$_1$-ATPase. Add solid (NH$_4$)$_2$SO$_4$ to a final concentration of 2.8 M. Incubate overnight at 4°C.

17. Centrifuge at 2500 g for 5−10 min.

18. Dissolve the pellet in 250 mM sucrose, 2 mM EDTA, 4 mM ATP, 50 mM Tris−HCl, pH 8 at 25°C. Centrifuge at 2500 g for 10 min to remove undissolved protein.

19. Add solid (NH$_4$)$_2$SO$_4$ to the supernatant to give a concentration of 2 M. Store at 4°C for up to several weeks.

---

for preparing submitochondrial particles including ETP$_H$ should be calibrated by varying the power output and determining the % yield of particles and the resultant enzyme activity of interest.

4.3.6 *Purification of oligomycin-sensitive ATP synthase and oligomycin-insensitive (F$_1$)ATPase*

Oligomycin-sensitive ATP synthase (H$^+$-ATPase, F$_0$F$_1$-ATPase) is prepared from ETP$_H$ using the method of Hughes *et al.* (66) (*Method Table 16*). Oligomycin-insensitive ATPase (F$_1$-ATPase) is prepared from the disrupted mitochondria, obtained as a by-product of ETP$_H$ preparation, using the method of Knowles and Penefsky (67) (*Method Table 17*). F$_1$-ATPase is cold-labile and therefore many of the steps throughout the purification are carried out at room temperature. In addition the enzyme should be stored as an ammonium sulphate precipitate at 4°C and not in solution.

Purity of the ATPases is assessed by SDS−PAGE and by enzyme assays. Oligomycin-insensitive ATPase should contain five subunits. Enzyme assays for adenylate kinase (57), succinate dehydrogenase (68) and cytochrome oxidase (69) should be negative. The activity of the ATPases is measured by hydrolysis of ATP (48,49) and ATP−^{32}P$_i$ exchange (50). Oligomycin-sensitive ATP synthase preparations usually have a specific activity of 500−800 nmol min^{-1} mg^{-1} protein for ATP−^{32}P$_i$ exchange and both the ATPase activity and ATP−^{32}P$_i$ are inhibited by oligomycin (50,70). Preparations of the oligomycin-insensitive ATPase normally have a specific activity of 100 μmol ATP hydrolysed min^{-1} mg^{-1} protein at 30°C; this activity should be insensitive to added oligomycin and there should be no ATP−^{32}P$_i$ exchange activity.

4.4 **Isolation of chloroplasts**—M.R.Hartley

Chloroplasts, in addition to carrying out photosynthesis and possessing a genetic system which encodes and synthesizes a limited number of proteins and RNA, are the site of biosynthesis of a multitude of compounds (e.g. starch, amino acids, lipids, terpenoids, phenolics and quinones) (71). There is no one source of plant material nor one all-embracing method of chloroplast isolation which will yield chloroplasts suitable for all purposes. For example chloroplasts isolated from mature leaves of spinach (*Spinacea oleracea*) show excellent rates of CO$_2$-dependent O$_2$ evolution, but low rates of protein and RNA synthesis. Also the composition of the medium in which the chloroplasts are suspended is critical for success. For purification of proteins, sources which give the highest yield of intact chloroplasts are preferable (e.g. spinach or pea). Limitations of isolated chloroplasts are their short life-span, the loss of cytoplasmic factors which may control chloroplast functions and the fact that for many species there are inherent unresolved problems in isolating active chloroplasts. A major problem is the presence of phenolics in the vacuoles of leaf cells which come into contact with phenol oxidases on disruption of the tissue and produce quinones and condensed tannins. These are very potent inhibitors of enzymes and irreversibly damage the chloroplast envelope (71). High levels of thiol reagents have been used to inhibit polyphenol oxidase (72), but these may also inhibit other enzymes (e.g. those involved in photophosphorylation). Alternatively, polyvinyl pyrollidone may be included in the medium to bind tannins (73).

4.4.1 *Classification of isolated chloroplasts*

The most widely adopted classification of isolated chloroplasts was introduced by Hall (74) and is summarized in *Table 11*. It should be emphasized that this classification is based on photosynthetic activities and structural criteria and may not be directly

Table 11. Types of chloroplast preparations and some of their properties, after Hall (74).

Chloroplast type	Description	Preparation	Envelope	Rate of CO_2 fixation (mol CO_2 mg^{-1} chlorophyll h^{-1})	Permeability properties of envelope
A	Whole, intact chloroplasts	Rapid grinding in isotonic sugar solution. Rapid centrifugation	Intact	50–250	Impermeable to NADP$^+$ and ferricyanide
B	Intact chloroplasts	Isotonic or hypertonic sugar or salt, often involving lengthy centrifugation	Appear intact by microscopy but have been damaged	50 depending on envelope damage and loss of stroma	Permeable to NADP$^+$ and ferricyanide

applicable to the macromolecular synthetic activities of chloroplasts. Type A chloroplasts show high rates of CO_2 fixation, but are unable to translocate ferricyanide (a commonly used electron acceptor), $NADP^+$ and ATP through their intact envelopes; however, they do translocate metabolites for which specific carrier systems exist. Type A chloroplasts isolated from young, expanding leaves are also able to carry out RNA, DNA and protein synthesis, the post-translational uptake and processing of the cytoplasmically-synthesized precursors of chloroplast proteins and the assembly of functional multi-subunit enzyme complexes (75,76). Type B chloroplasts appear intact by microscopy and possess a complete envelope; however, they are permeable to ferricyanide and $NADP^+$. The differences between type A and type B chloroplasts are the result of lengthy centrifugations, often involving sucrose gradients, which result in damage to the envelope and leakage of stroma. There is evidence that envelopes can rupture and reseal (77). Historically, type B chloroplasts have proved valuable for enzyme localization studies and most of the accumulated knowledge in this area has been gleaned with their use (78). Type A chloroplasts are preferable for protein isolation, therefore the remainder of this section will concentrate on methods which yield type A chloroplasts and their purification.

4.4.2 *Choice of plant material*

Spinach (*S. oleracea*) and garden pea (*Pisum sativum*) are among the few plants which consistently yield type A chloroplasts and are relatively easy to grow. Tobacco, lettuce and cereals (wheat, barley and maize) have advantages for some purposes (71). Preparations of chloroplasts derived by mechanical disruption of leaves (Section 4.4.3) of the majority of higher plants have little biochemical activity. Recently, the use of protoplasts derived from leaves by digestion of cell walls with cellulase and pectinase has greatly extended the range of plants from which active chloroplasts can be prepared, but has the limitation that only relatively small quantities of chloroplasts can be prepared (79). Spinach Hybrid 102 Yates (The Seed Centre, Withyfold Drive, Macclesfield, UK) and pea var. Feltham First (Brookes Seeds Limited, Sleaford, Lincs, UK) are recommended varieties. Market-purchased spinach is often a suitable source of chloroplasts for photosynthetic studies, but should be avoided for macromolecular synthesis studies because of spurious activity resulting from bacterial contamination. Be sure to check that market-bought spinach is 'true' spinach and not 'perpetual' spinach (*Beta vulgaris*), which is unsuitable.

(i) *Growth of plants*. Two conditions must be met if intact chloroplasts with good rates of light-driven macromolecular synthesis are to be routinely obtained. Firstly, young expanding leaves must be used, since chloroplasts in mature leaves synthesize little RNA and protein. Secondly, the leaves must be free of starch grains, otherwise these dense particles rupture the chloroplast envelope during leaf grinding and centrifugation. The former condition is relatively unimportant for chloroplasts for photosynthesis studies. Peas are much easier to grow than spinach. At the University of Warwick peas are grown in a windowless, air-conditioned room maintained at $20-24°C$. Seeds are sown at high density (~ 30 seeds per 100 cm^2) in Levingtons compost in seed trays and placed under a bank of warm white fluorescent tubes (~ 40 μmol photons m^{-2} s^{-1} at plant height) with a 12 h photoperiod for $7-10$ days. This relatively low light intensity

Table 12. Nutrient solution for spinach water culture.

No.	Solutions	Stock concentration	Volume (ml per 20 litres nutrient solution made up with distilled water)
1	KNO_3	1 M	120
2	$Ca(NO_3)_2$	1 M	80
3	$MgSO_4$	1 M	40
4	KH_2PO_4	1 M	20
5	$MgCl_2$	1 M	80
6	Trace elements	(see below)	20
7	NaFe−EDTA	40 mM	20

	Trace elements	Quantity (mg) in 250 ml H_2O	
	H_3BO_3	715	
	$MnCl_2 \cdot 4H_2O$	452	
	$ZnSO_4 \cdot 7H_2O$	55	
	$CuSO_4 \cdot 5H_2O$	20	
	$NaMoO_4 \cdot 2H_2O$	7.25	

does not induce starch formation. Spinach seeds are germinated in Levingtons compost for about 8 days, when the cotyledons have expanded. The seedlings are then pricked out and held in 2.5 cm apertures cut in the lid of an aquarium tank by means of foam rubber plugs. The tank is filled with nutrient solution (*Table 12*) and aerated with an aquarium pump. Spinach may be successfully grown under greenhouse conditions without supplementary lighting during the spring and autumn months. However, it flowers when the daily photoperiod exceeds about 13 h, producing little leaf material, thus in the summer months the photoperiod should be shortened to 12 h by an automatic blind. During the winter months supplementary illumination is necessary and at Warwick we use Growlux sodium vapour lamps delivering about 200 μmol m^{-2} sec^{-1} at plant height. Temperature is relatively unimportant in full daylight, but in poor winter light, 25°C should not be exceeded. Starch accumulation is not usually a problem in rapidly growing spinach, but it is advisable to check for its presence by I/KI staining or microscopic examination of a squashed leaf. If starch is detected, plants should be transferred into darkness for 24 h, then illuminated for 15−20 min immediately prior to chloroplast isolation.

4.4.3 *Isolation of crude chloroplasts*

The most active type A chloroplasts are obtained from leaves homogenized in isotonic medium at low temperature, followed by rapid separation of the chloroplasts from the homogenate. The method described below is suitable for chloroplasts to be used for both photosynthesis and macromolecular synthesis studies.

(i)   Harvest the apical leaves of pea plants using scissors, or remove spinach leaves by hand. Large spinach leaves should be de-ribbed and cut into 10 mm wide strips.

(ii)   Place the leaves (up to 90 g) in a pre-cooled square sided Perspex vessel (50 × 50 × 200 mm deep) and pour in 300 ml of semi-frozen SIM medium (0.35 M sucrose, 25 mM Hepes−NaOH, pH 7.6, 2 mM EDTA Na$_2$, 2 mM sodium isoascorbate). Autoclave at 10 p.s.i. for 15 min, minus isoascorbate, allow to cool, then add the isoascorbate (NB, isoascorbate is destroyed by autoclaving). Place at −20°C prior to use until an ice slurry forms). Homogenize for 4 sec using a Polytron PT20 with a PT35 head (Northern Media Supply) at speed setting 7. The top of the probe should be 2 cm above the bottom of the container. If a Polytron is not available domestic food blenders are also suitable.

(iii)   Using gloved hands, immediately squeeze the homogenate through four layers of hairy muslin and pour the filtrate through eight more layers. Pour the filtrate into polycarbonate centrifuge tubes and place them in a suitable swing-out centrifuge rotor pre-cooled to 0°C. It is important that the centrifuge has rapid acceleration and braking. We use the Mistral 6L (MSE) with a 4 × 100 ml rotor (radius 14.5 cm). The rotor is accelerated rapidly to 4000 r.p.m. (2500 *g*) and held at this speed for 60 sec, and then decelerated with the brake full on.

(iv)   Decant the supernatant, and keeping the tubes inverted, wipe their insides with tissue. Place the tubes on ice and add 5 ml of SIM to each pellet. Gently resuspend the pellets with a small piece of cotton wool and a glass rod. Alternatively, the cotton buds sold in pharmacies are ideal for this purpose. Add a further 45 ml SIM to each tube and repeat the centrifugation step above.

(v)   Gently resuspend the chloroplasts in the incubation medium of choice. For experiments involving measurements of CO_2 fixation or CO_2-dependent O_2 evolution the medium of Jensen and Bassham (80) is often used (0.33 M sorbitol, 2 mM EDTA Na$_2$, 1 mM MgCl$_2$, 50 mM Hepes, 5 mM Na$_2$P$_4$O$_7$, 0.5 mM Na$_2$HPO$_4$ and adjusted to pH 7.6 with KOH). Two incubation solutions have been used to study light-driven protein and RNA synthesis. Solution A (0.2 M KCl, 66 mM Tricine−KOH, pH 8.3, 6.6 mM MgCl$_2$) was devised by Ramirez *et al.* (81) and contains K^+ ions both as osmoticum and essential cofactor for protein synthesis. Solution B (0.33 M sorbitol, 66 mM Tricine−KOH, pH 8.3) contains K^+ ions from the KOH used to neutralize the buffer. The relative merits and limitations of these two solutions are discussed in detail by Ellis and Hartley (75) and briefly summarized here.

(1)   Pea and spinach chloroplasts show high rates of CO_2-dependent O_2 evolution in solution B but not in solution A.

(2)   The rates and products of light-driven protein synthesis are the same in both solutions, but the assembly of *in vitro* synthesized large subunit of ribulosebisphosphate carboxylase:oxygenase into the holoenzyme occurs only in solution B.

(3)   The rate of light-driven RNA synthesis by spinach chloroplasts is higher in solution A than solution B, but the reaction proceeds for longer times in solution B and the RNA products formed undergo more extensive processing.

For most purposes, a chlorophyll concentration of between 100−500 μg ml^{-1} is suitable.

Method Table 18. Solution required for silica-sol gradients.

1. 5 × GR medium stock: 1.65 M sorbitol, 0.25 M Hepes, 10 mM EDTA Na$_2$, 5 mM Na$_4$P$_2$O$_7$, 5 mM MgCl$_2$, 5 mM MnCl$_2$. Adjust to pH 6.8 with NaOH; autoclave for 15 min at 15 p.s.i.
2. Isoascorbate buffer: 0.5 M sodium isoascorbate, 50 mM Hepes. Adjust to pH 7.0 with NaOH, filter sterilize and store frozen.
3. 1 × GR medium (200 ml): 40 ml of 5 × GR medium stock, 2 ml of isoascorbate buffer. Dilute to 200 ml with sterile, distilled water.
4. PCFB medium: 25 ml of Percoll^R, 1.05 g of polyethylene glycol 6000 (BDH), 0.25 g of BSA (Sigma, fraction V), 0.35 g of Ficoll (Sigma, type 400). Stir for ~30 min until all solids dissolve.
5. 'Light' gradient solution: 7 ml of 5 × GR medium stock, 0.35 ml of isoascorbate buffer, 6 mg of solid glutathione, 3.5 ml of PCFB medium. Add sterile distilled water to 35 ml.
6. 'Heavy' gradient solution: 7 ml of 5 × GR medium stock, 0.35 ml of isoascorbate buffer, 6 mg of solid glutathione. Add PCBF medium to 35 ml.

Method Table 19. Preparation of gradients.

Gradients should be formed in 50 ml polycarbonate centrifuge tubes for appropriate wind shield swing-out rotors (e.g. DuPont/Sorval HB4 or MSE 43124-705). It is important that the centrifuge used has a slow acceleration device to avoid disruption of the gradient on starting.
1. Pipette 17 ml of 'light' gradient solution in the chamber of a parallel-sided, two chamber, gradient maker furthest from the outflow. Pipette 17 ml of 'heavy' gradient solution in the mixing chamber, add a small magnetic flea and open the valve connecting the chambers.
2. Using a peristallic pump, pump the solution into the centrifuge tube held at an acute angle such that the solution runs slowly down the wall of the tube.
3. Cool the gradients on ice, or place at 4°C until required. They are stable for at least 16 h.

(vi) Determine the percentage of intact chloroplasts. This is most conveniently done by examination in a phase-contrast microscope. Intact chloroplasts appear highly refractile whereas broken ones appear dark and granular. Preparations should contain at least 40% intact chloroplasts.

Crude chloroplast suspensions prepared by this rapid procedure are contaminated to varying extents by stripped thylakoid membranes, chromatin (from disrupted nuclei), mitochondria, peroxisomes, ribosomes and cytosolic enzymes. The amounts of these contaminants may be lessened by several cycles of resuspension and re-pelleting of chloroplasts, but differential centrifugation can only be used to separate particles whose sedimentation coefficients differ by a factor of 10 or more (84). It must also be borne in mind that only a relatively small proportion (typically 10−15%) of the chloroplasts

Method Table 20. Running the gradients.

---

1. Resuspend crude chloroplast pellets in $1 \times$ GR medium to give a chlorophyll concentration of ~ 0.5 mg ml^{-1}.
2. Using a 10 ml pipette, carefully layer up to 5 ml of chloroplast suspension on top of each gradient.
3. Place tubes in a rotor and accelerate slowly to 8500 g max at 4°C. Run for 20 min then decelerate with the brake 'off'.
4. The gradients should contain two clearly separated green bands (*Figure 9*). Aspirate the upper band (broken chloroplasts) and discard. Carefully remove the lower band (intact chloroplasts) using a wide bore 10 ml pipette and place in a 50 ml centrifuge tube. Dilute with three volumes of sorbitol−Tricine (see Section 4.4.3).
5. Centrifuge at 3000 g, at 4°C for 5 min in a swing-out bucket rotor.
6. Resuspend each pellet in 20 ml of sorbitol−Tricine. Centrifuge at 1000 g, 4°C for 5 min.
7. Resuspend the pellets in the medium of choice (see Section 4.4.3) to give a chlorophyll concentration of $0.1-0.5$ mg ml^{-1}. Typical yields from 100 g of leaves are ~ 6 mg chlorophyll, ~ 40 mg thylakoid protein and ~ 60 mg soluble protein.

Note: In the pelleting and washing procedures (5) and (6) above it is inadvisable to exceed the centrifugation times and g forces specified, as highly compacted chloroplast pellets are difficult to resuspend without appreciable chloroplast breakage.

---

in the leaf are recovered as intact, isolated chloroplasts. The remainder are either present in unbroken cells or rupture during isolation. For these reasons the detection of an enzymatic activity in crude type A chloroplasts does not unambiguously show that the enzyme in question is located within chloroplasts. This problem of enzyme localization has been thoroughly addressed by Halliwell (78).

4.4.4 *Purification of intact chloroplasts by silica-sol density-gradient centrifugation*

Although sucrose is commonly used as a medium for density-gradient centrifugation, isolated chloroplasts purified in this way show very low rates of photosynthesis and protein synthesis (83). The reason for this is that the high osmotic pressures encountered cause irreversible damage to chloroplast envelopes. In contrast, silica-sols have low osmotic pressures and low viscosities and thus are the gradient materials of choice for the preparation of pure, active chloroplasts. Percoll^R (Pharmacia: colloidal silica coated with polyvinyl pyrolidone) is the most convenient source of silica-sol for chloroplast isolations. It is advisable to aliquot in small volumes and autoclave as soon as the manufacturer's container is opened. The methods detailed in *Method Tables 18−20* are based on those of Price and Reardon (84) and are suitable for the purification of pea and spinach chloroplasts. Type A chloroplasts purified by silica-sol gradient centrifugation are essentially free of other membrane-bound organelles, cytoplasmic enzymes, ribosomes and nuclear DNA. They are the preferred source of chloroplasts

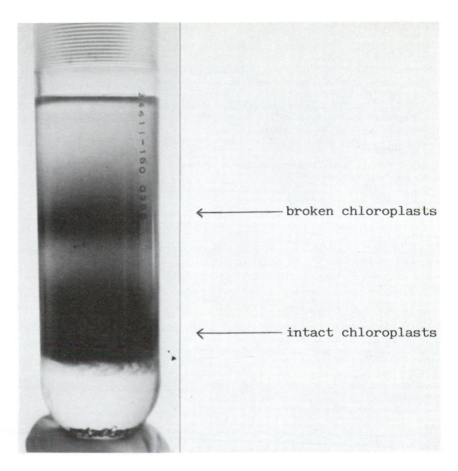

Figure 9. Separation of intact and broken chloroplasts by silica-sol density-gradient centrifugation. Crude chloroplast suspension containing 1 mg chlorophyll was applied to the gradient. The arrows indicate green bands of broken and intact chloroplasts.

for the subsequent purification of low-abundance soluble and thylakoid proteins (84) and chloroplast envelopes (85). Spinach and pea chloroplasts prepared in this way show consistently high rates of light-driven protein and RNA synthesis and are active in the post-translational uptake and processing of cytoplasmically-synthesized precursors of chloroplast proteins. With modifications, the methods described here can be successfully applied to the isolation and purification of type A chloroplasts from *Euglena* (84) and the mesophyll and bundle sheath of maize (86).

5. ACKNOWLEDGEMENTS

N.Lambert is greatly indebted to Drs R.B.Freedman, T.Hallinan and E.N.C.Mills for their critical appraisals and comments.

6. REFERENCES

1. Gray,G.L., Baldridge,J.S., McKeown,K.S., Heynekeer,H.L. and Chang,C.N. (1985) *Gene,* **39**, 247.
2. Zemel-Dreasen,O. and Zamir,A. (1984) *Gene,* **27**, 315.
3. Marston,F.A.O. (1987) In *DNA Cloning: A Practical Approach.* Glover, D.M. (ed.), Vol. III, p. 59.
4. Marston,F.A.O. (1986) *Biochem. J.,* **240**, 1.
5. Rickwood,D. (ed.) (1984) *Centrifugation: A Practical Approach.* IRL Press, Oxford, 2nd Edn.
6. Rickwood,D. (ed.) (1983) *Iodinated Density Gradient Media: A Practical Approach.* IRL Press, Oxford.
7. Hall,J.I. and Moore,A.L. (ed.) (1983) *Isolation of Membranes and Organelles from Plants Cells.* Academic Press, New York and London.
8. Albertsson,P.-A. (ed.) (1986) *Partition of Cell Particles and Macromolecules.* Wiley-Interscience, London and New York.
9. Findlay,J. (1989) In *Protein Purification Applications: A Practical Approach.* Harris,E.L.V. and Angal,S. (eds), IRL Press, Oxford.
10. Duance,V. (1989) In *Protein Purification Applications: A Practical Approach.* Harris,E.L.V. and Angal,S. (eds), IRL Press, Oxford.
11. Schoner,R.G., Ellis,L.F. and Schoner,B.E. (1985) *Bio/Technology,* **3**, 151.
12. Marston,F.A.O., Lowe,P.A., Doel,M.T., Schoemaker,J.M., White,S. and Angal,S. (1984) *Bio/Technology,* **2**, 800.
13. Brewer,S. and Sassenfeld,H. (1989) In *Protein Purification Applications: A Practical Approach.* Harris,E.L.V. and Angal,S. (eds), IRL Press, Oxford.

Clarification

14. Shimazaki,K. *et al.* (1986) *Ann N.Y. Acad. Sci.,* **469**, 63.
15. Svarovsky,L. (1981) In *Solid−Liquid Separation,* Butterworth, London.
16. Moks,T., Abrahmsén,L., Osterlöf,B., Josephson,S., Östling,M., Enfors,S.-O., Persson,I., Nilsson,B. and Uhlén,M. (1987) *Bio/Technology,* **5**, 379.
17. Bratby,J. (1980) In *Coagulation and Flocculation.* Uplands Press, Croydon.
18. Hanisch,W. (1986) In *Membrane Separations in Biotechnology.* McGregor,W.C. (ed.), Marcel Dekker Inc, New York, p. 61.
19. Porter,M.C. (1979) In *Membrane Filtration: Handbook of Separation Techniques for Chemical Engineers.* Schweitzer,P.A. (ed.) McGraw Hill, New York, pp. 2/3−2/103.
20. Gutman,R.G. (1987) *Membrane Filtration,* Adam Hilger, Bristol.

Disruption

21. Edebo,L. (1969) In *Fermentation Advances.* Perlman,D. (ed.), Academic Press, New York and London, p. 249.
22. Currie,J.A., Dunnill,P. and Lilly,M.D. (1972) *Biotechnol. Bioeng.,* **12**, 63.
23. Melling,J. and Phillips,B.W. (1975) In *Handbook for Enzyme Biotechnology.* Wiseman,A. (ed.), Ellis Horwood, London, p. 58.
24. Buckland,B.C. *et al.* (1974) In *Industrial Aspects of Microbiology.* Spencer,B. (ed.), FEBS, North Holland Publishing, Amsterdam, Vol. 30.

Isolation of organelles

25. Busch,H. and Daskal,Y. (1977) In *Methods in Cell Biology.* Stein,G., Stein,J. and Kleinsmith,J.L. (eds), Vol. XVI, p. 1.
26. Hewish,D.R. and Burgoyne,L.A. (1973) *Biochem. Biophys. Res. Commun.,* **52**, 504.
27. Dignam,J.P., Lebowitz,R.M. and Roeder,R.G. (1983) *Nucleic Acids Res.,* **11**, 1475.
28. Gorski,K., Carneiro,M. and Schibler,U. (1986) *Cell,* **47**, 767.
29. Conaway,J.W., Bond,M.W. and Conaway,R.C. (1987) *J. Biol. Chem.,* **262**, 8293.
30. Plumb,M.A., Lobanenkov,V.V., Nicolas,R.H., Wright,C.A., Zavou,S. and Goodwin,G.H. (1986) *Nucleic Acids Res.,* **14**, 7675.
31. DePierre,J.W. and Dallner,G. (1975) *Biochem. Biophys. Acta,* **415**, 411.
32. DePierre,J.W. and Dallner,G. (1976) In *Biochemical Analysis of Membranes.* Maddy,A.H. (ed.), Chapman and Hall, London, p. 79.
33. Various authors (1983) In *Methods in Enzymology.* Fleischer,S. and Fleischer,B. (eds), Academic Press, New York, Vol. 96.
34. Various authors (1978) In *Methods in Enzymology.* Fleischer,S. and Fleischer,B. (eds), Academic Press, New York, Vol. 52.

35. Various authors (1974) In *Methods in Enzymology*. Fleischer,S. and Fleischer,B. (eds), Academic Press, New York, Vol. 31.
36. Siekevitz,P. (1963) *Annu. Rev. Physiol.*, **25**, 15.
37. Lambert,N. and Freedman,R.B. (1985) *Biochem. J.*, **228**, 635.
38. Meier,P.J., Spycher,M.A. and Meyer,U.A. (1981) *Biochem. Biophys. Acta.*, **646**, 283.
39. Pickett,C.B., Rosenstein,N.R. and Jeter,R.L. (1981) *Exp. Cell Res.*, **132**, 225.
40. Tangen,O., Jonsson,J. and Orrenius,S. (1973) *Anal. Biochem.*, **54**, 597.
41. Kamath,S.A., Kummerow,F.A. and Narayan,K.A. (1971) *FEBS Lett.*, **17**, 90.
42. Schenkman,J.B. and Cinti,D.L. (1978) In *Methods in Enzymology*. Fleischer,S. and Packer,L. (eds.), Academic Press, New York, Vol. 52.
43. Kamath,S.A. and Narayan,K.A. (1972) *Anal. Biochem.*, **48**, 53.
44. Hallinan,T. and deBrito,A.E.R. (1981) *Hormone Cell Reg.*, **5**, 73.
45. Brown,D. and Hallinan,T. (1984) *Biochem. Soc. Trans.*, **12**, 680.
46. Fleischer,S. and Packer,L. (eds). (1979) In *Methods in Enzymology*. Academic Press, New York, Vol. 55.
47. Tzagoloff,A. (ed.) (1982) *Mitochondria*. Plenum Press, New York.
48. Taussky,H.H. and Schorr,E. (1953) *J. Biol. Chem.*, **202**, 675.
49. Pullman,H.E., Penefsky,H.S., Datta,A. and Racker,E. (1960) *J. Biol. Chem.*, **235**, 3322.
50. Joshi,S., Hughes,J.V., Shaikh,F. and Sanadi,D.R. (1979) *J. Biol. Chem.*, **254**, 10145.
51. Mende,P., Kolbe,V.J., Kadenback,B., Stipani,I. and Patmieri,F. (1982) *Eur. J. Biochem.*, **128**, 91.
52. Cadman,E., Hostwick,J. and Eichberg,J. (1979) *Anal. Biochem.*, **96**, 21.
53. Kaplan,R.S. and Pedersen,P.L. (1985) *J. Biol. Chem.*, **260**, 10293.
54. Jacobs,E.E., Jacobs,M., Sanadi,D.R. and Bradley,L.M. (1956) *J. Biol. Chem.*, **223**, 147.
55. Pedersen,P.L., Greenawalt,J.W., Reynafarje,B., Hullihen,J., Decker,G.L., Soper,J.W. and Bustamente,E. (1978) *Methods Cell Biol.*, **20**, 411.
56. Glew,R.H., Kayman,S.C., Kuhlenschmidt,M.S. (1973) *J. Biol. Chem.*, **248**, 3137.
57. Schnaitman,C.A. and Greenawalt,J.W. (1968) *J. Cell Biol.*, **38**, 158.
58. Kaplan,R.S., Pratt,R.D. and Pedersen,P.L. (1986) *J. Biol. Chem.*, **261**, 12767.
59. Wohlrab,H. (1980) *J. Biol. Chem.*, **255**, 8170.
60. Gordon,A.J. and Ford,R.A. (1972) In *The Chemists Companion*. J. Wiley & Son, London.
61. Azzone,F.G., Colonna,R., Ziche,B. (1979) In *Methods in Enzymology*. Fleischer,S. and Packer,L. (eds), Academic Press, New York, Vol. 55, p. 46.
62. Gomez-Puyon,T. de., Gomez-Puyon,M. and Beigel,M. (1976) *Arch. Biochem. Biophys.*, **173**, 326.
63. Joshi,S. and Sanadi,D.R. (1979) In *Methods in Enzymology*. Fleischer,S. and Packer,L. (eds), Academic Press, New York, Vol. 55, p. 384.
64. Godinot,C., Vial,C., Font,B. and Gautheron,D.C. (1986) *Eur. J. Biochem.*, **8**, 385.
65. Beyer,R.E. (1967) In *Methods in Enzymology*. Academic Press, New York, Vol. 10, p. 186.
66. Hughes,J., Joshi,S., Torok,K., Sanadi,D.R. (1982) *J. Bioenerget. Biomemb.*, **14**, 287.
67. Knowles,A.F. and Penefsky,H.S. (1972) *J. Biol. Chem.*, **247**, 6617.
68. Le Quoc,K., Le Quoc,D. and Gaudemer,Y. (1981) *Biochemistry*, **20**, 1705.
69. Tolbert,N.E., Oeser,A., Kisaki,T. and Hageman,R.H. and Yamazaki,R.K. (1968) *J. Biol. Chem.*, **243**, 5179.
70. Penin,F., Godinot,C. and Gautheron,D.C. (1979) *Biochim. Biophys. Acta.*, **548**, 63.
71. Kirk,J.T.O. and Tilney-Bassett,R.A.E. (1978) *The Plastids*. 2nd edn, Elsevier/North Holland.
72. Anderson,J.W. (1966) *Phytochemistry*, **7**, 1973.
73. Loomis,W.E. and Battaile,J. (1966) *Phytochemistry*, **5**, 423.
74. Hall,D.O. (1972) *Nature*, **235**, 125.
75. Ellis,R.J. and Hartley,M.R. (1982) In *Methods in Chloroplast Molecular Biology*. Edelman,M. *et al.* (eds), Elsevier Biomedical Press, p. 169.
76. Ellis,R.J. and Robinson,C. (1987) In *Advances in Botanical Research*. Academic Press, Vol. 14, p. 1.
77. Lilley,R.M.C., Fitzgerald,M.P., Rienits,K.G. and Walker,D.A. (1975) *New Phytol.*, **75**, 1.
78. Halliwell,B. (1981) *Chloroplast Metabolism*. Clarendon Press, Oxford.
79. Robinson,S.P., Edwards,G.E. and Walker,D.A. (1979) In *Plant Organelles*. Reid,E. (ed.), Ellis Harwood Publishers, Chichester, p. 13.
80. Jensen,R.G. and Bassham,J.A. (1966) *Proc. Natl. Acad. Sci. USA.*, **56**, 1095.
81. Ramirez,J.M., Del Campo,F.F. and Arnon,D.I. (1968) *Proc. Natl. Acad. Sci. USA.*, **59**, 606.
82. Hartley,M.R. and Ellis,R.J. (1973) *Biochem. J.*, **134**, 249.
83. Morgenthaler,J.-J., Price,C.A., Robinson,J.M. and Gibbs,M. (1974) *Plant Physiol.*, **54**, 532.

84. Price,C.A. and Reardon,E.M. (1982) In *Methods of Chloroplast Molecular Biology*. Edelman,M. *et al.* (eds), Elsevier Biomedical Press, p. 189.
85. Douce,R. and Joyard,J. (1982) In *Methods in Chloroplast Molecular Biology*. Edelman *et al.* (eds), Elsevier Biomedical Press, p. 239.
86. Walbot,V. and Hoisington,D.A. (1982) In *Methods in Chloroplast Molecular Biology*. Edelman,M. *et al.* (eds), Elsevier Biomedical Press, p. 211.

CHAPTER 3

Concentration of the extract

E.L.V.HARRIS

1. INTRODUCTION

A concentration step is frequently required after a clarified solution of the protein has been obtained, in order to aid subsequent purification steps. This is particularly important when the protein is obtained in culture medium from cells (e.g. bacteria or tissue culture cells). Concentration of the protein solution results in a decreased volume, as well as a higher protein concentration. Clearly a smaller volume of solution is easier to handle in subsequent steps, such as precipitation or loading onto a chromatography column. Higher protein concentrations minimize protein losses by non-specific adsorption to container walls or column matrices. In addition many subsequent purification steps require a minimum protein concentration to be effective, for example, precipitation is more efficient at concentrations above $100 \mu g$ ml^{-1}, whilst for adsorption chromatography (e.g. ion-exchange or affinity) the concentration of protein must be greater than the dissociation constant (see Chapters 4 and 5).

Concentration is achieved by removal of water and other small molecules:

(i)   by addition of a dry matrix polymer with pores that are too small to allow entry of the large protein molecules (Section 2);
(ii) by removal of the small molecules through a semi-permeable membrane which will not allow the large molecules through (i.e. ultrafiltration, Section 3); or
(iii) by removal of water *in vacuo* (i.e. lyophilization, Section 4).

Precipitation can also be used to concentrate proteins if the pellet is redissolved in a smaller volume, and in addition often results in some degree of purification of the protein of interest. However, as mentioned above precipitation is more effective if the total protein concentration is above $100 \mu g$ ml^{-1} (see Section 6). Two-phase aqueous extraction can also be used to concentrate the protein, with an associated degree of purification (see Section 7).

2. ADDITION OF A DRY MATRIX POLYMER

This is one of the simplest and quickest methods of concentrating solutions of proteins, requiring minimal apparatus. A dry inert matrix polymer, such as Sephadex, is added to the protein solution and allowed to absorb the water and other small molecules; the pores within the matrix are too small to allow the protein to be absorbed. When the matrix has swollen to its full extent the remaining protein solution is removed after the matrix has been settled by gravity, filtration or centrifugation. A method for this is given in *Method Table 1*. The degree of concentration obtained by this method is

Method Table 1. Concentration using dry Sephadex.

---

1. Add dry Sephadex G-25 to the solution to be concentrated at a ratio of 20 g Sephadex per 100 ml solution.
2. Allow the Sephadex to swell for ~15 min.
3. Remove the supernatant by either:
   (i) centrifuging at 2000 *g* for 10 min; or
   (ii) filtration through Whatman No. 1 filter paper. Typically the volume of supernatant will be ~30% of the original volume and contain ~80% of the protein.

---

low compared to some of the other methods. Another major disadvantage of this method is that not all the protein solution can be recovered, since some will be trapped in the matrix bed, between the matrix particles, resulting in a low yield (at best 80–90%, depending on the volume of the matrix bed). Thus, this method is not widely used unless concentration is of more importance than the yield of the protein.

Recently, matrices have been developed which have a temperature-dependent hydrated volume (1). These may prove useful in large-scale applications such as the biotechnology industry. The dry matrix is added at a carefully controlled temperature and the bulk of the protein solution removed after the water has been absorbed; the temperature is raised by as little as 1°C resulting in a rapid shrinking of the volume occupied by the matrix of up to 10-fold. The released solution can then be removed and the matrix more rapidly dried for subsequent use.

3. ULTRAFILTRATION

In ultrafiltration, water and other small molecules are driven through a semi-permeable membrane by a transmembrane force such as centrifugation or high pressure. For ultrafiltration, the membrane pores range in diameter from 1 to 20 nm; the diameter is chosen such that the protein of interest is too large to pass through. Pore sizes of microfiltration membranes range from 0.1 to 10 μm diameter and allow proteins and other macromolecules to pass through, whilst retaining larger particles such as cells (see Chapter 2). Rather than quote the pore size of the ultrafiltration membranes it is more common to quote a nominal molecular weight cut-off (NMWC) for the membrane. The NMWC is defined as the minimum molecular weight globular molecule which will not pass through the membrane. It is important to remember that the shape of the molecule will affect whether it can pass through the pores, thus whilst a globular protein of 100 000 mol. wt will not pass through a 100 000 NMWC membrane, linear molecules of 1 000 000 mol. wt have been shown to pass through under certain conditions (2). The pore sizes are not uniform and will show a normal distribution around the mean pore size (*Figure 1*). The range of this distribution will vary with the method of manufacture of the membrane, and therefore also between manufacturers. A 10-fold change in pore size will result in a 100-fold change in the NMWC. Thus, the NMWC of the membrane used should usually be significantly less than the molecular weight of the protein of interest (usually ≥20% less). If too small a NMWC is chosen the flow rates through the membrane will be reduced, resulting in longer process times or a requirement for a higher driving force.

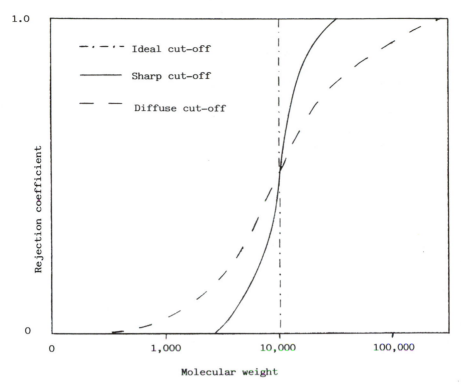

Figure 1. Rejection characteristics of a membrane with a NMWC of 10 000. Ideally all molecules of ≥ 10 000 molecular weight will not pass through the membrane whilst those of < 10 000 will. However, due to the distribution of pore sizes in membranes there is also a distribution in molecular weight of the molecules able to pass through the membrane. Thus, for a particular membrane the filtrate will contain a certain percentage of molecules with molecular weight less than the NMWC and a similar percentage with a higher molecular weight.

Concentration by ultrafiltration offers several advantages over alternative methods.

(i)   Precipitation followed by centrifugation requires a minimum concentration of 100 μg ml^{-1} and often results in poor recoveries due to the phase change. In addition the volumes which can be handled easily are limited (particularly if continuous flow centrifuges are unavailable), and aerosols are hard to contain.

(ii)   Concentration by dialysis requires longer processing times and volumes are severely limited by ease of handling.

(iii) Freeze-drying requires longer processing times and can result in poor recoveries due to the phase change. In addition to concentrating the protein, salts are also concentrated.

Equations which fully describe the observed behaviour of microfiltration and ultrafiltration processes have not yet been devised. The reader is referred to references 3−5 for discussion of the mathematical models; the qualitative principles are discussed below. The flux rate across the membrane (i.e. the volume of filtrate per unit area per unit of time) is the main factor requiring optimization, since a higher flux rate results in shorter processing time. Flux rate is directly proportional to the transmembrane pressure (up to a certain limit) and indirectly proportional to the resistance against passage

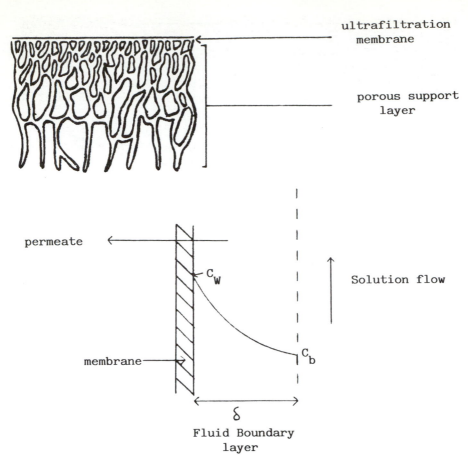

Figure 2. (**a**) Asymmetric ultrafiltration membranes consist of a thin upper ultrafiltration layer and a deep porous support layer. (**b**) Concentration polarization is caused by a build up of solute molecules at the surface of the membrane. A solution of concentration C_b flows across the membrane. Water and small solute molecules are forced across the fluid boundary layer (thickness δ) and pass through the membrane. Larger solute molecules, which cannot pass through the membrane, concentrate at the membrane surface. A concentration gradient of these solute molecules therefore occurs within the fluid boundary layer. This concentration gradient is termed concentration polarization.

of molecules across the membrane. This resistance is the sum of three factors: the membrane resistance, and the resistances caused by concentration polarization and fouling (see below).

The resistance of the membrane is minimized by the following.

(i)   Increased pore size, hence the maximum pore size which does not let the protein of interest pass through should be chosen.

(ii)   Increased pore density, which will vary from manufacturer to manufacturer, and often batch to batch.

(iii)   Minimal thickness of the membrane, hence most ultrafiltration membranes are asymmetric (or anisotropic). These ultrafiltration membranes consist of a thin upper layer (~ 0.5 μm thick) of small defined pore size (i.e. the ultrafiltration

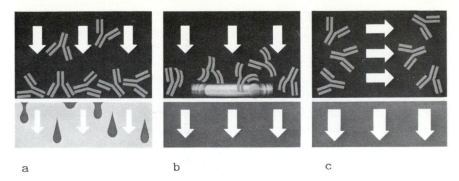

a   b   c

Figure 3. With dead-end filtration a gel-like layer rapidly builds up at the membrane surface (**a**) and severely reduces flux rates. The build up of this layer is minimized by stirring as in stirred cells (**b**) or by tangential or cross-flow (**c**). (Courtesy of Millipore.)

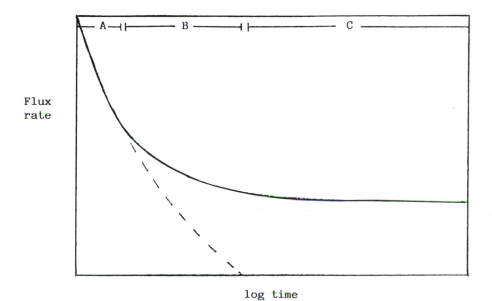

Figure 4. Decay of flux rate across the membrane with time. (**A**) The initial flux rate decreases as concentration polarization occurs. (**B**) As solute concentration increases the viscosity increases and causes a decrease in flux rate. (**C**) Finally a steady state flux is achieved. When fouling occurs (- - - -) this steady state flux is never achieved. Typically phases A−B occur within 5−10 min.

membrane itself) with a deep, large pore, spongy lower layer (i.e. a support layer) about 150 μm thick (*Figure 2a*).

(iv)   Maximum wettability of the membrane.

(v)   Minimum viscosity of the solution. Viscosity is indirectly proportional to temperature and directly proportional to the concentration of solutes.

Concentration polarization is caused by a build-up of molecules at the surface of the membrane (*Figure 2b*) (6−8). A gel-like layer forms and acts as a second ultrafiltration layer decreasing the flux rate and preventing passage of some molecules which would normally pass through the membrane. The build-up of this layer is minimized by allowing the molecules to diffuse back into the solution by agitating the solution close to the membrane. This is achieved either by stirring as in stirred cells or by tangential or cross-flow (*Figure 3*).

Fouling of the membrane is caused by particles, or macromolecules becoming adsorbed to the membrane, or physically embedded in the pores. Once a membrane is fouled, cleaning will not normally return the flux rate to its initial value, whilst in contrast cleaning will remove the layer caused by concentration polarization. A simple way to distinguish between whether the decrease in flux rate is caused by concentration polarization or by fouling is to observe the effect of increasing the tangential flow on the flux rate. If increasing the tangential flow increases the flux rate then concentration polarization was the problem; if increasing the tangential flow decreases the flux rate then it was fouling. Observation of the decay in flux rate with time will also distinguish between concentration polarization and fouling. The flux rates through a membrane will follow a sigmoidal decrease (*Figure 4*). With a new or clean ultrafiltration membrane the initial flux is relatively high and will rapidly decrease as concentration polarization occurs. An equilibrium will be reached and maintained until the concentration, and hence the viscosity increases; at equilibrium the cleaning caused by tangential flow equals the rate of concentration polarization. If fouling occurs, the flux rate will continue to

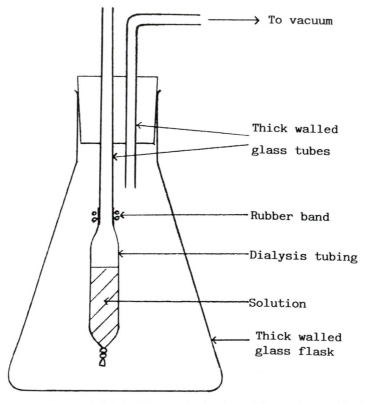

Figure 5. Apparatus for vacuum dialysis. NB ensure that the tubes and flask used are capable of withstanding a vacuum.

decrease with time and an equilibrium will not be achieved. Fouling can be minimized by removing particulate matter prior to ultrafiltration and by choosing conditions where the macromolecules in the solution do not interact with the membrane (thus pH, ionic strength and membrane type are important things to consider).

3.1 Equipment

3.1.1 *Small scale concentration*

Although not strictly defined as ultrafiltration, the principles of concentration by dialysis are similar and are therefore covered here. The protein solution to be concentrated is placed in a bag of dialysis tubing (see Section 5.1), which is placed in a solution or powder that draws water though the dialysis membrane. Solutions of polyethylene glycol (PEG) (mol. wt $\geq$ 20 000) at a concentration of 20% w/v are frequently used.

Alternatively, a dry matrix polymer such as Sephadex can be used. The former method has the advantage that the concentration can be left unattended without fear of complete removal of the water, which could result in loss of the protein by irreversible adsorption to the dialysis membrane. However, PEG may contain small molecular weight impurities that can inactivate enzymes. Water can also be 'sucked' through the dialysis membrane by applying a vacuum to the outside of the dialysis bag whilst the inside of the bag is maintained at atmospheric pressure. For this technique the apparatus shown in *Figure 5* is used. These methods of concentration are very suitable for small volumes ($\leq$ 50 ml), practical details are given in *Method Table 2*.

Several types of apparatus are available commercially for ultrafiltration on a small scale, which use either an absorbent pad (e.g. Minicon, Amicon) or centrifugal force to 'suck' or drive the small molecules through a semi-permeable membrane (e.g. Centricon, Amicon, and Centrisart 1, Sartorius, see *Figure 6*). The semi-permeable membrane is available in a range of pore sizes, with molecular weight cut-offs ranging from 5 to 30 kd. Care should be taken to note the chemical compatibility indicated by the manufacturer, particularly with regard to pH and organic solvents. Both of these

Method Table 2. Concentration by dialysis against PEG or dry Sephadex.

---

1. Treat the dialysis tubing as described in *Method Table 5*.
2. Place the solution to be concentrated in a dialysis bag sealed at one end with a knot or plastic clip.
3. Expel the air from the dialysis bag and seal the remaining end of the bag. Place the bag in a container of either:
   (i) 20% w/v PEG 2000 solution, 5–10× the volume of the solution to be concentrated; or
   (ii) dry Sephadex G-25, ~20 g per 100 ml of solution.
4. Stir the PEG solution using a magnetic follower and stirrer, or occasionally remove the swelling Sephadex from the dialysis bag and agitate.
5. Incubate until the desired concentration is reached. With the Sephadex it is important to ensure that the water is not completely removed, and therefore the volume in the dialysis bag should be checked about every 30 min. With the PEG solution the concentration may be left unattended overnight.

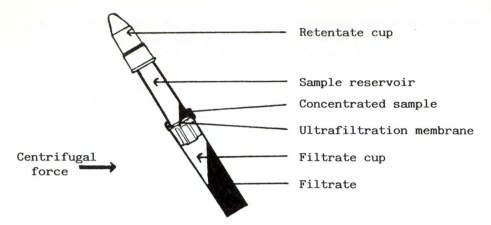

Retentate cup

Sample reservoir

Concentrated sample

Ultrafiltration membrane

Centrifugal force

Filtrate cup

Filtrate

Figure 6. Centricon shown after concentration. To recover the retentate, the sample reservoir and retentate cup can be removed from the filtrate cup, inverted and centrifuged, thus the retentate is obtained in the retentate cup.

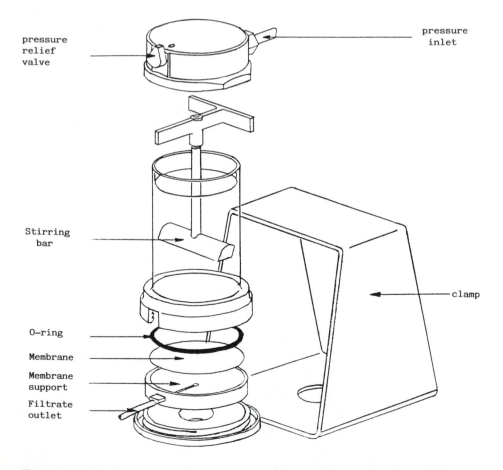

pressure relief valve

pressure inlet

Stirring bar

clamp

O–ring

Membrane

Membrane support

Filtrate outlet

Figure 7. Stirred cell.

types of apparatus are convenient for small volumes ($\leq$ 5 ml) of multiple samples and are therefore more often used for concentration, or removal of small molecular weight contaminants prior to analysis of proteins, rather than for purification.

For larger scale applications high flow rates are desirable to minimize process time. These are achieved (i) by the use of asymmetric (or anisotropic) membranes; and (ii) by agitating the solution close to the membrane by stirring or tangential flow to minimize membrane fouling and concentration polarization.

3.1.2 *Stirred cells*

Stirred cells are available to cover the range 1−400 ml (e.g. Amicon, Filtron, Sartorius) (*Figure 7*). Several types of membranes are available (*Table 1*) with NMWCs from 500 to 1 000 000. YM membranes (Amicon) and Omega membranes (Filtron) are particularly useful for concentrating dilute solutions of proteins due to their low non-specific binding properties. To minimize losses the smallest surface area of membrane should be used which will still allow a reasonable flow rate. It is advisable not to allow the solution to dry out completely, since this can cause irreversible loss of the protein onto the membrane surface. After use, membranes can be cleaned with dilute NaOH, 1−2 M NaCl or a dilute surfactant (check manufacturer's recommendations), washed with water and stored in 10% ethanol at 4°C. A protocol for using stirred cells is given in *Method Table 3*.

3.1.3 *Tangential or cross-flow systems*

Stirred cells are easy and convenient to use at the laboratory scale, but cannot be used at process scale because of the large surface areas of membrane required to achieve suitable flow rates. There are several types of ultrafiltration systems and membranes suitable for large-scale use. These fall into four categories.

(i) *Flat plate (Figure 8a).* Flat sheets of membrane are stacked between stainless steel or acrylic plates. The membranes may be individual or stacked in cassettes. The solution is pumped tangentially across the membrane or membrane stacks and is recycled (retentate), whilst filtrate passes through the membrane and is channelled along a separate flow path to the collection vessel. Scale-up is easily achieved by increasing the number of membranes in a stack, or by connecting stacks together, thus increasing the membrane surface area. These systems can be used to process volumes of 200 ml or more.

(ii) *Spirals (Figure 8b).* Several flat sheets of membrane are sandwiched between spacer screens and then the whole stack is wound spirally around a hollow, perforated cylinder. The solution is pumped parallel to the long axis of the cylinder, the filtrate (permeate) passes through into a collection channel which is connected to the central hollow cylinder. The retentate exits from the spiral cartridge and is recirculated until the desired concentration has been achieved. Spiral membrane systems can be used for volumes of a few litres upwards.

(iii) *Hollow fibres (Figure 8c).* The membrane is produced as a self-supporting hollow fibre with an internal diameter of 0.5−3.0 mm. The fibres are assembled in bundles in a cylindrical cartridge. The ultrafiltration membrane is usually on the inside of the fibres, in which case the solution is pumped through the fibres

Table 1. Membranes for stirred cells and their properties.

Membrane	Manufacturer	NMWC	Sterilization	Not compatible with:	Other properties
PM Polysulphone Hydrophobic	Amicon	10 000 30 000	5% formalin; 25% ethanol	Organic solvents Strong acids Some detergents >100°C	Absorbs hydrophobic macromolecules, but not ionic molecules High flux rates
Xm Acrylic Moderately hydrophilic	Amicon	50 000 100 000 300 000	5% formalin; 70% ethanol	Most organic solvents Strong acids >70°C	Low absorption of macromolecules. Resistant to phosphates and detergents
YM Cellulosic Hydrophilic	Amicon	1000; 5000 10 000 30 000 100 000	70% ethanol; autoclave (submerged in buffer or water)	Some organic solvents Formalin >5% ammonium	Exceptionally low absorption of proteins
YC	Amicon	500	5% formalin	Most organic solvents Strong acids and bases >50°C	Large hydrated salts (e.g. phosphates and sulphates) retained, thus causing low flux rates
Nova series Polyethersulphone	Filtron	1000; 3000 5000; 8000 10 000 30 000 50 000 100 000	70% ethanol	Esters Ethers Ketones Halogenated or aromatic hydrocarbons	

Membrane	Manufacturer	Molecular weight cut-off	Sanitizing agent	Chemical incompatibilities	Comments
Omega series Polyethersulphone	Filtron	As Nova series plus 300 000 1000 000	70% ethanol	As Nova series	Very low absorption of proteins
Alpha series Polyethersulphone	Filtron	3000 10 000 30 000 100 000	70% ethanol	As Nova series <pH 3	Modified to minimize decrease in flux rate caused by anti-form
Cellulose triacetate	Sartorius	5000 10 000 20 000	3% formaldehyde 5% hydrogen peroxide	Strong acids and bases Polar solvents Ketones Halogenated hydrocarbons >80°C	
Polysulphone	Sartorius	100 000	As above	As above	
Polysulphone	Millipore	10 000 30 000 100 000 300 000	70% ethanol	Esters Ethers Ketones Halogenated hydrocarbons >80°C	
Cellulosic	Millipore	1000 10 000	70% ethanol	Most organic solvents Strong acids and bases	Low protein binding

Method Table 3. Ultrafiltration using stirred cells.

---

1. Pre-treat the ultrafiltration membrane according to the manufacturer's instructions to remove preservatives. Typically soak the membrane in three changes of water for ~1 h total.

2. Assemble the stirred cell according to the manufacturer's instructions. Ensure the membrane is placed the correct way up (shiny surface uppermost).

3. Pre-filter or centrifuge the solution to remove particulate matter and gently pour into the stirred cell. Attach the cap assembly and place in the retaining stand.

4. Ensure the pressure relief cap is closed and attach to a regulated pressure source, typically a nitrogen cylinder. Place the assembly on a magnetic stirrer.

5. Apply the minimum pressure required to give an acceptable flow rate; higher pressures will lead to increased concentration polarization and fouling, and therefore a reduced flux rate. Do not exceed the maximum pressure recommended by the manufacturer (e.g. 75 p.s.i. recommended by Amicon). Place the filtrate tubing into a collection vessel.

6. Turn on the magnetic stirrer and adjust the stirring rate so that the vortex is no more than one-third the depth of the solution. Excess stirring will denature the protein by shear forces and foaming; insufficient stirring will cause increased concentration polarization.

7. When the desired concentration has been achieved turn off the pressure and open the pressure relief valve. Continue stirring for 5 min to resuspend the polarized layer.

8. Remove the cap assembly and gently pour out the concentrate.

---

i

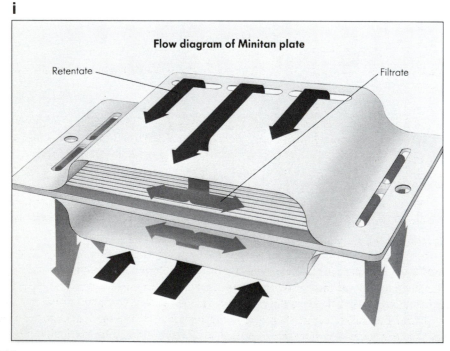

Flow diagram of Minitan plate

Retentate

Filtrate

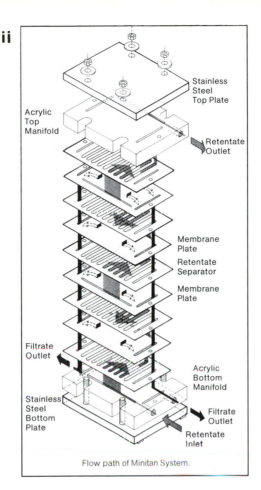

Figure 8a. Flat plate ultrafiltration system showing an individual plate (i) and the complete Minitan system (ii); (courtesy of Millipore).

and the filtrate (permeate) passes out into the cylindrical cartridge. Hollow fibre systems are available to cover a wide range of volumes from 25 ml upwards by using different cartridge sizes and linking several cartridges together.

(iv) *Tubes*. These are similar to hollow fibres but have much larger diameters (typically 2–3 cm) and consequently much larger internal volume to surface area ratios. Because of these larger ratios the flux through the membranes is lower than with hollow fibres, and therefore tube systems are not as widely used.

Several factors influence the choice of membrane configuration: hold-up volumes; flux rates; sensitivity to fouling; ease of cleaning; and ease of scale-up (*Table 2*). Systems based on the flat sheet configuration have minimal hold-up volumes and are therefore particularly useful for lower volume work, and where minimal losses are required. The flux rate is proportional to the ratio of surface area to volume, thus tube systems give poorer flux rates and are therefore less economical due to the longer processing times. Tube systems are, however, the least sensitive to fouling by particulate matter in the retentate stream and are therefore particularly useful for concentrating turbid solutions, or for clarification by microfiltration. Spirally wound membranes are not recommended for use with sample streams containing particulate matter. Hollow fibre membranes are the most effectively and easily cleaned, since they can be back-flushed to remove fouling material. The most widely available systems are those based on hollow fibre membranes or flat sheets, followed by spirally wound membranes (*Table 3* lists many of the manufacturers of ultrafiltration systems together with their membrane configurations).

With all these membrane configurations a pump is required to recirculate the retentate through the system until the desired concentration has been reached. Suitable pumps are supplied by the manufacturers of the ultrafiltration system. Different types of pump are available: peristallic, rotary piston, rotary lobe, diaphragm, progressing cavity, piston and centrifugal. Choice will be influenced by several factors, such as cost, pressure required, hold-up volume, shear sensitivity of protein, period of continuous use and requirement for steam sterilization. Peristallic pumps are relatively cheap, have minimal hold-up volumes and give minimal shear. However, they are only suitable for use at less than 2 bar and, unless heavy duty, can only be used for continuous periods of less than 2 h. The hoses must be carefully watched as they are prone to failure. Centrifugal pumps can be used at less than 5 bar, but are the most likely pumps to cause shear-induced denaturation and have high hold-up volumes. Diaphragm pumps are less likely to cause shear-induced denaturation, but can be hard to scale-up. For pressures of more than 10 bar piston, diaphragm or multistage centrifugal pumps are suitable. Most

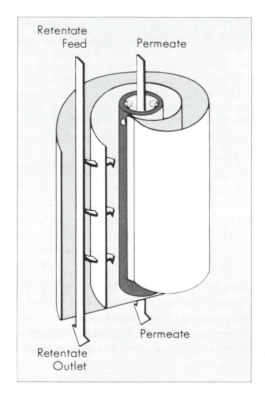

i

Retentate Feed

Permeate

Permeate

Retentate Outlet

ii

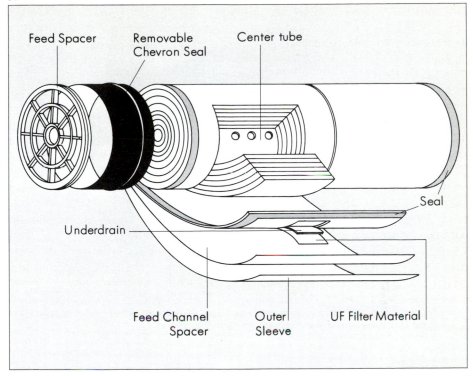

Feed Spacer

Removable
Chevron Seal

Center tube

Seal

Underdrain

Feed Channel
Spacer

Outer
Sleeve

UF Filter Material

Figure 8b. Spiral ultrafiltration system showing a flow schematic (i) and the complete cartridge Type SS-50 (ii); (courtesy of Millipore).

manufacturers tend to prefer positive deplacement rotary lobe pumps running at well below their maximum rated speeds.

3.1.4 *Membrane composition*

Traditionally all ultrafiltration membranes have been made from cellulosic materials (e.g. cellulose acetate). However, these materials have limited chemical and thermal stability, and therefore cannot be used for all applications. Recently several plastic membrane types have become available, such as polycarbonate, polyamides, polysulphone, polyvinyl chloride and acrylonitrile polymers. These membranes are compatible with a wider range of chemical conditions than the cellulosic membranes, and can be used over the pH range 1 − 13 and with a variety of organic solvents. They are also more stable to heat and can be used at up to 80°C. A few membranes can be autoclaved, but this often results in an increase in the NMWC. Many manufacturers give little information about the composition of their non-cellulosic membranes, however it appears that polysulphone is currently the most commonly used material. There are a few inorganic-based membranes available, for example Carbosep from the French company SFEC, which have a support layer of carbon with an ultrafiltration layer of zirconium oxide. These inorganic membranes have extremely good chemical and thermal resistance, and can be operated at 100°C and sterilized by autoclaving. *Table 3* lists the properties of some of the commercially available ultrafiltration membranes.

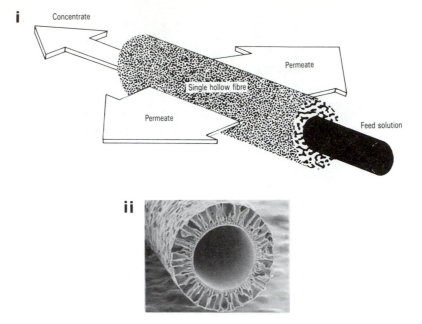

Figure 8c. Hollow fibre ultrafiltration system showing a flow schematic (i) and an electron micrograph of an individual fibre (ii) (reproduced with permission from Romicon).

3.2 **Operation**

3.2.1 *Optimization*

One of the key parameters requiring optimization is the flux rate, particularly for large-scale applications. In addition denaturation and loss of the protein must be minimized and, if appropriate, selectivity should be maximized to achieve purification in addition to concentration.

Although the choice of factors to consider during optimization (see below) can be minimized by careful theoretical consideration, final optimization can only be determined by experimental studies. An ideal optimization would follow the protocol given in *Method Table 4*.

Important factors to consider in optimizing a process were briefly discussed in Section 3 and are discussed in more detail here.

(i) *Transmembrane pressure.* Maximum transmembrane pressure will give maximum flux rate, but above a certain limit concentration polarization will limit the flux rate and fouling may be enhanced. Pumping costs will also increase.

(ii) *Tangential flow.* Maximum tangential flow will minimize concentration polarization and fouling but may cause denaturation of the protein by shear-stress (10,11) and will increase pumping costs.

(iii) *Viscosity of the solution.* High viscosity causes decreased flux rate (12). Viscosity can be minimized by increasing the temperature, however, since most proteins are denatured by heat the practical limit is usually ~40°C. DNA is often a cause of high viscosity, therefore prior treatment with DNase will often increase the flux rate. As

Table 2. Properties of different ultrafiltration membrane configurations.

	Pre-filtration required	Hold-up volume	Power consumption	Compactness	Other properties
Flat plate	Yes	Low	Medium	Medium	Withstands high pressure differentials; suitable for highly viscous solutions; steam sterilizable *in situ*
Spiral cartridge	Yes	Low	Low	High	Not suitable for particulate or viscous solutions
Hollow fibre	Yes	Low	Low	High	Can be back-flushed to ease cleaning; will not withstand high pressure differentials
Tubular	No	High	High	Low	Poor flux rates; require high pumping rates; suitable for particulate solutions

Table 3. Availability of ultrafiltration system configurations and membrane types.

Manufacturer	System configuration	NMWC range	Membrane type
Abcor	Spiral cartridge Tubular	5000−35 000	Polysulphone Cellulosic Non-cellulosic
Amicon[a]	Hollow fibre Flat plate	1000−100 000	Polysulphone Cellulosic
DDS (De Danske Sukkerfabrikker)	Flat plate	1500−500 000	Cellulosic Polysulphone Polyvinyldifluoride
Dorr-Oliver	Flat plate	1000−300 000	Cellulosic Polysulphone Non-cellulosic
Filtron	Flat plate	1000−1 000 000	Polyethersulphone
Gelman[a]	Flat plate	10 000−300 000	Cellulosic Polysulphone Non-cellulosic
Millipore[a]	Flat plate Spiral cartridge	1000−1 000 000	Polysulphone Cellulosic
Nitto	Hollow fibre	6000−100 000	Non-cellulosic
Nucleopore	Flat plate Hollow fibre	500−300 000	Non-cellulosic
Rhone Poulenc	Flat plate	10 000−20 000	Acrylic Polysulphone
Romicon	Hollow fibre	1000−100 000	Non-cellulosic Polysulphone
Sartorius[a]	Flat plate	5000−50 000	Cellulosic Polysulphone
SFEC	Tubular	20 000−50 000	Zirconium oxide
Whatman	Hollow fibre	1000−20 000	Polysulphone

[a]Laboratory-scale equipment available.

the concentration of the protein increases the viscosity will increase and therefore decrease the flux rate.

(iv) *Composition of the solution.* Particulate matter should usually be removed prior to ultrafiltration. This is best done by centrifugation, flocculation, coarse filtration or

Method Table 4. Optimization of ultrafiltration.

1. Assemble the system according to the manufacturer's instructions. Ideally use a new membrane. Check for leaks.

2. Pump clean water through the system at an appropriate transmembrane pressure and tangential flow rate. Measure the flux rate whilst recycling the filtrate, either with an in-line flow meter on the filtrate line or by measuring the volume of filtrate collected over a given period of time (e.g. 1 min). It is important to use water purified by reverse osmosis, since fouling may occur with inferior quality water (9). Continue measuring flux rate for at least 15 min and plot against time. Flux rate should not decrease significantly with time.

3. Check membrane integrity by either of the following methods. Use a solution ($\sim 1\%$) of pure protein which will not pass through the pores. Pump this solution through the system and check the filtrate for absence of protein by a standard protein assay (see Chapter 1). Before continuing to step 4 clean the membrane as described in steps 8 and 9. Alternatively, attach a supply of nitrogen to the system and pressurize to ~ 10 p.s.i. Monitor the decrease in pressure with time; a slow decrease is expected due to diffusion. If a leak is present in the membrane or system the pressure will not be maintained. Alternatively measure the volume of water displaced into the filtrate stream; typically this will be $1-2$ ml min^{-1} ft^{-2} membrane.

4. Pump the solution of interest through the system allowing the permeate to recycle into the retentate. Initially use the highest tangential flow achievable. Measure the flux rate with time at various transmembrane pressures. Repeat at several different rates of tangential flow. Typical plots are shown in *Figure 9a*. Determine the optimum transmembrane pressure and tangential flow rate. Predict the expected duration of the final process from the flux rate at the plateau point. Continue the test for this length of time.

5. Continue pumping the solution at the optimum transmembrane pressure and tangential flow rate but do not recycle the permeate. Measure the flux rate. This part of the test will determine the effect of increased concentration on the flux rate (*Figure 9b*).

6. Clean the system by pumping through hot water (e.g. 50°C; check the manufacturer's recommendations).

7. Measure the flux rate of clean water as in step 1. Ensure that the temperature is identical to that used in step 1. If fouling has occurred the flux rate will not be restored to close to that observed in step 1.

8. Clean the system with one of the solutions suggested by the manufacturer (see Section 3.2.3).

9. Measure the flux rate as in step 1. Unless considerable fouling has occurred this should be $\geq 95\%$ flux rate measured in step 1.

10. Repeat steps 4, 5, 8 and 9 to optimize each parameter (see *Figure 9* for typical results).

11. Finally concentrate several batches under optimum conditions to ensure batch to batch reproducibility. Also measure yields using an assay specific for the protein of interest to determine whether losses are being incurred, e.g. by absorption to the membrane or shear-induced denaturation.

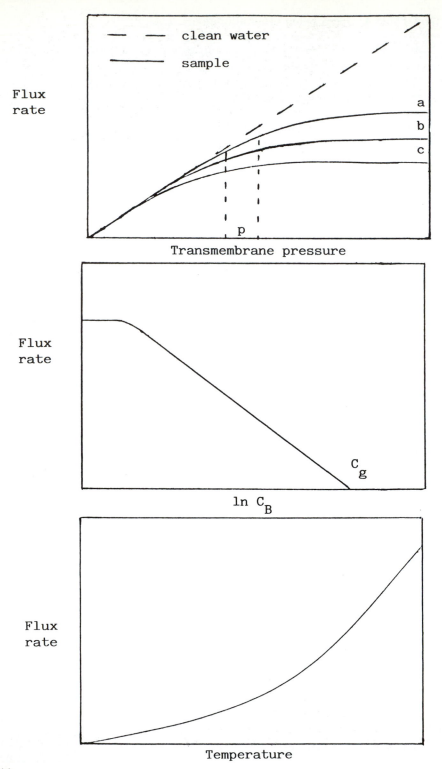

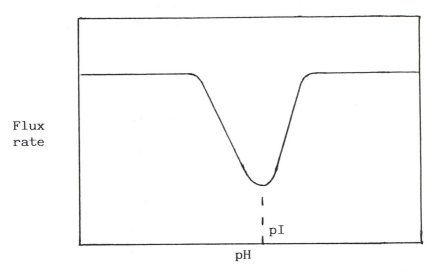

Figure 9. Flux rate as a function of transmembrane pressure with clean water or sample. **a**, **b** and **c** are obtained at decreasing rates of tangential flow. Above **a**, the optimum rate of tangential flow, further increases do not increase the flux rate. p, optimum transmembrane pressure. Flux rate as a function of the protein concentration in solution (C_B). Extrapolation to C_g gives the concentration of protein in the gel layer. For diafiltration the optimum protein concentration in solution is $C_g/2.7$) Flux rate as a function of temperature. In practice the optimum temperature is limited to $\leq 40\,°C$ by the stability of the protein. Flux rate as a function of the pH of the solution. At a pH approximately equal to the pI of the protein a minimum flux rate is usually obtained due to precipitation of the protein.

microfiltration (see Chapter 2). Anti-foam, often included in fermentation media, increases concentration polarization and fouling (2). The magnitude of the effect varies from type to type; the silicone emulsions have the greatest effect. Thus the anti-foam should be chosen with care and its concentration minimized.

pH and ionic strength can also have a profound effect on the flux rate. The buffer conditions (i.e. pH and ionic strength) should be chosen to minimize precipitation of any of the sample components in order to minimize membrane fouling. This is especially important for the protein of interest, since precipitation onto or into the membrane will cause irreversible losses. The pH should usually not be close or equal to the pI of the protein (13), since the flux rate will be minimal. Ionic strength may affect the aggregation of a protein and may therefore also affect the observed rejection characteristics (9).

(v) *Choice of membrane and system.* Flux rate increases with size and number of pores. In order to choose an appropriate membrane the flux rate should be measured experimentally for each application. The experimentally observed NMWC may differ significantly from the manufacturer's claimed NMWC and both size distribution and numbers of pores may vary from batch to batch (9). The susceptibility to fouling also varies with membrane type and manufacturer (9). The degree of non-specific adsorption of the protein to the membrane varies with membrane type, from protein to protein, and with buffer conditions and protein concentration. As an example, polysulphone membranes have been reported to show higher adsorption of bovine serum albumin (BSA) than cellulose acetate membranes (14); polyamide based membranes showed even higher adsorption. pH and ionic strength can both affect the degree of adsorption, thus optimum conditions should be determined for each application. Phosphate buffers are also reported to cause higher adsorptive losses than Tris or succinate buffers (literature from Amicon). Adsorption losses will be more marked at lower protein concentrations, therefore under these conditions membranes with very low non-specific protein binding should be chosen. The choice of membrane and system may also be dictated by the desired chemical or thermal resistance. For example, cellulose acetate membranes can generally only be used between pH 4−8 whereas polysulphone membranes can be used over the pH range 1−13. Also, not all membranes are resistant to organic solvents (polysulphone membranes are more resistant than cellulose acetate membranes).

Increasing the surface area of membrane will increase the flux rate, but this will also increase the cost, hold-up volume, power consumption and protein loss. System configuration may also affect flux rate, thus tubular systems give lower flux rates than other systems.

3.2.2 *Pre-treatment and sterilization of the system and the membranes*

Membranes are often supplied by the manufacturer impregnated with glycerin, to prevent the membrane drying out, and a bacteriostat. These can be removed prior to use by soaking in water or by connecting to the ultrafiltration system and passing water through the system. For some applications the ultrafiltration system and the membrane must be sterilized. Before sterilizing an ultrafiltration system or the membranes consult the manufacturer's literature for appropriate conditions, since these will be limited by the thermal and chemical resistance of both the system components and the membranes. Autoclaving can only be used to sterilize a few systems (i.e. those made of stainless steel, e.g. Millipore Pellicon, Sartorius Sartocon Mini) and a few membranes (e.g. SFEC Carbosep and Amicon spiral wound cartridges and YM membranes—autoclaving of these Amicon membranes leads to an increase in their NMWC). Chemical methods of sterilization for example, with 1−5% formaldehyde, 25−75% ethanol or 50−200 p.p.m. sodium hypochlorite, are therefore more widely applicable.

Before use the system and membrane should be tested for integrity. A protein of known molecular weight above the NMWC may be used, and the filtrate checked for absence of the protein. This method may not always be desirable since cross-contamination of the solution with the test protein may occur. In these cases the rate of diffusion of an inert gas, such as nitrogen, across the wetted membrane can be measured. A slow rate is to be expected; if the membrane is damaged the gas pressure will rapidly drop.

Membranes can be depyrogenated prior to use by flushing with 0.2% sodium hydroxide followed by 0.1% hydrochloric acid (9).

3.2.3 *Cleaning and storage of membranes*

Membrane fouling and concentration polarization will eventually affect the performance of the membrane by decreasing the flux rate to an unacceptable level. The gel layer formed by concentration polarization can be removed by cleaning, but removal of material on the membrane is less easily achieved. Cleaning may also be necessary for hygiene reasons (e.g. prior to storage or if the membrane is to be used for a different application). Cleaning can be done *in situ* with many systems, or after dismantling. The manufacturer's recommendations for cleaning should be followed to avoid damage of the membrane or system. Some recommended cleaning solutions are 5% NaCl, 1 M HCl or 1 M NaOH; some manufacturers supply proprietary cleaning agents (e.g. Sartorius). Use of elevated temperatures may enhance the cleaning procedure. Hollow fibre configurations can be back-flushed to maximize the efficiency of cleaning; particularly useful if the membrane has been fouled. After cleaning, the system and/or membrane should be thoroughly rinsed with water to remove the cleaning agent. The efficiency of cleaning can be assessed by measuring the flux rate, which should be restored to 95% or more of the initial flux rate of the new membrane (2,15).

All the plastic membranes are hydrophobic and therefore to prevent irreversible damage they must be stored wet and never allowed to dry out. To prevent microbial contamination the membranes should be stored in the presence of a bacteriostat according to the manufacturer's recommendation (e.g. 0.1% sodium azide or 10% ethanol). This should normally be removed prior to use.

3.2.4 *Operating procedures*

The reader is referred to the operating instructions supplied by the manufacturer for details of how to set up the system, and what operating pressures to use, etc.

3.3 **Other applications of ultrafiltration**

3.3.1 *Diafiltration*

In addition to concentration, ultrafiltration can be used to remove salts from protein solutions (diafiltration). Prior to diafiltration the protein solution is usually concentrated by ultrafiltration. Water or buffer is then added to the retentate and ultrafiltration continued until the filtrate reaches the desired ionic strength and pH. This technique will be described in more detail in a later section in this Chapter.

3.3.2 *Purification*

(i) *Separation by size*. Ultrafiltration can to some extent be used to purify proteins on the basis of size (16). Although in principle this appears to be an attractive method of purification, in practice the resolution of the technique is poor due to the following causes.

(1)   The distribution of pore sizes is not tight, resulting in some proteins larger than the NMWC passing through the membrane. Since a 10-fold change in pore size results in a 100-fold change in NMWC a small change in pore size results in significantly larger proteins passing through.

(2)   Linear molecules pass more readily through the membrane than globular molecules. Thus, linear proteins larger than that predicted by the NMWC can be deformed and squeezed through the pores. In addition pH, ionic strength and the presence of polyelectrolytes influence the effective size of the protein, since a charged protein has an effectively larger size.

(3)   Concentration polarization effectively lowers the NMWC of the membrane, thus inhibiting the passage of molecules smaller than the NMWC.

(4)   Fouling also causes a lowering of the effective NMWC.

In practice the proteins to be separated should differ in molecular weight by a factor of 100 before purification by ultrafiltration can be used effectively.

(ii) *Depyrogenation.* A frequently used application of ultrafiltration is depyrogenation of solutions, such as water for preparation of injectables, antibiotic and sugar solutions. Most pyrogens are bacterial lipopolysaccharides (LPS), which range in size from subunits of 20 000 molecular weight to aggregates of greater than 0.1 μm in diameter. Ultra-filtration membranes with NMWCs of 10 000 can be used economically and effectively to depyrogenate the filtrate. Thus, using these conditions only molecules with molecular weights less than 10 000 can be depyrogenated. This technique is therefore of limited application for proteins, but could be used for small molecular weight peptides (say <80 amino acids long). Addition of calcium and/or magnesium ions can induce aggregation of LPS allowing use of membranes with 0.025 μm pores to retain LPS and allow smaller molecules to pass through. For depyrogenating water, reverse osmosis membranes are more commonly used.

(iii) *Affinity purification.* Ultrafiltration can be exploited to achieve affinity purification by using a large molecular weight affinity ligand which is retained by the ultrafiltration membrane (17). The protein to be purified will therefore be retained by the membrane, even though it is smaller than the NMWC. The ligand, such as Cibacron blue or *p*-aminobenzamidine, is covalently coupled to a large molecular weight polymer, such as dextran or starch. The protein mixture to be purified is mixed with the ligand−polymer complex and then ultrafiltered. The protein of interest plus ligand−polymer are retained and can then be separated by dissociation and further ultrafiltration.

An exciting new advance in filtration is the development of membranes for ion-exchange or affinity purification (18,19). In this case the membrane is modified by covalent attachment of an ion-exchange group or an affinity ligand. These membranes offer the advantage of higher flow rates over conventional column matrices.

4. FREEZE-DRYING OR LYOPHILIZATION

In contrast to ultrafiltration, lyophilization also results in concentration of any salts present in the initial solution; in addition lyophilization may cause greater losses in enzyme activity. Lyophilization is, however, an invaluable method both for concentrating small molecular weight peptides which are not retained by ultrafiltration membranes, and for obtaining a dry powder of protein. Once obtained a dry powder of enzyme is more stable than an aqueous preparation of enzyme, since many degradation processes require the presence of water. Hence many commercially available proteins are obtained as freeze-dried powders.

For laboratory-scale operations, the solution to be lyophilized is placed in a freeze-drying flask and 'shell-frozen' by slowly rotating the flask in a bath of dry ice and methanol, or liquid nitrogen (NB These solutions are extremely cold and will cause frost-bite); this results in a film of frozen solution around the outside of the flask. Freeze-drying flasks come in a variety of shapes, pear-shaped flasks are more convenient for small volumes, whilst flat-bottomed flasks are best used for larger volumes. The freeze-drying flask is then rapidly attached to a mechanical vacuum pump (e.g. Edwards), ensuring that all the solution remains frozen prior to applying the vacuum. Any thawed liquid will rapidly degas and 'bump' with possible loss of solution out of the flask. To preserve the life of the pump a cold-trap must be placed between the pump and the frozen solution, this traps the water drawn off by the vacuum and prevents it entering the pump and causing rusting. A simple cold-trap consists of a glass vessel placed in a solution of dry-ice and methanol. Alternatively, purpose built freeze-driers are available from several manufacturers (e.g. Edwards); these consist of a cold-trap which is kept at $-60°C$ by a compressor, and an adaptor for either attaching several freeze-drying flasks, or using several vials. As water is removed from the solution the outside of the flask will become cold, due to evaporation; this effect also ensures that the solution remains frozen. When the outside of the flask warms up to room temperature this indicates that the freeze-drying process is completed; if a vacuum gauge is fitted into the system (e.g. when a purpose built freeze-dryer is used) another indication that the freeze-drying process is completed is given when the gauge reaches a minimum value (approximately equal to the reading obtained from the pump when run in isolation).

For larger scale freeze-drying, purpose designed automated freeze-driers are available. The product is placed on shallow trays, either loose or in vials. Freezing is often done within the freeze-drier prior to application of the vacuum. The shelves are heated during the drying cycle to speed up the process.

If the solution thaws out during the process this may result in greater losses of activity, and will also result in a glassy residue which is difficult to redissolve rather than the light, fluffy powder usually obtained by freeze-drying. Many buffers are suitable for lyophilization. However, phosphate buffers are not ideal since the pH will drop on freezing with subsequent denaturation of the protein. Also buffers with one volatile component should be avoided since again the pH may change dramatically during the lyophilization and on redissolving. Volatile buffers such as ammonium bicarbonate, or water may be preferable to minimize interference with subsequent steps. Buffer concentrations should be minimized to prevent losses in recovery of activity. Additives, such as lactose, trehalose or mannitol may be added to aid recovery of activity; suitable concentrations are $1-5\%$. Solutions containing azide should not be lyophilized, particularly using equipment fitted with a condenser, since the equipment becomes potentially explosive.

5. REMOVAL OF SALTS AND EXCHANGE OF BUFFER

5.1 **Dialysis**

Frequently it is necessary to remove salts or change the buffer after one step in the purification for the next step to work efficiently (e.g. for ion-exchange chromatography, the pH and/or the ionic strength may need to be changed to ensure that the protein will bind to the matrix). This is often achieved by dialysis; the protein solution is placed in a bag of semi-permeable membrane and placed in the required buffer, small molecules can pass freely across the membrane whilst large molecules are retained. The semi-permeable dialysis tubing is usually made of cellulose acetate, with pores of between $1-20$ nm in diameter. The size of these pores determine the minimum molecular weight of molecules which will be retained by the membrane (NMWC). Dialysis tubing often requires pre-treatment to ensure a more uniform pore size and removal of heavy metal

Method Table 5. Dialysis.

---

1. Select dialysis tubing of suitable diameter and cut into suitable lengths to contain the volume required.
2. Submerge in a solution of 2% sodium bicarbonate and 0.05% EDTA. Ensure sufficient volume is used to amply cover the dialysis tubing. Boil for 10 min. Ensure the dialysis tubing remains submerged by placing a conical flask partially filled with water on top of the tubing.
3. Discard the solution and boil for 10 min in distilled water. Repeat once more.
4. Cool and place into a suitable solution to prevent microbial growth (e.g. 20% v/v ethanol or 0.1% sodium azide). Store at 4°C for up to 3 months.

Note Wear gloves for the following steps.

5. Prior to use rinse the dialysis tubing inside and outside with distilled water or buffer.
6. Seal one end of the tubing with a double knot or dialysis clip (e.g. Pierce). Dialysis clips are easy to use, allow easy labelling of each dialysis bag and float.
7. Pour in the solution to be dialysed (a small funnel may be helpful). It is a good idea to support the bottom end of the bag on the bench. Do not overfill the bag since the volume may increase during dialysis of solutions with higher osmolarity than that used for the dialysis. For solutions of high osmolarity (e.g. 4 M $MgCl_2$ or supernatants from ammonium sulphate precipitations) the volume can increase by 2-fold or more.
8. Expel the air from the bag and seal the top end with a double knot or dialysis clip.
9. Place the bag in a large volume of buffer and agitate gently with a magnetic bar and stirrer motor. Ensure the bag is not knocked by the magnetic bar, to prevent rupture.
10. Leave to reach equilibrium, usually ≥ 3 h, preferably at 4°C.

---

contaminants; a method for pre-treatment is given in *Method Table 5*. Some dialysis membranes manufactured by Spectropor and available from several suppliers (e.g. Pierce) only require wetting in an appropriate buffer and do not require pre-treatment, except for the most sensitive enzymes. Spectropor also make various NMWC dialysis membranes ranging from 1000 to 50 000; the traditional visking tubing has a NMWC of about 15 000−20 000.

Small molecules pass freely through the membrane until the osmotic pressure is equalized, thus complete exchange from one buffer requires several changes of the dialysis buffer. For example, if a 20 ml sample containing 1 M NaCl is placed in 1 litre of water the concentration of salt at equilibrium will be:

$$\frac{20}{1020} \times 1 = 19.6 \text{ mM}$$

After changing into a further 1 litre of buffer the concentration would reach 384 nM. Equilibrium is usually reached after approximately 3 h with efficient stirring and 15 000 NMWC membranes; the time taken increases with decreasing NMWC. Dialysis is often carried out overnight, usually at 4°C to minimize losses in activity (*Method Table 5*). The Multidialyser manufactured by BRL is convenient for dialysing multiple samples of less than 2 ml.

5.2 Diafiltration

A quicker, alternative method for desalting or buffer exchange is diafiltration. This method is also more applicable to larger scale applications (i.e. >100 ml). Ultrafiltration equipment is used for diafiltration (see Section 3). With some types of equipment, water or buffer is added to the protein solution, which is then concentrated by ultrafiltration; this process is repeated until the ionic strength of the filtrate reaches that of the added buffer or water. With other types or equipment (e.g. Amicon's hollow fibre systems) the plumbing is altered to allow uptake of water or buffer at a rate equal to the flux through the membrane; this method has the advantage that it can be left unattended whilst equilibrium is achieved. The time taken to achieve equilibrium depends on the volume of sample and the flux rate (as an indication 1 litre can be diafiltered in 2−4 h). The optimum concentration for diafiltration can be determined as described in *Method Table 4* and *Figure 9b*. Higher concentrations will decrease the flux rate and hence increase the processing time, whilst lower concentrations and hence larger volumes will require larger volumes of buffer. The reader is referred to the instructions supplied by the system manufacturer for details on how to carry out diafiltration.

5.3 Gel filtration

Another quicker alternative to dialysis is gel filtration. This method is only applicable to small volumes. The maximum sample volume should not exceed 25−30% of the volume of the column to ensure adequate resolution between the protein and salt. A gel filtration matrix with a small pore size (e.g. Sephadex G-25, Pharmacia; for more detail on gel filtration media see Chapter 6) is poured into a column to give a bed volume of approximately five times the volume of sample to be desalted. A syringe plugged with glass wool or a glass fibre disc can be used, or small disposable columns (e.g. Bio-Rad's Disposocolumns). Pre-packed columns for desalting and buffer exchange are available from Pharmacia (PD-10 columns). Methods for using these columns are given in *Method Table 6*. Unfortunately, this method of desalting and buffer exchange results in dilution of the sample.

6. PURIFICATION AND CONCENTRATION BY PRECIPITATION

Many of the early protein purification procedures used only precipitation methods as a means of separating one protein (or class of proteins) from another. For example, the core histones (H2a, H2b, H3 and H4) were purified by ethanol and/or acetone precipitation (20). Differences in solubility have also been used to purify albumins and globulins from serum (21) (the globulins are precipitated by diluting serum with water, whilst the albumins remain soluble). An example of purification by precipitation is given in *Protein Purification Applications: A Practical Approach* (22), for the purification of mammalian cytochrome oxidase. Nowadays precipitation is usually only used as a fairly crude separation step often during the early stages of a purification procedure, and this is then followed by chromatographic separations. Precipitation can also be used

Method Table 6. Desalting columns.

---

1. Select a suitable size column. This may be a chromatography column or a plastic syringe plugged with a small amount of glass wool.
2. Pack the pre-swollen matrix (e.g. Sephadex G-25) into the column (the length of the column should be > twice its diameter to ensure adequate resolution). To do this make a slurry, pour into the column and allow to settle. If required top up the volume with additional slurry. (A perfectly packed column is not necessary since high resolution is not required).
3. To equilibrate the column in the required buffer pass more than a column volume through by gravity or using a pump (if using a pump ensure the maximum recommended flow rate is not exceeded). Alternatively, matrices packed in syringes can be centrifuged as follows. Place the syringe in a tube, supported off the bottom of the tube by the flanges on the syringe. Place approximately a column volume of buffer on top of the matrix and centrifuge at 1600 *g* for 4 min.
4. Place the sample to be desalted on top of the matrix and allow it to pass into the matrix by the method used in step 3. The volume of sample must be ≤30% of the volume of matrix. Discard the flow-through.
5. Continue passing through buffer until one column volume has been used and collect suitable size fractions (e.g. one-tenth of the total column volume). Larger fractions will cause higher dilution of the sample, and will decrease the resolution.
6. Determine which fractions contain the protein by measuring absorbance at 280 nm or by using another protein assay. The protein will usually elute between 0.2 and 0.7 times the column volume.

---

as a method of concentrating proteins prior to analysis or a subsequent purification step.

The solubility of a protein molecule in an aqueous solvent is determined by the distribution of charged hydrophilic and hydrophobic groups on its surface. The charged groups on the surface will interact with ionic groups in the solution (*Figure 10*). Protein precipitates are formed by aggregation of the protein molecules, induced by changing pH or ionic strength, or by addition of organic miscible solvents or other inert solutes or polymers. Temperature will also affect the degree of aggregation achieved. Precipitates can be recovered by filtration or centrifugation, washed and redissolved in an appropriate buffer, if required. To remove final traces of the precipitating agent, which might interfere with subsequent purification steps, the redissolved precipitate should be dialysed, diafiltered or desalted on a gel filtration column (see Section 5).

6.1 Precipitation by alteration of the pH

One of the easiest methods of precipitating a protein and achieving a degree of purification is by adjusting the pH of the solution to close or equal to the pI of the protein (termed isoelectric precipitation). The surface of protein molecules is covered by both negatively and positively charged groups. Above the pI the surface is predominantly negatively charged, and therefore like-charged molecules will be repelled from one another; conversely below the pI the overall charge will be positive and again

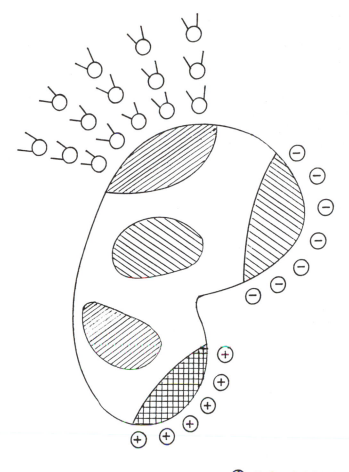

Figure 10. Schematic representation of a protein showing negatively and positively charged areas on the protein interacting with ions in the solution. The hydrophobic areas on the protein interact with water molecules causing an ordered matrix of water molecules to form over these areas.

like-charged molecules will repel one another. However, at the pI of the protein the negative and positive charges on the surface of a molecule cancel one another out, electrostatic repulsion between individual molecules no longer occurs and electrostatic attraction between molecules may occur, resulting in formation of a precipitate.

Isoelectric precipitation is often used to precipitate unwanted proteins, rather than to precipitate the protein of interest, since denaturation and inactivation can occur on precipitation (see Section 6.6).

6.2 **Precipitation by decreasing the ionic strength**

Some proteins can be precipitated by lowering the ionic strength. This can rarely be achieved with crude extracts, since the ionic strength can only be lowered by addition of water, which will also lead to a decrease in the concentration and hence an increased solubility (a notable exception is the serum globulins). However, this form of precipitation can often occur at later stages of a purification, for example, when removing salts by diafiltration, dialysis or gel filtration. This may not always be a welcome occurrence, for example, when using gel filtration a precipitated protein will be trapped and may block the matrix. Precipitation at low ionic strength is more likely to occur at or close to the pI of the protein, since the causes of precipitation are similar and therefore additive.

6.3 **Precipitation by increasing the ionic strength (salting-out)**

Precipitation by addition of neutral salts is probably the most commonly used method for fractionating proteins by precipitation. The precipitated protein is usually not denatured and activity is recovered upon redissolving the pellet. In addition these salts can stabilize proteins against denaturation, proteolysis or bacterial contamination. Thus, a salting-out step is an ideal step at which to store an extract overnight, either before or after centrifugation. The cause of precipitation is different from that for isoelectric precipitation, and therefore the two are often used sequentially to obtain differential purification. Salting-out is dependent on the hydrophobic nature of the surface of the protein. Hydrophobic groups predominate in the interior of the protein, but some are located at the surface, often in patches. Water is forced into contact with these groups, and in so doing becomes ordered (*Figure 10*). When salts are added to the system, water solvates the salt ions and as the salt concentration increases water is removed from around the protein, eventually exposing the hydrophobic patches. Hydrophobic patches on one protein molecule can interact with those on another, resulting in aggregation. Thus, proteins with larger or more hydrophobic patches will aggregate and precipitate before those with smaller or fewer patches, resulting in fractionation. The aggregates formed are a mixture of several proteins, and like isoelectric precipitation the nature of the extract will affect the concentration of salt required to precipitate the protein of interest. In contrast to isoelectric precipitation, increasing the temperature increases the amount of precipitation; however, salting-out is usually performed at 4°C to decrease the risk of inactivation (by, e.g. proteases).

Several aspects of the salt used should be considered. The effectiveness of the salt is mainly determined by the nature of the anion, multi-charged anions being the most effective; the order of effectiveness is phosphate > sulphate > acetate > chloride > (and follows the Hofmeister series). Although phosphate is more effective than sulphate, in practice phosphate consists of mainly HPO_4^{2-} and $H_2PO_4^-$ ions at neutral pH, rather than the more effective PO_4^{3-}. Monovalent cations are most effective, with $NH_4^+ > K^+ > Na^+$. The salt must be relatively cheap with few impurities present. The solubility is also an important consideration, since concentrations of several molar are required; thus, many potassium salts are not suitable. Because of the risk of possible denaturation, or changes in solubility, there should be little increase in heat caused by the salt dissolving. The final consideration is the density of the resultant solution, since

Table 4. The amount of solid ammonium sulphate to be added to a solution to give the desired final saturation at 0°C.

Initial concentration of ammonium sulphate

g solid ammonium sulphate to add to 100 ml of solution

Initial concentration of ammonium sulphate	Final concentration of ammonium sulphate, % saturation at 0°C																
	20	25	30	35	40	45	50	55	60	65	70	75	80	85	90	95	100
0	10.7	13.6	16.6	19.7	22.9	26.2	29.5	33.1	36.6	40.4	44.2	48.3	52.3	56.7	61.1	65.9	70.7
5	8.0	10.9	13.9	16.8	20.0	23.2	26.6	30.0	33.6	37.3	41.1	45.0	49.1	53.3	57.8	62.4	67.1
10	5.4	8.2	11.1	14.1	17.1	20.3	23.6	27.0	30.5	34.2	37.9	41.8	45.8	50.0	54.5	58.9	63.6
15	2.6	5.5	8.3	11.3	14.3	17.4	20.7	24.0	27.5	31.0	34.8	38.6	42.6	46.6	51.0	55.5	60.0
20	0	2.7	5.6	8.4	11.5	14.5	17.7	21.0	24.4	28.0	31.6	35.4	39.2	43.3	47.6	51.9	56.5
25		0	2.7	5.7	8.5	11.7	14.8	18.2	21.4	24.8	28.4	32.1	36.0	40.1	44.2	48.5	52.9
30			0	2.8	5.7	8.7	11.9	15.0	18.4	21.7	25.3	28.9	32.8	36.7	40.8	45.1	49.5
35				0	2.8	5.8	8.8	12.0	15.3	18.7	22.1	25.8	29.5	33.4	37.4	41.6	45.9
40					0	2.9	5.9	9.0	12.2	15.5	19.0	22.5	26.2	30.0	34.0	38.1	42.4
45						0	2.9	6.0	9.1	12.5	15.8	19.3	22.9	26.7	30.6	34.7	38.8
50							0	3.0	6.1	9.3	12.7	16.1	19.7	23.3	27.2	31.2	35.3
55								0	3.0	6.2	9.4	12.9	16.3	20.0	23.8	27.7	31.7
60									0	3.1	6.3	9.6	13.1	16.6	20.4	24.2	28.3
65										0	3.1	6.4	9.8	13.4	17.0	20.8	24.7
70											0	3.2	6.6	10.0	13.6	17.3	21.2
75												0	3.2	6.7	10.2	13.9	17.6
80													0	3.3	6.8	10.4	14.1
85														0	3.4	6.9	10.6
90															0	3.4	7.1
95																0	3.5
100																	0

the difference between the densities of the aggregate and the solution determines the ease of separation by centrifugation (see Chapter 2).

In practice ammonium sulphate is the most commonly used salt (other salts which have been used in particular applications are ammonium acetate, sodium sulphate, and sodium citrate). Ammonium sulphate is cheap, and sufficiently soluble; a saturated ammonium sulphate solution in pure water is approximately 4 M. The density of a saturated solution is 1.235 g ml^{-1}, compared to 1.29 g ml^{-1} for a protein aggregate in this solution. In practice the density of a 75 − 100% saturated solution may be higher than 1.235 g ml^{-1}, due to the presence of other salts and compounds in the extract, therefore making recovery of a protein aggregate by centrifugation difficult.

The ammonium sulphate concentration is usually quoted as percent saturation, assuming that the extract will dissolve the same amount of ammonium sulphate as pure water. To calculate the number of grams of ammonium sulphate (g) to add to one litre at 20°C to give a desired concentration use the following equation:

$$g = \frac{533\ (S_2 - S_1)}{100 - 0.3\ S_2}$$

where S_1 is the starting concentration, and S_2 is the final concentration. This equation allows for the increase in volume that occurs on addition of the salt. Alternatively, the amount of ammonium sulphate to add can be read from *Table 4*.

Ammonium sulphate will slightly acidify the extract, therefore a buffer of about 50 mM should be used to maintain a pH between 6.0 − 7.5. If a higher pH is preferred then sodium citrate should be used instead of ammonium sulphate. Although ammonium sulphate is sufficiently pure for most applications, if the enzyme of interest is sensitive to heavy metals, EDTA should be included in the buffer (even if the level of contamination is only 1 part per million, with an ammonium sulphate concentration of 75% saturation, the concentration of contaminant will be 3 μM). Usually an ammonium sulphate cut is taken in order to obtain a higher degree of purification. Thus, the extract is taken to a percent saturation where the protein of interest does not precipitate and the precipitate is removed by centrifugation, More ammonium sulphate is added to the supernatant to give a percent saturation where the protein of interest does precipitate, and can therefore be obtained by centrifugation. In order to determine the appropriate percent saturations follow the method given in *Method Table 7*. As mentioned before the composition of the extract will influence the precipitation of the protein, as will its concentration. These initial trial studies must therefore be carried out on an extract obtained by the same conditions that will be used in the purification procedure. The temperature at which the precipitation is carried out is also important; the higher the temperature the lower the solubility of the protein. Slow addition of the ammonium sulphate and efficient stirring are important, particularly as the desired saturation is approached. Dissolved air may come out of solution and cause frothing; this is not deleterious, but frothing caused by over-vigorous stirring may cause denaturation of the protein. The precipitate is usually removed by centrifugation, though filtration can be used, particularly if the density of the protein aggregate is similar to or lower than that of the solution. For centrifugation 100 000 *g* min^{-1} is normally sufficient, although the higher salt concentrations may require more; thus, 10 000 *g* for 10 min

Method Table 7. Optimization of ammonium sulphate precipitation.

---

1. Take aliquots of the extract, ideally 20–50 ml, place in beakers and pre-chill to 4°C.
2. Calculate the amount of ammonium sulphate required to give 20, 30, 40, 50, 60, 70 and 80% saturation from the equation given in the text or *Table 4*. Weigh out the required amounts of ammonium sulphate and ensure all lumps are removed (use a pestle and mortar).
3. Slowly add the ammonium sulphate to each aliquot whilst stirring (use a magnetic follower and stirrer). Leave each aliquot stirring for 1 h at 4°C.
4. Centrifuge each aliquot at 3000 *g* for 40 min.
5. Remove the supernatants and drain the pellets. Dissolve the pellets in buffer (e.g. PBS or 50 mM Tris–HCl, pH 8). Use the same volume for each pellet, approximately twice the volume of the largest pellet.
6. If undissolved material remains, centrifuge at 3000 *g* for 15 min.
7. Assay the supernatants for total protein using the Bradford assay (see Chapter 1) and by an assay specific for the protein of interest. (Dialyse the supernatants if ammonium sulphate interferes with the assay.)
8. Plot the concentration of total protein and the protein of interest against % saturation of ammonium sulphate. For an ammonium sulphate cut choose the maximum % saturation which does not precipitate the protein of interest, and the minimum % saturation that precipitates all the protein of interest (see *Figure 11*).
9. Repeat the procedure with one aliquot. First add ammonium sulphate to give the lowest % saturation chosen. After centrifuging, add further ammonium sulphate to give the highest % saturation chosen. Check that none of the protein of interest is in the first pellet or in the supernatant from the higher % saturation. Since the extent of the precipitation depends on the nature of the extract, the higher % saturation may require further optimization to achieve maximum purification.

---

or 3000 *g* for 35 min may be used. After centrifugation the pellet should dissolve in a volume of buffer equal to one or two times the volume of the pellet; any material which does not dissolve is particulate or denatured protein, and should be removed by centrifugation. The redissolved pellets will contain ammonium sulphate, therefore any assays used should not be sensitive to ammonium sulphate; the Bradford dye-binding assay (see Chapter 1) is a suitable assay for total protein. Desalting (Section 5) prior to assaying may be necessary. The choice of appropriate ammonium sulphate concentrations for the cut will be a compromise between the yield and the degree of purification; thus, a narrow cut will result in a lower yield with a higher degree of purification.

6.4 Precipitation by organic solvents

Many proteins can be precipitated by addition of water-miscible organic solvents, such as acetone and ethanol. The factors which influence the precipitation behaviour of a

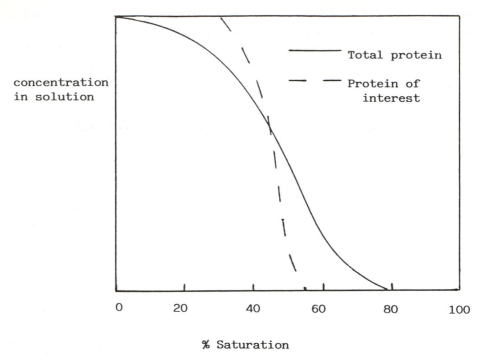

concentration
in solution

─────── Total protein

─ ─ ─ Protein of
            interest

0   20   40   60   80   100

% Saturation

Figure 11. Typical profile for ammonium sulphate precipitation. The optimum % saturations for an ammonium sulphate cut would be 30% and 55% for purification of this protein.

protein are similar to those involved in isoelectric precipitation and different from those involved in salting-out; thus, this method can be used as an alternative to isoelectric precipitation in a purification sequence, perhaps in conjunction with salting-out. Addition of the organic solvent lowers the dielectric constant of the solution, and hence its solvating power. Thus, the solubility of a protein is decreased and aggregation through electrostatic attraction can occur. Precipitation occurs more readily when the pH is close to the pI of the protein. The size of the protein also influences its precipitation behaviour; thus, a larger protein will precipitate in lower concentrations of organic solvent than a smaller protein with otherwise similar properties. However, some hydrophobic proteins, particularly those which are located in the cellular membranes are not precipitated by organic solvents, and in fact can be solubilized from the membranes by addition of organic solvents (see ref. 23). With these proteins the organic solvent will displace the water molecules from around the hydrophobic patches of the protein, resulting in an increased solubility.

   To minimize denaturation, precipitation with organic solvents should be carried out at or below 0°C. At higher temperatures the protein conformation will be rapidly changing, thus enabling molecules of the organic solvent to gain access to the interior of the protein, where they can disrupt the hydrophobic interactions and cause denaturation. Fortunately mixtures of aqueous solutions and organic solvents freeze at well below 0°C. However, as the organic solvent is added the temperature of the solution will rise due to the negative $\Delta H°$ of hydration of the solvent molecules. Thus, care must be taken to prevent the temperature rising above 10°C, by using pre-cooled organic solvent, and adding it slowly, with mixing, to the solution contained in a vessel in an ice-salt bath (or other low temperature bath). Lower temperatures will also promote the degree of precipitation. The ionic strength of the solution should be between 0.05 − 0.2; higher ionic strengths require higher amounts of organic solvent with an increased risk of denaturation, whilst lower ionic strengths can result in the precipitate being very fine and difficult to sediment. Prior to adjustment of the pH close to the pI of the protein will result in precipitation at a lower concentration of organic solvent.

Acetone and ethanol are the most commonly used solvents; others which have been used are methanol, propan-1-ol, and propan-2-ol. Safety aspects should be considered, particularly when working on a large scale; thus, the solvent should be relatively non-toxic and have a relatively high flashpoint, above 20°C (thus dioxans and ethers should be avoided). The longer chain alcohols, such as butanol, cause a higher degree of denaturation than ethanol. In many cases acetone is preferable, since lower concentrations are required, and therefore less denaturation occurs. Most proteins will be precipitated by addition of an equal volume of acetone, or four volumes of ethanol. Clearly addition of such large volumes will result in dilution of the protein, thus the starting concentration of protein should be more than 1 mg ml^{-1}. The concentration of organic solvent is usually expressed as a percentage, assuming that the volumes are additive. Thus, if an equal volume of acetone is added this corresponds to 50% v/v; however, there will be a small decrease in volume, approximately 5% with acetone, which is due to the formation of hydrated solvent molecules that occupy a smaller volume than their constituent components.

As with salting-out, an organic solvent cut is often taken to increase the degree of purification; the optimum amount of organic solvent to add can be determined using a similar method to that described in *Method Table 7*. The following equation is used to calculate the volume, in ml, of organic solvent to add to one litre:

$$v = \frac{1000 \ (P_2 - P_1)}{100 - P_2}$$

where P_1 is the starting percentage of organic solvent, and P_2 is the desired percentage of organic solvent. Since the density of the organic solvent mix is often less than that of water, the precipitates will sediment more readily than, for example, those formed with ammonium sulphate. Thus, the precipitate may be left to settle by gravity and the supernatant decanted, or centrifugation may be used (e.g. 10 000 g for 5 min). If centrifugation is used, the rotor should be pre-chilled and the centrifuge should be maintained at 0°C, otherwise if the solution is allowed to warm up denaturation may occur, and/or some of the precipitate may redissolve. The resultant precipitate will normally redissolve in one or two times its own volume; any undissolved material will probably be denatured protein and can be removed by centrifugation. Organic solvent will be present in the redissolved pellet, therefore it is important to ensure that this will not interfere with the assays, or any subsequent purification steps. The solvent can be removed by evaporation at reduced pressure, or by dialysis or gel filtration, preferably at 4°C. In the case of subsequent purification steps only those involving hydrophobic interactions will be affected (precipitation with ammonium sulphate is usually possible in the presence of low concentrations of organic solvents, but a higher concentration of salt will be required).

6.5 **Precipitation by organic polymers**

PEG is the most commonly used organic polymer (24). The mechanism of precipitation is similar to that of precipitation by organic solvents, however, lower concentrations are required, usually below 20%. Higher concentrations result in viscous solutions, making recovery of the precipitate difficult. The molecular weight of the polymer should be greater than 4000; the most commonly used molecular weights are 6000 and 20 000. PEG can be removed by ultrafiltration, provided its molecular weight differs significantly from that of the protein of interest. However, PEG does not interfere with many of the possible subsequent purification steps (e.g. ion-exchange or affinity chromatography).

6.6 **Precipitation by denaturation**

Precipitation by denaturation can be used as a purification step if the protein of interest is not denatured by the treatment, whilst many of the contaminant proteins are. This method can also be used to concentrate the proteins in a solution prior to analysis (e.g. for gel electrophoresis or amino acid analysis). Denaturation can be caused by changes in temperature, pH or addition of organic solvents. The tertiary structure of the proteins is disrupted during denaturation, resulting in the formation of random coil structures. In solution these random coils become entangled with one another, thus forming aggregates. Aggregate formation is influenced by pH and ionic strength, occurring more readily close to the pI of the protein, and at lower ionic strength.

6.6.1 *Denaturation by high temperatures*

High temperatures induce denaturation by breaking many of the bonds holding the protein in its native conformation (e.g. van der Waal's forces, ionic interactions, and even the peptide bond itself at the higher extremes), and also causing the release of associated solvent molecules. Different proteins are denatured at different temperatures, thus achieving purification; for example, human tumour necrosis factor has been partially purified in this way. Small scale trials using 1 ml of extract can be carried out to determine a temperature (usually between 45 and 65°C) and incubation time which will cause maximum precipitation of contaminating proteins, with minimum loss of activity of the protein of interest. For these small scale trials the extract used should be obtained in exactly the same way as it will in the purification itself, since pH, ionic strength and composition all affect the precipitation obtained. Although on a small scale short times (e.g. 1 min) may be practical, due to the increased time a larger volume will take to reach the temperature, and cool down, a time of about 30 min or longer is usually more practical. Unfortunately many proteases are relatively stable to high temperatures, and can in fact be more active under these conditions, thus protease inhibitors should be added prior to this step.

6.6.2 *Denaturation by extremes of pH*

Extremes of pH cause internal electrostatic repulsion, or loss of internal electrostatic attraction by changing the charges on the side chains of the amino acids. Thus, the protein then opens up and bound solvent is lost, resulting in denaturation. This method of purification is often extremely useful as a preliminary purification step of recombinant proteins expressed in prokaryotes, since many bacterial proteins have pIs in the region of 5.0 and can therefore often be removed by adjusting the pH of the extract to around 5.0. In general, many proteins can be precipitated by adjusting the pH to 5 or below, fewer proteins precipitate at neutral and alkaline pHs. Before using this treatment as a purification step, the stability of the protein of interest under these conditions should be established. Use of strong acids and bases should be avoided; thus use acetic acid for a pH of 4 or above, or citric acid for a pH of 3 or more and diethanolamine or sodium carbonate for a pH of 8 or more. Small-scale trials should be carried out to determine the optimum pH. Remember when determining the optimum pH that the aggregate formed is made up of many proteins, as well as particulate material such as ribosomes and membrane fragments; thus the composition of the aggregate will depend on the composition of the extract. The extract used to determine the optimum pH must therefore be the same as that to be used in the purification itself. Also, if the earlier purification procedures of an established protocol are changed (e.g. the method of extraction) the optimum pH should be redetermined; or the protein of interest might end up in the wrong fraction and be inadvertently discarded! The trials should also be carried out at the same temperature as will be used in the purification; increased temperature will cause increased precipitation.

Strong acids, such as perchloric acid or trichloroacetic acid (TCA), can be used to concentrate proteins for analysis when activity is not important, or occasionally for purification (e.g. perchloric acid is used to extract histones and HMG proteins as a first step in their purification). Care should be taken with these acids, since they are highly corrosive (a method is given in the Appendix for making a stock solution of TCA which avoids weighing out the solid). In addition perchloric acid can form explosive compounds, particularly when in contact with wood. Most proteins will be precipitated by addition of TCA to 10% v/v; smaller proteins, of molecular weight less than 20 000, may require up to 20%. Excess TCA can be removed from the protein pellets by washing with buffer.

6.6.3 *Denaturation by organic solvents*

Selective denaturation by organic solvents is often carried out at $20-30°C$ to promote denaturation. Small-scale trials should be used to establish optimum conditions, remembering that pH, temperature, ionic strength and extract composition will affect the precipitation behaviour. For analysis when activity is not required, to ensure maximum precipitation, the organic solvent should be added at room temperature and then the mixture cooled to $-20°C$ for at least an hour prior to centrifugation. Four volumes of acetone or nine volumes of ethanol may be required to achieve maximum precipitation.

7. AQUEOUS TWO-PHASE PARTITIONING—B.A.Andrews and J.A.Asenjo

Aqueous two-phase partitioning exploits the incompatibility between aqueous solutions of two polymers, or a polymer and a salt at high ionic strength. This incompatibility arises from the inability of the polymer coils to penetrate into each other. Hence, as the polymers are mixed large aggregates form and the two polymers will tend to separate due to steric exclusion. The most commonly used polymers are polyethylene glycol (PEG) and dextran (25,26). A similar exclusion phenomenon can be observed between a polymer and a high concentration of salt (e.g. PEG and phosphate, 27).

Two-phase partitioning can be used to separate proteins from cell debris, to purify proteins from other proteins, or to concentrate proteins. Most soluble and particulate matter will partition to the lower, more polar phase, whilst proteins will partition to the top, less polar phase (usually PEG). Separation of proteins from one another is achieved by manipulating the partition coefficient by altering the average molecular weight of the polymers, the type of ions in the system, the ionic strength, or the presence of hydrophobic groups (28,29). To achieve a higher degree of purification several sequential partitioning steps can be carried out, or alternatively polymers can be mixed to yield more than two phases. Affinity partitioning can also be used to increase the degree of purification; in this case affinity ligands, such as triazine dyes (e.g. Cibacron blue) are covalently attached to one or both polymers. Alternatively, one polymer may be modified with hydrophobic groups to alter the partitioning of the protein (26). Two-phase aqueous partitioning is a very mild method of protein purification, and denaturation or loss of biological activity are not usually seen. This is probably due to the high water content and low interfacial tension of the systems which will therefore protect the proteins. The polymers themselves may also have a stabilizing effect.

Phase diagrams can be constructed for each polymer system and are used to predict the volume and compositions of the top phases. The curve T−C−B in *Figure 12* is called the binodial curve. It separates the area to the left of TCB where there is only one phase present, with both PEG and dextran in solution, and the area to the right of TCB where two phases form. The overall composition of any particular system is given by the ordinate (PEG) and the abcissa (dextran) values. For the region to the left of TCB (one phase) this is straightforward. For the region to the right of TCB (two phases) a mixture of composition given by point M will separate into two phases of composition T (top) and B (bottom), where T and B are on the equilibrium tie line through point M and on the binodial curve. Other tie lines are also shown. Tie lines are defined as the straight lines that join the composition values of two phases in equilibrium. They will be on different sides of the binodial curve. Tie lines will never cross each other. All points on one tie line will have identical compositions in the top (T) and bottom (B) phases, but the relative phase volumes and hence overall composition will be different. The ratio of the volumes of the upper to lower phases is given by the ratio of the length of the line sections MB/MT. Thus, to concentrate a protein the phase into which the protein partitions should have a small relative volume (i.e. M will be nearer the bottom of the tie line). At the critical point (C) the two phases theoretically have identical compositions, are of equal volumes and have a partition coefficient of 1.

The partition coefficient (K) is defined as

$$K = C_T/C_B$$

where C_T and C_B represent the equilibrium concentrations of the partitioned protein in the top and bottom phases, respectively. Systems represented by points on the same tie line have identical partition coefficients.

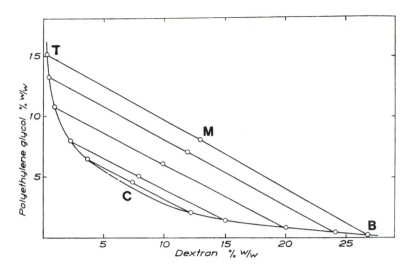

Figure 12. Phase diagram and phase compositions of the dextran/PEG system D24−PEG 6000 at 20°C (26).

The mechanism governing partition is largely unknown; qualitatively it can be described as follows (26). A protein interacts with the surrounding molecules within a phase via various bonds, such as hydrogen, ionic and hydrophobic interactions, together with other weak forces. The net effect of these interactions is likely to be different in the two phases and therefore the protein will partition into one phase where the energy is more favourable. If the energy needed to move a protein from one phase to the other is ΔE, at equilibrium the relationship between the partition coefficient (K) and ΔE can be expressed as:

$$K = \exp(\Delta E/kT)$$

Where k is the Boltzman constant and T the absolute temperature. E will depend on the size of the partitioned molecule, since the larger it is, the greater the number of atoms exposed which can interact with the surrounding phase. Thus, the Bronsted theory gives an exponential relationship between the partition coefficient and molecular weight (M):

$$K = \exp(\lambda M/kT)$$

where λ is a factor which depends on properties other than molecular weight, such as charge and density. The amount of protein extracted into the top phase (Y_t) can be determined using the following equation:

$$Y_t\ (\%) = \frac{100}{1 + \dfrac{V_b}{V_t.K}}$$

where V_t and V_b are the volumes of the top and bottom phases, respectively.

The following properties of partitioning can be exploited individually or in conjunction to achieve an effective separation of a particular protein (26).

163

(i)   Size-dependent partition where molecular size of the proteins or surface area of the particles is the dominating factor.

(ii)   Electrochemical, where the electrical potential between the phases is used to separate molecules or particles according to their charge.

(iii)   Hydrophobic affinity, where the hydrophobic properties of a phase system are used for separation according to the hydrophobicity of proteins.

(iv)   Biospecific affinity, where the affinity between sites on the proteins and ligands attached to one of the phase polymers is exploited for separation.

(v)   Conformation-dependent, where the conformation of the proteins is the determining factor.

Thus, the overall partition coefficient can be expressed in terms of all these individual factors:

$$\ln K = \ln K_0 + \ln K_{el} + \ln K_{hfob} + \ln K_{biosp} + \ln K_{size} + \ln K_{conf}$$

where el, hfob, biosp, size and conf stand for electrochemical hydrophobic, biospecific, size, and conformational contributions to the partition coefficient and $\ln K^0$ includes other factors.

The factors which influence partitioning of a protein in aqueous two-phase systems as follows.

(i)   the polymer used;

(ii)   molecular weights and size of polymers;

(iii)   concentration of polymer;

(iv)   ionic strength;

(v)   pH;

(vi)   purity of protein solution.

Generally, the higher the molecular weight of the polymers the lower the concentration needed for the formation of two phases, and the larger the difference in molecular weights between the polymers the more asymmetrical is the curve of the phase diagram. Also, the larger the molecular weight of the PEG, the lower the value of K, whereas the molecular weight of dextran does not have such a strong effect on K. A high pH can also result in a several-fold increase in the value of K (27), as can a high phosphate concentration [e.g. $0.3-0.4$ M (25)].

7.1 Equipment and materials

Polymers used for two-phase partitioning are listed in *Table 5*, together with their properties, and suppliers. For large-scale use the highly purified dextrans are prohibitively expensive; thus, a high molecular weight (5 000 000) crude dextran and a crude dextran after limited acid hydrolysis have been used (30). Starch polymers have recently been used as substitutes for dextran (31) (Reppal PES, Aquaphase PPT).

At a laboratory scale, equipment will mainly consist of plastic centrifuge tubes provided with lids, pipettes, a centrifuge and a small ultrafiltration unit with a low molecular weight cut-off membrane (e.g. 2000). For a large-scale operation usually a liquid−liquid separator (e.g. nozzle centrifuge), or a mixer settler will be required. This will mainly depend on the density difference between the two phases and their

Table 5. Polymers used in two-phase systems.

Polymer	Composition	Molecular weight	Supplier	Price (US $ kg^{-1})
Polyethylene glycol (PEG)	Polymer of ethylene glycol	600–20 000; usually 4000–8000	Union Carbide Corp.	1.4–1.6
Dextran, technical grade	(1–6) linked glucose polymer	500 000	Pharmacia-LKB; Pfeiffer Langen	80–100
Ficoll	Copolymer of sucrose and epichlorohydrin	400 000	Pharmacia-LKB	
Reppal PES	Hydroxypropyl derivative of starch	100 000 or 200 000	Reppe Glykos AB	17
Aquaphase PPT	Hydroxypropyl derivative of starch		Perstorp Biolytica AB	

165

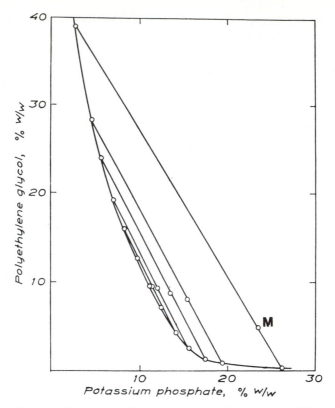

Figure 13. Phase diagram and phase compositions of the potassium phosphate−PEG 4000 system at 20°C (26).

viscosity. Generally, in a PEG/phosphate system more rapid separation is obtained than in a PEG/Dextran (or a PEG/starch polymer) system (10−60 min as opposed to 30−180 min using gravitational settling).

7.2 Optimization

To develop a suitable two-phase system for the concentration of a specific protein all the factors which can influence the system must be taken into account. These parameters will also influence each other (27).

From the phase diagram a system can be constructed to allow concentration of the protein into the top phase. For example, using *Figure 13*, a system with an overall composition at point M would have a large potassium phosphate bottom phase and a small PEG-rich top phase into which the protein will partition.

When developing methods at the laboratory scale 1−10 ml systems are used (32). The method given below is suitable for this scale.

(i)　Prepare stock solutions on a weight to weight basis in water (e.g. 20−30% w/w for dextran or Ficoll, 30−40% w/w for PEG, and for salts at least four times the concentration required in the phase system). Add sodium azide to 0.05% if the solutions are to be stored for more than 1 week. Store at 4°C.

(ii) Weigh the required amounts of polymer stock solutions in graduated glass tubes. Ensure the temperature of the stock solutions has been equilibrated to that at which the partitioning will be carried out (usually $\sim 20\,^\circ C$).

(iii) Add the sample and water to give the appropriate concentration of polymers (e.g., a system of 15% PEG/15% potassium sulphate contains the given % w/w in the final overall composition of the system).

(iv) Mix for ~ 30 sec on a vortex mixer.

(v) Separate the two phases by:

    (1) centrifuging at $100-2000$ *g* for 5 min using a swing-out rotor; or

    (2) gravitational settling, typically for 10 min to 2 h. The settling time varies with the system type, the volumes used, and the ratio of the volumes. Also sedimentation time decreases with increasing length of the tie line due to higher interfacial tension, and larger density differences (thus, systems close to the critical point require longer settling times). With increasing polymer concentration this effect may be counteracted by the increase in viscosity (26).

(vi) Measure the volumes of the two phases and remove samples from both with a pasteur pipette.

(vii) Measure the total protein concentration and/or enzyme activity in each phase. Ensure that the assay is not affected by the polymers (e.g. PEG strongly affects the Folin protein assay).

At the laboratory scale the following parameters should be determined to assist optimization (32).

    Concentration of protein of interest in both phases.

    Volume of each phase.

    Material balance of protein of interest.

    Concentration of contaminants in each phase (e.g. total protein, nucleic acids).

    Material balance of contaminants.

    pH of each phase.

These parameters will be influenced by a variety of factors $(33-36)$, such as polymer molecular weight and concentration, pH, ionic strength and salt type. Thus, for each purification, these factors should be optimized.

The following general rules can be applied to optimize partition. To increase the partition coefficient of a protein in the PEG phase of PEG/dextran systems:

(i) lower the molecular weight of PEG;

(ii) increase the molecular weight of dextran;

(iii) increase the pH in the presence of phosphate; or

(iv) increase the phosphate concentration.

If the pI of the protein is known then the influence of salt type (used to change the electrical potential difference between both phases) on the partition coefficient can be more readily predicted.

For separating insoluble material, such as cell debris, from proteins by partition, conditions must be established so that solids are collected in the bottom phase and the protein in the top phase (with an enhanced enrichment). A screening for suitable phase systems may start with the following system compositions: 9% PEG 4000; 2% dextran

Table 6. Extraction of enzymes from cell homogenates (30).

Organism	Enzyme	Constituent of the phase system	K_{enzyme}	Yield (%)
Candida boidinii	Catalase	PEG 4000/crude dextran	2.95	81
	Formaldehyde dehydrogenase	PEG 4000/crude dextran	10.8	94
	Formate dehydrogenase	PEG 4000/crude dextran	7.0	91
	Formate dehydrogenase	PEG 1000/potassium phosphate	4.9	94
	Isopropanol dehydrogenase	PEG 1000/potassium phosphate	18.8	98
Saccharomyces carlsbergensis	α-Glucosidase	PEG 4000/dextran T-500	1.5	75
Saccharomyces cerevisiae	α-Glucosidase	PEG 4000/dextran T-500	2.5	86
	Glucose-6-phosphate dehydrogenase	PEG 1000/potassium phosphate	4.1	91
Streptomyces sp.	Glucose isomerase	PEG 1550/potassium phosphate	3.0	86
Klebsiella pneumoniae	Pullulanase	PEG 4000/dextran T-500	2.96	91
	Phosphorylase	PEG 1550/dextran T-500	1.4	85

Organism	Enzyme	System		
Escherichia coli	Isoleucyl tRNA synthetase	PEG 6000/potassium phosphate	3.6	93
	Leucyl tRNA synthetase	PEG 6000/potassium phosphate	0.8	75
	Phenylalanyl tRNA synthetase	PEG 6000/potassium phosphate	1.7	86
	Fumarase	PEG 1550/potassium phosphate	3.2	93
	Aspartase	PEG 1550/potassium phosphate	5.7	96
	Penicillin acylase	PEG 4000/crude dextran	1.7	90
Bacillus sp.	Leucine dehydrogenase	PEG 4000 crude dextran	9.5	98
Bacillus species	Glucose dehydrogenase	PEG 4000 crude dextran	3.2	95
Brevibacterium ammoniagenes	Fumarase	PEG 1550 potassium phosphate	0.24	90
Lactobacillus cellobiosus	β-glucosidase	PEG 1550/potassium phosphate	2.2	98
Lactobacillus sp.	Lactate dehydrogenase	PEG 4000/dextran PL-500	6.3	95

Table 7. Purification of leucine dehydrogenase from *B.sphaericus* (37).

Step	Overall yield (%)	Specific activity (U mg^{-1})	Purification (fold)
1. Cell disruption	100	0.4	—
2. Heat denaturation	100	2.1	5.3
3. PEG/dextran[a] system	97	5.2	2.5
4. PEG/salt system	83	6.5	1.3
5. Diafiltration	70	11.0	1.7
Overall yield:	70%		
Overall purification:	27.5 times		

[a]Crude dextran PL, Pfeiffer and Langen.

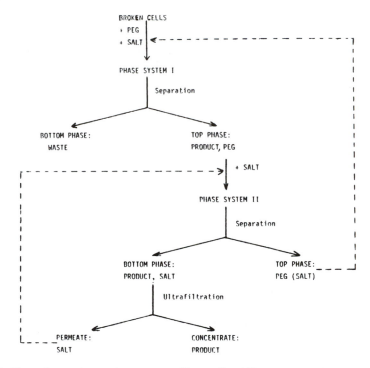

Figure 14. Flow scheme of extractive recovery with recycling (42).

500 (or 1% crude dextran); 0.2 M potassium phosphate, pH 7 to 8; biomass content approximately 20% (wet weight/v). PEG and dextran concentrations may then be varied in both directions and the partitioning of the protein should be optimized as described above.

7.3 Using phase partitioning in a purification procedure

Partitioning using aqueous systems will be predominantly used for cell debris removal and the primary purification steps. As a method of cell debris removal, aqueous phase

Table 8. Examples of affinity extraction of proteins with PEG/dextran systems

Protein	Ligand	Reference
Trypsin	Trypsin inhibitor	49
Phosphofructokinase	Triazine-dye	50
Formate dehydrogenase	Triazine-dye	51
Glucose-6-phosphate dehydrogenase	Triazine-dye	52
Serum albumin and α-lactalbumin	Fatty acids	53
Albumin and α-fetoprotein	Triazine-dye	54
Thaumatin	Glutathione	55

systems offer a number of important advantages (36). In addition to cell debris removal, the protein of interest is enriched and other contaminants, such as proteins, nucleic acids, polysaccharides and coloured by-products from fermentation can also be removed. The polymers present in the aqueous phases also enhance stability of the protein and little or no protein denaturation is observed. The process can also be carried out at room temperature without cooling.

For economic reasons the amount of cell mass processed within the phase system should be as high as possible. However, with increased cell mass the volume ratio of the phases is decreased and thus conditions must be defined to give a sufficiently high partition coefficient for efficient extraction. *Table 6* shows examples of extractive cell debris removal experiments in PEG/dextran systems. In most cases the partition coefficients of these enzymes are between 2 and 10, and the overall yields are 90% or more.

After the first extraction further purification of the protein collected in the top phase is generally accomplished by addition of salts or dextran, thus generating a new phase system. Depending on the requirements for protein purity, or of the following purification procedures, the product can be partitioned either to the new top or bottom phase (32). If further removal of nucleic acids and/or polysaccharides and higher enrichment factors are required the product should be extracted into the top phase and a third separation should be carried out.

For the separation of contaminating proteins no general rules have been defined; conditions are verified experimentally. *Table 7* summarizes an example where the extraction has been coupled to other purification procedures for the recovery of leucine dehydrogenase from *Bacillus sphaericus*. For the large-scale purification of β-galactosidase a conventional procedure can be compared with the use of aqueous two-phases. Higgins *et al.* (38) outlined a process of successive centrifugation steps to remove cell debris, precipitated nucleic acids and precipitated protein product. Veide *et al.* (39) showed that a single aqueous extraction with PEG and salt in which β-galactosidase partitions to the PEG-rich upper phase could be used. Cells, nucleic acids and most contaminating proteins partition to the salt-rich phase.

After partitioning the polymer can be removed from the protein of interest by ultra-filtration through a membrane of the appropriate type (40). Alternatively, for a PEG-rich phase, salt can be added to establish a new phase system where the protein will partition into the salt phase (26, 41). Other methods include adsorption and subsequent elution, or precipitation with salt (26). Recently, an enzyme extraction method has been devised using a PEG/phosphate system with two sequential extractions which allows recycling of the phase-forming chemicals (see *Figure 14*). The overall product yield was 77% and a 13-fold purification was obtained (42).

7.4 Affinity partitioning

Affinity partitioning, where a biospecific ligand is either present free in the system, or bound to the less polar polymer, offers increased selectivity. Most proteins, nucleic acids and cell debris will partition to the more polar phase, whilst proteins which bind to the ligand will partition into the less polar phase. Ligands used can be specific for one protein [e.g. substrates, inhibitors, antibodies and protein A (43,44)] or specific for a group of proteins [e.g. cofactors (45), or triazine dyes (46−48)]. Some examples are given in *Table 8*; for a more comprehensive list the reader is referred to ref. 26.

Several methods are available for covalent attachment of the ligand to the polymer, usually PEG (28,56−58). The trichloro-s-triazine method (58) has several disadvantages: it is complex to carry out, uses highly toxic chemicals, and non-specific adsorption to the activated groups may occur. Alternatives currently being developed are bisepoxy-oxirane activation (55,56), periodate and epichlorohydrin (55, authors' unpublished results). After activation, the polymer is incubated with ligand. Unbound ligand is removed by ultrafiltration, gel filtration or two-phase partitioning. Unreacted groups are then blocked.

Factors which must be considered during optimization include the partition coefficient of the protein, the ligand concentration in the phase system, the partition coefficient of the ligand, the dissociation coefficient of the protein−ligand complex and the number of binding sites that the protein has for the ligand. A potential complicating factor is non-specific binding of contaminants to the ligand-polymer, often due to interactions with the activating group. In affinity partitioning recovery and recycle of the usually costly PEG−ligand complex is a crucial issue. However, the PEG−ligand complex will preferably partition to the PEG-rich phase making the ligand losses in the bottom-phase negligible (27).

8. ACKNOWLEDGEMENTS

E.L.V.Harris is grateful to J.Fletcher (Millipore), P.Reardon (Romicon) and J.Hickling (Sartorius) for their help in providing information and in particular to Millipore for providing figures for the ultrafiltration section.

9. REFERENCES

1. Freitas,R.D.S. and Cussler,E.L. (1987) *Chem. Eng. Sci.*, **42**, 97.
2. Docksey,S.J. (1986) In *Bioactive Microbial products III. Downstream Processing*. Academic Press, New York and London, p. 161.
3. Tutunjian,R.S. (1985) *Bio/Technology*, **3**, 615.
4. Anon. (1987) *Bio/Technology*, **5**, 915.
5. Hanisch,W. (1986) In *Membrane Separations in Biotechnology*. McGregor,W.C. (ed.), Marcel Dekker, New York and Basel, Vol. 1, p. 61.
6. Blatt,W.F., Dravid,A., Michaels,A.S. and Nelsen,L. (1970) In *Membrane Science and Technology*. Flinn,J.E. (ed.), Plenum, New York, p. 47.
7. de Fillipi,R.P. and Goldsmith,R.L. (1970) In *Membrane Science and Technology*. Flinn,J.E. (ed.), Plenum, New York, p. 33.

8. Porter,M.C. (1972) *Ind. Eng. Chem. Prod. Res. Dev.*, **11**, 234.
9. McGregor,W.C. (1986) In *Membrane Separations in Biotechnology*. McGregor,W.C. (ed.), Marcel Dekker, New York and Basel, Vol. 1, p. 1.
10. Dunnill,P. (1983) *Process Biochem.*, **18**, 9.
11. Narendranathan,T.F. and Dunnill,P. (1982) *Biotechnol. Bioeng.*, **24**, 2103.
12. Pace,G.W., Scherin,M.J., Archer,M.C. and Goldstein,D.J. (1976) *Sep. Sci.*, **11**, 65.
13. Swaminathan,T., Chandhuri,M. and Sirkar,K.K. (1981) *Biotechnol. Bioeng.*, **23**, 1873.
14. Matthiasson,E. (1983) *J. Membr. Sci.*, **16**, 23.
15. Sirkar,K.K. and Prasad,R. (1986) In *Membrane Separations in Biotechnology*. McGregor,W.C. (ed.), Marcel Dekker, New York and Basel, Vol. 1, p. 37.
16. Ingham,K.C., Busby,T.F., Sahlestrom,Y. and Castino,F. (1980) In *Ultrafiltration Membranes and Applications*. Cooper,A.R. (ed.), Plenum, New York, p. 141.
17. Luong,J.H.T., Nguyen,A.L. and Male,K.B. (1987) *TIBTECH*, **5**, 281.
18. Mandaro,R.M., Roy,S. and Hou,K.C. (1987) *Bio/Technology*, **5**, 928.
19. Brandt,S., Goffe,R.A., Kessler,S.B., O'Connor,J.L.and Zale,S.E. (1988) *Bio/Technology*, **6**, 779.
20. Johns,E.W. (1964) *Biochem. J.*, **92**, 55.
21. Cohn,E.J., Strong,L.L., Hughes,W.L., Mulford,D.L., Ashworth,J.N. and Taylor,H.L. (1946) *J. Am. Chem. Soc.*, **68**, 459.
22. Froud,S.J. (1989) In *Protein Purification Applications: A Practical Approach*. Harris,E.L.V. and Angal,S. (eds), IRL Press, Oxford.
23. Findlay,J. (1989) In *Protein Purification Applications: A Practical Approach*. Harris,E.L.V. and Angal,S. (eds), IRL Press, Oxford.
24. Ingham,K.C. (1984) In *Methods in Enzymology*. Jakoby,W.B. (ed.), Academic Press, New York and London, Vol. 104, p. 351.

Aqueous two-phase partitioning

25. Kula,M.R. (1979) *Appl. Biochem. Bioeng.*, **2**, p. 71.
26. Albertsson,P.A. (1986) *Partition of Cell Particles and Macromolecules*. 3rd edition, Wiley & Sons, New York.
27. Dove,G.B. and Mitra,G. (1986) In *Separation, Recovery and Purification in Biotechnology: Recent Advances and Mathematical Modeling*. Asenjo,J.A., and Hong,J. (eds), American Chem. Soc., Washington, p. 93.
28. Birkenmeier,G., Kopperschlager,G., Albertsson,P.A., Johansson,G., Tjerneld,F., Akerlund,H.E., Berner,S. and Wickstroem,H. (1987) *J. Biotechnol.*, **5**, p. 115.
29. Mattiasson,B. and Kaul,R. (1986) In *Separation, Recovery and Purification of Biotechnology: Recent Advances and Mathematical Modeling*. Asenjo,J.A. and Hong,J. (eds), American Chem. Soc., Washington, p. 78.
30. Kula,M.R., Kroner,K.H. and Hustedt,H. (1982) *Adv. Biochem. Eng.* **24**, p. 73.
31. Skuse,D.R., Norris-Jones,R., Brooks,D.E., Abdel-Malik,M.M. and Yalpani,M. (1987) Paper presented at the International Conference on Partitioning in Aqueous Two Phase Systems, Oxford, August, 1987.
32. Pfeiffer and Langen (1986) *Aqueous Phase Systems on the Basis of Dextran*. Pfeiffer and Langen, Dormagen, FRG.
33. Bamberger,S., Brooks,D.E., Sharp,K.A., Van Alstine,J.M. and Weber,T.J. (1985) In *Partitioning in Aqueous Two-Phase Systems*. Walter,H., Brooks,D.E. and Fisher,D. (eds), Academic Press, New York and London, p. 85.
34. Hustedt,H., Kroner,K.H. and Kula,M.R. (1985) In *Partitioning in Aqueous Two-Phase Systems*. Walter,H., Brooks,D.E. and Fisher,D. (eds), Academic Press, New York and London, p. 529.
35. Johansson,G. (1984) *Acta. Chem. Scand. Ser. B.*, **28**, 873.
36. Kula,M.R., Kroner,K.H., Hustedt,H., Grandja,S. and Stach,W. (1976) German patent No. 26.39.129; US patent No. 4,144,130.
37. Schutte,H., Kroner,K.H., Hummel,W. and Kula,M.R. (1983) *Ann. N.Y. Acad. Sci.*, **413**, 270.
38. Higgins,J.J., Lewis,D.J., Daly,W.H., Mosqueira,F.G., Dunnill,P. and Lilly,M.D. (1978) *Biotechnol. Bioeng.*, **20**, 159.
39. Veide,A., Smeds,A.L. and Enfors,S.O. (1983) *Biotechnol. Bioeng.*, **25**, 1789.
40. Hustedt,H., Kroner,K.H., Menge,U. and Kula,M.R. (1978) *Abst. 1st. Eur. Congr. Biotechnol.*, Interlaken.
41. Hustedt,H., Kroner,K.H., Stach,W. and Kula,M.R. (1978) *Biotechnol. Bioeng.*, **20**, 1989.
42. Hustedt,H. (1986) *Biotechnol. Lett.*, **8**, 791.
43. Mattiasson,B. (1980) *J. Immunol. Methods*, **35**, 137.
44. Ling,T.G.I. and Mattiasson,B. (1982) *J. Chromatogr.*, **252**, 159.
45. Patton,J.S., Albertsson,P.-A., Erlanson,C. and Borgstrom,B. (1978) *J. Biol. Chem.*, **253**, 4195.
46. Lowe,C.R.,Small,D.A.P. and Atkinson,A. (1981) *Int. J. Biochem.*, **13**, 33.

47. Johansson,G., Kopperschlager,G. and Albertsson,P.-A. (1983) *Eur. J. Biochem.,* **3**, 589.
48. Johansson,G. and Joelsson,M. (1985) *Biotechnol. Bioeng.,* **27**, 621.
49. Segard,E., Takerkart,G., Monsigny,M. and Oblin,A. (1973) French patent No. 7342320.
50. Kopperschlager,G. and Johansson,G. (1982) *Anal. Biochem.,* **124**, 117.
51. Cordes,A. (1985) Dissertation,T.U. Braunschweig, FRG.
52. Kroner,K.H., Cordes,A., Schelper,A., Moor,M., Buckmann,A.F. and Kula,M.R. (1982) In *Affinity Chromatography and Related Techniques*. Gribnau,T.C.J., Visser,J. and Nivard,R.J.F. (eds), Elsevier, Amsterdam, p. 491.
53. Johansson,G. and Shanbhag,V. (1984) *J. Chromatography,* **284**, 63.
54. Birkenmeier,G., Usbeck,E. and Kopperschlager,G. (1984) *Anal. Biochem.,* **136**, 264.
55. Dunthorne,P. (1988) B.Sc. project in Biotechnology, University of Reading.
56. Head,D.M., Andrews,B.A. and Asenjo,J.A. (1989) *Biotechnol. Tech.,* **3**, 27.
57. Janson,H.-C. (1984) *Trends Biotechnol.,* **2**, 31.
58. Abuchowski,A., Van Es., T., Palczuk,C. and Davis,F. (1977) *J. Biol. Chem.,* **252**, 3578.

CHAPTER 4

Separation based on structure

S.ROE

1. INTRODUCTION

This chapter aims to provide guidance in the selection of techniques and operating conditions for the purification of proteins based on some of their structural properties. Each technique uses a single parameter (e.g. charge or hydrophobicity) to effect adsorption and separation. In contrast, separation based on biological interaction, which is mostly mediated through a combination of such interactions is covered in the next chapter.

Some general concepts are described in the first instance followed by sections on matrices and equipment which are common to all the techniques. This is followed by description of the theory and practice of individual techniques. The chapter is aimed at those not familiar with these techniques and who will be considering protein purification on a laboratory scale. For further information on this subject the reader is directed to references 1−11.

2. GENERAL CONCEPTS

2.1 **Chromatography**

Chromatography is the differential separation of sample components between a mobile phase and a stationary phase. In the majority of applications the stationary phase consists of spherical particles which are packed into a column. A mixture of proteins to be separated is introduced into the mobile phase and allowed to migrate through the column. Those proteins having a greater attraction for the solid phase migrate slower than proteins more attracted to the mobile phase, thus effecting resolution (*Figure 1*).

2.2 **Adsorption/desorption**

While purification by chromatography may provide excellent resolution of protein mixtures adsorption/desorption is of equal importance, particularly for preparative work. This technique also separates proteins according to their relative distribution between a liquid and a solid phase, but is not based on speed of migration. It is generally used in a batch mode to provide a rapid method of concentration and purification, most usefully at an early stage of a process.

2.3 **Compatibility**

Table 1 shows that each method of purification places certain requirements on the sample which is to be purified. In the design of a purification procedure it is important that

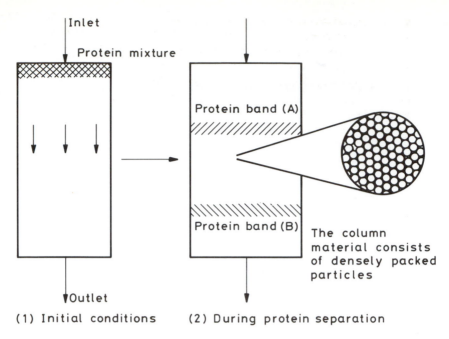

Figure 1. The principle of protein purification using chromatography. A protein mixture is applied to a column in the mobile phase. As it travels down the column proteins are separated depending on their degree of attraction for the column packing. Protein band (**A**) is more attracted to the stationary phase and moves slower than protein band (**B**).

consecutively used techniques are therefore compatible. After a salt precipitation step using ammonium sulphate, for example, the salt concentration of the protein-containing solution is high and therefore unsuitable for the effective use of ion exchange without prior desalting or dilution to reduce the ionic strength. Hydrophobic interaction, however, requires a high salt concentration to promote protein adsorption and may therefore be used after ammonium sulphate precipitation with minimal necessary adjustments to the sample. Elution of proteins in hydrophobic interaction is frequently accomplished using a reduced ionic strength, allowing the subsequent use of an ion exchange step. Thus the selection of technique is clearly dependent on its position in a purification scheme.

2.4 Capacity

Capacity is a measure of the amount of protein which can be adsorbed from solution onto a unit volume or weight of the stationary phase. High capacity techniques can remove proteins from large volumes of solution and are therefore frequently used early on in a purification strategy where both the total protein level and solution volume are high. Ion exchange is an example of a high capacity technique often used to concentrate and purify proteins at an early stage of a purification.

2.5 Selectivity

Selectivity measures the ability of a purification technique to adsorb a protein from

Table 1. Characteristics of the purification techniques discussed in this chapter.

Separation techniques	Characteristics	Sample composition		Approx matrix cost £/litre	Stage of purification
		Before	After		
Ion exchange	High capacity; used in batch or column mode	Low ionic strength; correct pH	Change in pH; high ionic strength	200	Variable but especially early
Hydrophobic interaction	High capacity; useful after salt precipitation	High ionic strength	Change in pH and/or low ionic strength	400	Variable but especially early
Hydroxylapatite	Mild separation conditions	Neutral pH; low phosphate	Neutral pH; high phosphate concentration	280	Variable
Metal chelate	Not sensitive to ionic strength	Absence of chelators	Altered pH, presence of chelators	1000	Variable
Covalent	Matrix requires lengthy regeneration	Neutral at mildly acidic pH	Presence of low molecular weight thiols	2000	Late
Chromatofocusing	High resolution	Low ionic strength	Presence of amphoteric buffer. pH close to protein pI.	430	Late

solution with a high degree of rejection of contaminants. High selectivity methods are usually reserved for the last stages in a purification, since they are more able to separate the very similar proteins remaining after earlier general processing. Examples of highly selective techniques include affinity, covalent and metal chelate chromatography.

2.6 Resolution

Resolution of proteins is the aim of each purification step. It is a measure of the degree of separation of a required protein from contaminants. Methods which have a high selectivity can therefore separate proteins with a high degree of resolution. Chromatography achieves better resolution than adsorption/desorption since the degree of separation is improved by multiple opportunities for re-equilibration as the developing front migrates down a column. Chromatography is therefore ideally suited to the separation of similar proteins while adsorption allows protein concentration. High resolution techniques include chromatofocusing and HPLC. The latter technique uses fast flow rates and small diameter particles to provide rapid high resolution separations.

2.7 The efficiency of a separation

The aim of a protein separation is to purify the maximum amount of protein in the minimum amount of time. This is called the throughput. The throughput of a separation can be improved in two ways.

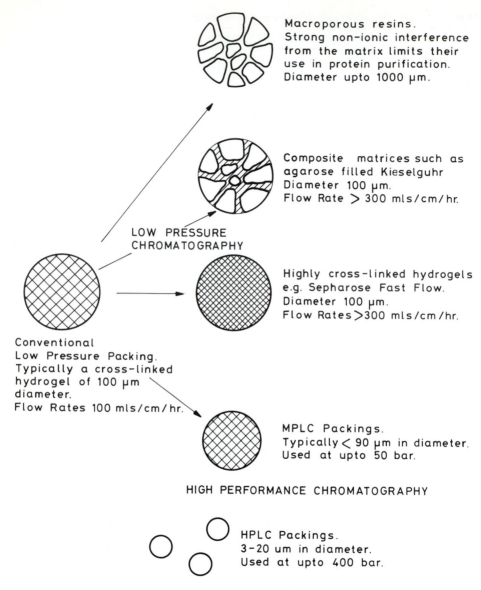

Macroporous resins.
Strong non-ionic interference
from the matrix limits their
use in protein purification.
Diameter upto 1000 μm.

Composite matrices such as
agarose filled Kieselguhr
Diameter 100 μm.
Flow Rate > 300 mls/cm/hr.

LOW PRESSURE
CHROMATOGRAPHY

Highly cross-linked hydrogels
e.g. Sepharose Fast Flow.
Diameter 100 μm.
Flow Rates >300 mls/cm/hr.

Conventional
Low Pressure Packing.
Typically a cross-linked
hydrogel of 100 μm
diameter.
Flow Rates 100 mls/cm/hr.

MPLC Packings.
Typically < 90 μm in diameter.
Used at upto 50 bar.

HIGH PERFORMANCE CHROMATOGRAPHY

HPLC Packings.
3-20 um in diameter.
Used at upto 400 bar.

Figure 2. Improvements in matrix design. The variety of stationary phases used in protein purification may be broadly divided between high performance and low pressure matrices.

(i)   By the use of high capacity matrices which allow a large amount of protein to be purified at a time.
(ii)   By using matrices which can be packed into a column and used at a high flow rate.

Improvements in matrix design have been achieved by using two different methods (*Figure 2*). Firstly, more rigid matrices have been designed by improving the flow

Table 2. Typical characteristics of low and high pressure chromatography.

Characteristic	Low pressure	High pressure
Particle size (μm)	100	10
Flow rate (ml cm^{-2} h^{-1})	10−30	100−300
Operating pressure (bar)	<5	>50
Separation time (h)	Up to 24	1−3
Sample volume	Variable (ml−litres)	Small (μl−ml)
Purification stage	Variable	Usually late
Resolution	Good	May be excellent
Drawbacks	Long process time; potential need for cold room facilities	High cost of equipment. Risk of denaturation from high pressure

characteristics of existing chromatographic materials. This has been accomplished using a higher degree of matrix cross-linking and by the design of composite materials with a rigid incompressible structure. These materials allow higher flow rates to be used without causing bed compression.

Secondly, an alternative approach to improving both throughput and resolution has been in the design of more rigid matrices of smaller particle size (3−20 μm) in high performance (pressure) liquid chromatography (HPLC). The improved efficiency results, in part, from the reduction of diffusion effects which are responsible for the lower resolution achievable in low pressure liquid chromatography (LPLC). Diffusion is also reduced since high flow rates (100−300 ml cm^{-2} h^{-1}) at high pressure (>50 bar) are used in the very evenly packed columns usually supplied ready packed by manufacturers. A comparison of the characteristics of LPLC and HPLC is given in *Table 2*.

Purifications in which the operating pressure lies between those used in LPLC or HPLC (i.e. 6−50 bar) are termed medium pressure liquid chromatography (MPLC). These techniques will be discussed in detail in subsequent sections.

2.8 Cost

The cost of each matrix and its re-usability are important considerations in its use. Generally those techniques providing a greater degree of selectivity are more expensive to use (*Table 1*). Consequently high selectivity techniques tend to be used later in the purification when sample volume is low and many contaminants have already been removed, thereby requiring smaller matrix volumes. Implications of matrix, equipment and reagent costs in large-scale use are discussed in more detail in the companion to this volume *Protein Purification Applications: A Practical Approach* (see ref. 12).

3. MATRIX MATERIALS

The matrix is the solid substrate of the stationary phase. It is modified to confer specific properties to the support by the chemical attachment of various functional groups. As the same functionality may be available on a wide range of matrices it is important to understand the physicochemical properties of the matrices and consider how these influence the selection of a stationary phase (*Table 3*).

Table 3. Some characteristics of commonly used matrices.

Matrix	Manufacturers	pH stability	Typical cross-linker	Drawbacks of use
Cellulose	Whatman Pharmacia-LKB Bio-Rad, Serva	3−10 2−12 if cross-linked	Epichlorohydrin	Cycling is required for dry material
Dextran	Pharmacia-LKB	2−12	Epichlorohydrin	Swells and shrinks depending on ionic strength
Agarose	Bio-Rad Pharmacia-LKB	4−9 3−14 if cross-linked	2,3-dibromopropanol	Must not dry out
Polyacrylamide	IBF Bio-Rad Pharmacia-LKB	2−11	*N-N'*-Methylene bisacrylamide	*N-N'* bis is toxic
Silica	Whatman Waters DuPont Merck	3−8	By poly-condensation	Unstable above pH8
Polystyrene	Rohm & Haas Dowex Bio-Rad	Any pH	Divinyl benzene	Non-ionic interaction

Important criteria in the choice of a matrix are as follows.

(i)   High mechanical stability: allows maximization of flow rate, minimizes the pressure drop and results in low abrasion.

(ii)   Good chemical stability: necessary to maintain the bed structure, allow sterilization and minimize contamination of the protein mixture.

(iii)   High capacity: for minimizing the bed volume thus maximizing speed of operation. The capacity is dependent on the density of groups available for functionalization.

(iv)   In porous matrices pore size must be sufficiently large to allow access of proteins to the internal surface.

(v)   Pore shape is also important since blocking of pore entrances can dramatically reduce available surface areas.

(vi)   The surface of the matrix should be inert in order to minimize non-specific adsorption which can eventually result in fouling of the matrix.

(vii)   Matrix density should be suitable for the application; e.g. fluidized beds require densities to be sufficiently high so as to maintain a stable bed.

(viii)   Particle size influences the choice of equipment required to carry out a separation. The mass transfer of solutes from the mobile phase onto the stationary phase is largely determined by the diffusion distance (6,7). If smaller particles are used specialized chromatographic equipment must be available to deliver an even flow of solvent at high back pressures.

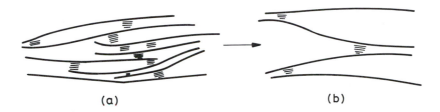

Figure 3. The chemical structure of cellulose: β1−4 linked glucose residues.

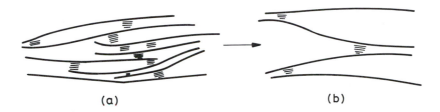

(a)   (b)

Figure 4. The result of cycling dry cellulose (**a**) in 0.5 M NaOH leads to an opening up of the fibrous structure and widening of the matrix pores (**b**).

3.1 Cellulose

This is a linear polysaccharide of β1−4 linked glucose monomers (*Figure 3*). Each glucose residue has three hydroxyl groups which makes the matrix very hydrophilic and easily derivatized. Cellulose itself is relatively inert towards proteins and unstable towards mineral acids, alkalis and oxidants. Cellulosic ion exchangers have an operating pH range of 3−10 but can withstand 0.5 M alkali for up to 2 h. The pores in cellulose are formed at regions of the polymer structure where hydrogen bonding between chains is reduced. The nature of these pores depends on the degree of solvation and the batch of cellulose. It is necessary to pre-treat dry cellulose in 0.5 M alkali for 30 min in order to swell the matrix and open up the pores (*Figure 4*).

Cellulose is available in fibrous, microgranular or bead form. The fibrous matrix has good hydrodynamic properties (400 ml cm^{-2} h^{-1}) and is recommended for use in early purification steps. Microgranular cellulose provides better resolution since matrix rigidity and porosity are increased by chemical cross-linking and partial acid hydrolysis. Both matrices are unstable at high flow rates and may collapse. Beaded cellulose is a further improvement in rigidity and chemical stability giving the best hydrodynamic properties.

3.2 Agarose

Agarose is a polysaccharide obtained by purifying agar, a component of seaweed. It is composed of polymeric chains of the disaccharide agarobiose (D-galactose and 3,6-anhydro-1-galactose) (*Figure 5*), and forms a very porous hydrophilic gel structure upon cooling hot solutions containing agarose at concentrations as low as 2% w/v (*Figure 6*) (13).

The matrix is highly porous with minimal non-specific adsorption which can be

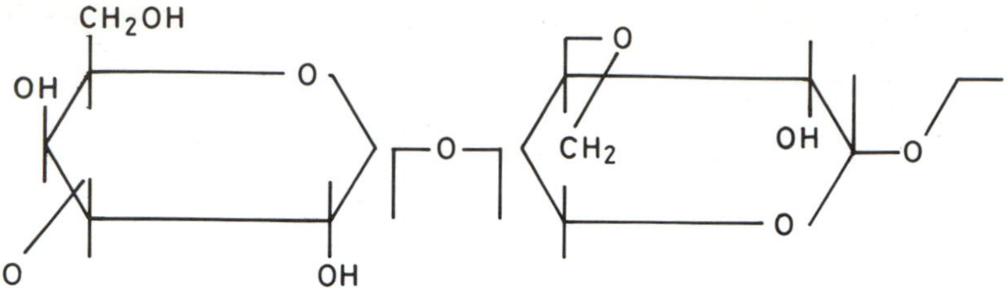

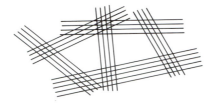

Figure 5. The chemical structure of agarose: the disaccharide repeating unit.

Figure 6. The macrostructure of agarose: a highly porous hydrophilic matrix.

reduced even further in the presence of low concentrations of salt (20 mM). While simple agarose gels have limited stability at extremes of pH, this can be extended by cross-linking with dibromopropanol to provide a gel which can be autoclaved at 120°C and is stable over a wider pH range (3−14). The level of cross-linking and amount of agarose used control the pore size of the gel. Agarose matrices should never be allowed to dry out since they undergo irreversible changes in structure once the structural water is lost.

Cross-linked agarose also has good hydrodynamic properties; flow rates of 10−30 ml cm^{-2} h^{-1} are common. However, a new generation of more highly cross-linked agaroses are now available (e.g. Sepharose Fast Flow) with flow rates up to 300 ml cm^{-2} h^{-1}. In practice, optimum resolution is obtained at 30−60 ml cm^{-2} h^{-1}.

3.3 Dextran

Dextran (4) is an extracellular polysaccharide produced by the bacterium *Leuconostoc mesenteroides* and consists of glucose residues linked by $\alpha1−6$ bonds (*Figure 7*). Each residue has six hydroxyl groups, providing a very hydrophilic matrix which is relatively chemically inert and easily derivatized. Dextran matrices are less stable than cellulose to acid hydrolysis but can withstand 0.1 M HCl for up to 2 h. The matrix is stable over a wide pH range (2−12) and is cross-linked with epichlorohydrin (*Figure 8*) for use as a gel matrix in column chromatography. As with agarose, the amount of cross-linker used controls the pore size of the matrix. The gels are rather soft, easily compressed and swell considerably in aqueous solutions.

The exclusion limit of commercially used dextran matrices is, at 300 kd (for a globular protein), rather less than agarose. Unlike cellulose, the pore size is relatively homogenous and does not require any pre-treatment other than soaking the matrix to allow swelling. Cross-linked dextrans are autoclavable but biodegradable. Dextrans can be repeatedly dried and swollen without alteration of chromatographic properties. Dextran-based ion exchangers suffer from a strong dependence of bed volume on pH and ionic strength.

182

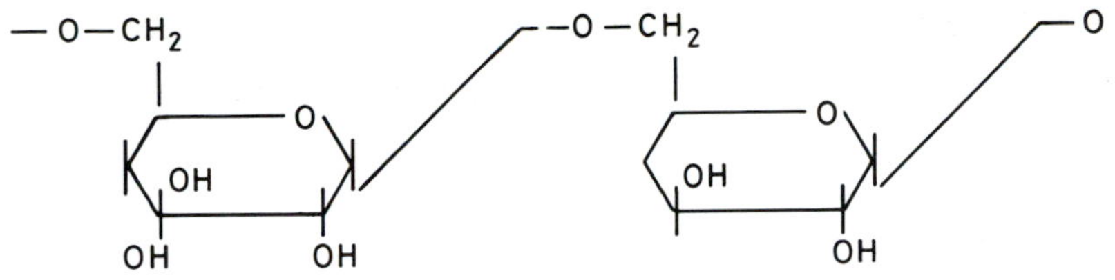

Figure 7. The chemical structure of dextran: $\alpha 1 - 6$ linked glucose residues.

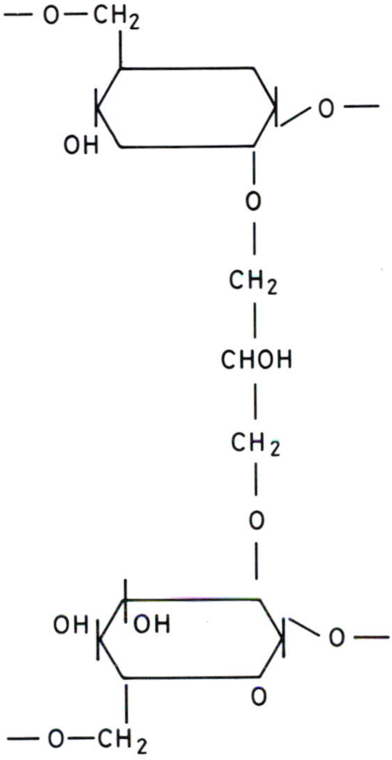

Figure 8. Epichlorhydrin cross-linked dextran (e.g. Sephadex).

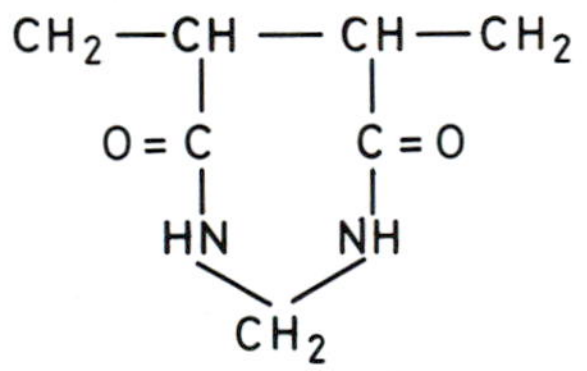

Figure 9. The repeating unit of polyacrylamide.

3.4 Polyacrylamide

Polyacrylamide gels (3) are produced by the polymerization of acrylamide in the presence of a cross-linking agent, *N,N'*-methylene bisacrylamide (*Figure 9*). The polymerization is carried out in the presence of a catalyst (e.g. persulphate) and in the absence of oxygen.

The hydrophilic compressible gels produced are used in gel permeation and modified to form ion exchangers. Gels are autoclavable and, although chemically stable, toxic acrylamides may slowly leach and may therefore limit their use in food and pharmaceutical applications. Manufacturers of these materials include IBF and Bio-Rad.

3.5 Natural earths

This group of matrices includes Kieselguhr, Fullers Earth, Bentonite and Alumina. These materials may serve without derivatization for the adsorption of proteins although the exact mechanisms involved are not well understood. Kieselguhr, derived from the silica shells of diatoms, has been used most as a matrix for both HPLC and LPLC. In the latter example, inorganic particles are fabricated into macroporous particulate matrices of size $250-500$ μm with very large pores (up to 7 μm) and highly stable characteristics. These granules may serve as adsorbents (e.g. Titania) or provide an inert matrix for subsequent use in combination with hydrogels as composite ion exchangers (14). Matrices are incompressible, allowing high flow rates with minimal back pressure in packed beds.

Alumina is formed by the precipitation of sodium aluminate to form aluminium hydroxide which, through thermal dehydration and ageing yields porous alumina aggregates (3). Matrices may be slightly acidic, neutral or basic in water depending on the treatment. Alumina is used in HPLC as porous or pellicular particles (Merck-Lichrosorb, Whatman-Pellumina) or as a batch adsorbent (Bio-Rad, BDH, Sigma).

3.6 Silica

Silica matrices (15), commonly used in HPLC, are formed by the acidification of sodium silicate to form a sol of orthosilicic acid which is subject to polycondensation during ageing to yield silica particles (*Figure 10*). The silanol (Si-OH) groups present at the silica surface make them very hydrophilic and easily derivatized. Silica matrices are incompressible and thus well suited to HPLC. They are unaffected by organic solvents and mechanically stable but gradually dissolve above pH 8. In derivatized silica, unreacted silanol groups provide the matrix with a weak cation exchange nature which is usually masked by end-capping with, for example, trimethylchlorosilane. Silicas are used in HPLC as porous particles or as coatings on glass beads in pellicular packings. Serva also produce silica matrices for low pressure chromatography (Daltosil) using a particle size of up to 200 μm and pore sizes up to 30 nm.

3.7 Porous glass

Controlled pore glass (CPG) is formed from the chemical treatment of alkaline borosilicate glass (16). CPG in bead form has been used in HPLC as a chemically inert, thermally stable matrix. Like silica, it is a negatively-charged aerogel which dissolves in alkaline conditions. CPG has a narrow pore size distribution and although used in HPLC, is a less common matrix than silica. Manufacturers include Serva, BDH and Sigma.

3.8 Polystyrene and phenol-formaldehyde

These organic synthetic polymers (*Figure 11*) are commercially available as $1-2$ mm

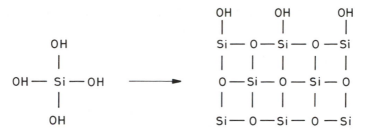

Figure 10. The polycondensation of silicic acid to produce silica gel. At the silica surface, silanol groups remain, giving silica gel its desiccant properties.

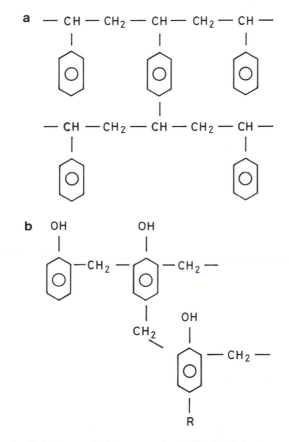

Figure 11. (a) The chemical structure of polystyrene resins (b) The chemical structure of phenol-formaldehyde.

beads, called resins. The pore size of the resins is controlled by the level of cross-linking which in turn is determined by the amount of divinyl benzene used in the manufacturing process (4). Although originally used in water treatment and essentially non-ionic in nature, the hydrophilic nature of the resins has been increased by attaching ionic groups so that they can be used in protein recovery and immobilization (17).

185

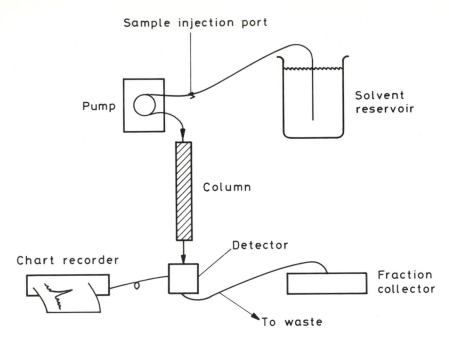

Figure 12. Typical components of low pressure chromatography. Notice that the tube length between the column outlet and the detector is kept small so as to minimize loss of resolution.

4. CHROMATOGRAPHIC EQUIPMENT AND BASIC PROCEDURES

Although proteins are separated using a variety of matrices the basic components of a chromatographic system are similar and consist of the following.

(i)   A pump to give an even flow of liquid.
(ii)   A column in which the protein separation occurs.
(iii) A detector to provide a continual measurement of a physical parameter of the eluent.
(iv)   A recorder to give a continuous visual read-out of the detector output.
(v)   A fraction collector to separate the column eluent into samples.

These are typically set up as shown in *Figure 12*. The requirements of each unit, in particular the pump and column, differ depending on the operating pressure. Consequently the equipment and procedures for protein separation will be subdivided as follows:

LPLC: <5 bar
MPLC: 6−50 bar
HPLC: >50 bar

Pressure units can be interconverted using the formula:

$$1 \text{ bar} = 15 \text{ p.s.i.} = 0.1 \text{ MPa}$$

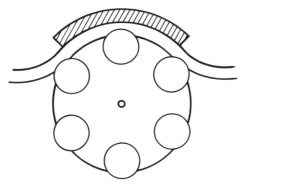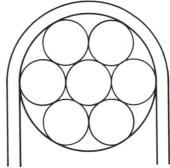

Figure 13. Roller (**left**) and planetary (**right**) pump designs commonly used in low pressure chromatography.

Table 4. Some commonly used peristaltic pumps.

Manufacturer	Pump name	Description
Bio-Rad	Econo-column	2 channel $0.6-325$ ml h^{-1}
Pharmacia-LKB	P1	1 channel $0.6-500$ ml h^{-1}
	P3	3 channel $0.6-400$ ml h^{-1}
	P500	Dual piston, high precision
Watson-Marlow	–	Wide variety available
Whatman	Peristaltic	4 channel $0.03-2280$ ml h^{-1}

4.1 Low pressure equipment

4.1.1 *Pumps*

A variable speed pump capable of directing solvent at a low flow rate is suitable. This is commonly a peristaltic pump which uses silicon, PVC or fluoro-rubber tubing and may be of two designs: (i) roller; (ii) planetary as shown in *Figure 13*.

A roller pump directs solvent by tube compression using rollers at the periphery of a revolving disc, the amount of compression being controlled by tightening a screw against a tube guiding plate. A drawback of this type of pump is that the tubing will gradually wear.

A planetary pump also directs solvent by tube compression using rollers but these rub against each other and gradually wear.

Examples of the most commonly available pumps and their specifications are given in *Table 4*. Many pumps can induce flow along several tubes of varying diameters simultaneously. Pumps must be calibrated for a specific diameter of tubing.

4.1.2 *Columns*

Low pressure chromatography columns are available from a large number of manufacturers in different lengths, sizes, degrees of chemical resistance and mechanical strength (*Table 5*). For simple small-scale laboratory experiments, however, columns may be made in the laboratory and, although less exact in their design, are suitable for some separations. A glass column with a fritted glass filter at its base can be used

Table 5. List of some small scale chromatography columns.

Manufacturer or supplier	Trade name	Description
Alltech	Cheminert	
	MB	Glass, fixed bed, 15−100 cm length
	LC	Glass, adjustable, 33−109 cm length
Amicon	Lab Columns	Glass + polyamide, polypropylene
		4−1520 ml volume
Beckman	−	For low−medium pressure
		Glass and fluorocarbon
		Adjustable, solvent resistant
Bio-Rad	Econo-columns	Glass, adjustable
		Up to 590 ml volume
	Bio-Rex MP	Solvent resistant
		Pressure resistant up to 500 p.s.i.
		15−2000 ml volume
Pharmacia-LKB	C-Series	Glass and polypropylene, polyamide, polyethylene and fluoro-rubber
		Up to 520 ml volume
		Adjustable, autoclavable
	K-Series	For labile biopolymers
		Available at different levels of solvent resistance
		Up to 1800 ml volume
		Adjustable
	SR Series	Minimal heavy metal contamination
Pierce	Chromato Flo	Glass and polypropylene or teflon
		Adjustable 19−1176 ml volume
Whatman	IEC	Glass and polypropylene
		volume up to −500 ml

for separations provided a tube can be fitted to the column top to supply a solvent inlet. The glass frit must retain the adsorbent particles but must not restrict flow.

Commercially available columns used in small-scale separations are usually made of glass with a variety of internal diameters and column lengths to give a packed bed volume of up to 2 litres. The volume of most columns may be adjusted using adaptors mounted at the inlet and/or outlet. All these columns are of modular design so that their end fittings can be interchanged for cleaning. Many columns are jacketed for temperature control and are autoclavable. Where columns cannot be autoclaved, sterilization with ethylene oxide is recommended. The sorbent bed is held in place by a mesh (usually 20 μm) while the inlet may be fitted with a reversed funnel design to provide an even solvent distribution over the packed bed.

4.1.3 *Detectors*

Detectors may measure UV absorbance, fluorescence, conductivity, radioactivity or optical density. The strong absorption of light by proteins at 206−215 nm and 280 nm is due to peptide bonds and aromatic amino acids, respectively. UV absorption is therefore the most common method of protein detection used and although fluorescence detectors may provide improved sensitivity, the necessity for sample derivatization prior to detection means that this method of detection is less common.

Ultraviolet detectors used in low pressure chromatography commonly have mercury or zinc lamps. The UV light from the lamp passes through the stream of an eluent liquid and then through an optically clear silica continuous flow cell. The diminution in light intensity due to absorption by the proteins is detected by a highly sensitive photocell. This generates a signal (0−10 mV) which is passed to a chart recorder. The sensitivity of the detector can be adjusted to alter the output signal.

The Pharmacia UV1 detector monitors at 254, 280 or 405 nm using a low pressure mercury lamp while the UV2 has two independent monitoring systems using two optical pathlengths (20 mm and 1 mm) on the same flow cell. This provides two levels of sensitivity from the same detector allowing the measurement of low levels of protein.

Bio-Rad market two detectors. The 1306 is a variable wavelength low noise UV detector which can be used on any scale of chromatography from microbore to preparative. The 1740 model monitors at 254 and 280 nm by changing filters using a mercury lamp.

4.1.4 *Fraction collectors*

Fraction collectors allow the automatic collection of samples taken from the column after they pass through the detector. The number of fractions taken and the time or volume of each sample can be pre-set. Many fraction collectors (*Table 6*) are programmable, allowing the collection of only the desired protein peak or the control of washing and elution cycles.

4.2 **Basic procedures in low pressure chromatography**

The basic procedures in establishing any LPLC purification step are discussed below.

4.2.1 *Choice of column dimensions*

The optimum size and dimension of a chromatography column will depend on several factors which include the sample size, the protein concentration, the adsorbent capacity and the type of matrix used.

Traditionally chromatography columns are long and thin (i.e. high aspect ratio). This leads to a low flow rate and long separation times. However, short, wide columns are required in order to minimize gel compression; flow rates are consequently higher. Both column designs have their drawbacks. Long, thin columns are excellent for maximizing peak resolution, the longer process time may, however, lead to loss of protein activity in some applications (8). Even a well distributed flow is less easy to control with wide bore columns so that channelling of solvent flow is more likely, leading to loss of resolution. Hydrogel matrices are commonly used in stacked column arrange-

Table 6. List of some commonly used fraction collectors.

Manufacturer	Name	Details
LKB	UltroRac	200 tubes maximum Three collection modes (time, volume, drop)
	SuperRac	312 tubes maximum. Time, volume, drop modes. Programmable
	HeliRac RediRac	Compact design, external control, 160 tubes maximum. Time, volume or drop mode
Pharmacia	Frac 100	Up to 95 tubes. Automated or manual
	Frac 200 Frac 300	Up to 300 tubes. Microprocessor controlled

Table 7. Sample loading requirements for different methods of protein purification.

Purification method	% of bed volume required for loading
Adsorption (concentration)	Up to 100
Stepwise gradient	Up to 50
Continuous gradient	5−10
Isocratic	1−5

ments in large-scale purification. More rigid matrices, typified by the new series of Macrosorb type adsorbents, can be used at a high flow rate in long, narrow columns without pressure build-up and matrix compression. The choice of optimum dimension is therefore clearly dependent on matrix type; this becomes more critical in large-scale process chromatography where back pressures are more significant.

The total volume of matrix required will depend on the capacity of the adsorbent and the protein content of the sample. A method for calculating adsorbent capacity is given in Section 4.2.2. Manufacturers' data on adsorbent protein capacity should always be treated as an upper limit since they are calculated under conditions of excess protein. Working capacities are frequently much less and dependent on the protein size and sample characteristics (pH, ionic strength, presence of contaminants). Working capacities of ion exchangers are usually around 30 mg ml^{-1} of matrix when using complex protein mixtures (8).

For stepwise elution the sample should be loaded onto approximately half the packed bed volume leaving the other half to effect separation during elution. The bed volume for elution is kept small to minimize band spreading through diffusion during stepwise elution. Columns are consequently short. In gradient elution, however, longer columns are required and best resolution is usually obtained if the top 10% of the bed is used for initial sample loading. A longer column can be used since band spreading is minimized due to the continuously changing eluting conditions.

In some purifications the primary objective of the step is concentration rather than

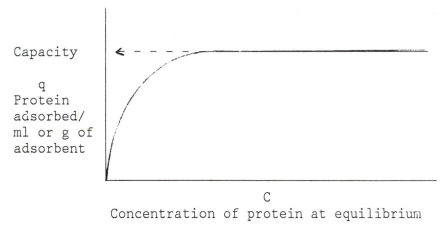

Figure 14. Adsorption isotherm used for the calculation of adsorbent capacity.

fractionation. In this technique the process of adsorption rather than chromatography is common. The total bed capacity may therefore be used for the adsorption step, allowing the concentration of protein from a large volume of sample onto a packed bed. The required protein is then eluted in a small volume.

The calculation of the required bed volume therefore depends on the method of purification: these are summarized in *Table 7*.

4.2.2 *Calculation of adsorbent capacity*

The capacity of an adsorbent is dependent on the molecular weight of the required protein, the adsorption conditions and the nature of the sample. Generally more porous matrices should be used for high molecular weight proteins (>100 kd). For the adsorption of very large proteins ($>4 \times 10^3$ kd) surface adsorption is usually preferable. A matrix should therefore be selected with an adequate degree of porosity for the protein to be purified.

The capacity should always be calculated using the conditions used for adsorption. In particular the pH, ionic strength and level of contaminants should be identical. An approximate indication of an adsorbent capacity can be obtained by loading a sample onto a small packed bed under the adsorption conditions until the desired protein appears in the eluent (i.e. breakthrough). Alternatively, the adsorption isotherm of the adsorbent for the protein may be determined. This can be calculated by measuring the batch adsorption of protein from a fixed volume of sample onto variable volumes of adsorbent. The adsorption can be carried out in test tubes in which the adsorbent and sample are gently mixed until equilibrium is reached. The amount of the protein left in each solution at equilibrium is then measured. From this the amount adsorbed per unit volume or weight of adsorbent (q) is calculated. The adsorption isotherm for the protein can then be drawn by plotting q against c (*Figure 14*) from which the capacity Q_{max} (expressed in mg of protein adsorbed per unit volume or weight of adsorbent) is calculated. The maximum volume or concentration of sample which can be applied per unit volume of adsorbent can then be determined. As mentioned earlier, the total bed volume required then depends on the method of purification (*Table 7*).

Method Table 1. General procedure for column packing.

---

1. If necessary, pre-treat the matrix to allow removal of contaminants or to ensure pore swelling. Fibrous celluloses, for example, require pre-soaking in 0.5 M NaOH for 30 min while synthetic resins may need an alkali/acid cycle. Manufacturers' literature should be consulted.
2. Remove any fines from the matrix. Mix the adsorbent with 5−6 times its volume of buffer, allow to settle and decant off the cloudy supernatant. Repeat 3−5 times.
3. Remove the buffer until its volume is half that of the adsorbent; this ratio is generally most suitable for column packing. If necessary degas the slurry under suction.
4. Clean the column, clamp the outlet and pour the matrix suspension in slowly with the column slightly tilted. An extension tube may be used if necessary. Packing can be speeded by opening the column outlet.
5. Ensure the column is vertical, and filled to the top with buffer so that an upward meniscus appears, eliminating any air. Fit a column adaptor at the inlet and pump through at least two bed volumes of buffer prior to sample application.

4.2.3 *Column packing*

A general procedure for column packing is shown in *Method Table 1*. It is essential to remove any fines from the matrix prior to use. If these are not removed they will fill the voids between matrix particles and block the support thus reducing the flow rate and causing a non-uniform flow. The time required for matrix settling in fines removal depends on the adsorbent and can vary from a few minutes (e.g. resins) to half an hour in the case of cellulose.

The ratio of buffer to matrix used in column packing should provide a suspension which is sufficiently dilute to allow the escape of trapped bubbles, but not so thin that turbulence is promoted during packing, leading to particle distribution according to size. This is critical when using hydrogels but with more dense packings it is frequently possible to fluidize the bed by back-washing after packing in order to achieve an even bed.

An extension tube is supplied by some companies (e.g. Pharmacia) and is fitted to the column inlet during packing to allow all the matrix slurry to be poured at once. Extension tubes are usually of the same diameter as the column but half the length. Once a column is packed it should never be allowed to dry; this is particularly important in the use of hydrogels such as agarose where the matrix is irreversibly damaged upon drying.

4.2.4 *Sample application*

The introduction of the sample in low pressure separations can be achieved by stopping the flow of buffer and layering the sample directly onto the matrix bed. This method is described in *Method Table 2*.

The protein sample should be equilibrated in the same buffer used for column equilibration. This minimizes localized pH or ionic strength fluctuations which may

Method Table 2. Procedure for open column sample application.

1.	Drain off excess buffer from the column top until its level just reaches the top of the bed.
2.	Close the outlet and layer on the sample using a syringe via the column wall so as to minimize disturbing the bed.
3.	Open the outlet and drain the sample onto the bed.
4.	Apply elution buffer until it is $1-2$ cm above the bed. The column wall should be washed in the process. Connect the column inlet and begin elution.

impair resolution. Samples should also be freed of particulates, including precipitate, using centrifugation or precipitation.

The sample may alternatively be layered onto the column bed under a layer of eluent; to achieve this the sample density is increased, usually with sucrose (3).

The volume of sample applied will vary. In ion-exchange chromatography it is dependent on the protein concentration of the sample; if too high (>50 mg ml^{-1}), the rapid adsorption of protein to the column may provide a localized pH and ionic strength fluctuation through rapid counter-ion release resulting in loss of activity. Reproducible column separations are usually best achieved with dilute protein samples. In ion exchange a protein concentration of $10-20$ mg ml^{-1} is recommended (8). Large volumes of samples can be pumped directly onto the column.

4.2.5 *Elution*

After sample application the column may be washed in equilibration buffer to remove unbound contaminants. For proteins with partition coefficients less than 1 the washing volume should be minimized in order to limit loss of protein in the column wash.

In column chromatography the protein of interest may be separated from contaminants by adsorbing the protein followed by selective desorption or by adsorption of contaminants, allowing the protein to pass through the column without retention. The former option is most commonly used in protein purification since it allows a greater degree of protein fractionation during the elution step.

Elution of proteins from columns is achieved using three possible methods.

(i)   Isocratic elution—the composition of the eluent does not change during the course of elution.

(ii) Stepwise elution—the eluent composition is changed at least once to provide conditions more favourable to elution.

(iii) Gradient elution—the eluent composition is changed continuously to conditions favouring protein dissociation from the packing.

In isocratic elution the degree of resolution is invariable and fixed by the eluent composition. A weak eluent may therefore never effectively elute bound protein while an excessively strong eluent would lead to loss of resolution. Trial and error is therefore required for its success. Its chief advantage is its simplicity and the minimal requirements for equipment and handling. It is seldom used in LPLC due to the improved resolution

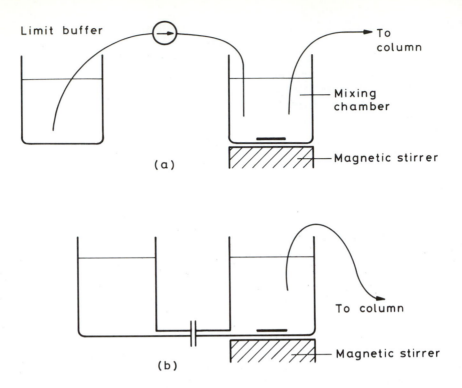

Figure 15. Simple apparatus for use in gradient elution consisting of two beakers connected via (**a**) a peristaltic pump or (**b**) a tube at their bases.

achievable using stepwise or gradient elution but is common in more rapid methods of protein resolution such as HPLC.

Stepwise elution is generally more reproducible than gradient methods (8) using laboratory apparatus. Two problems however arise in its use.

(i)   Following a step change in elution conditions, a large number of proteins may be co-eluted with the consequence that selective desorption is lost.

(ii)   The increase in partition coefficient which occurs as a protein is eluted may cause a fraction of the protein to remain bound to the column packing. Consequently in batch elution a protein may occur in several eluted fractions, while in column elution peak tailing may result.

Nevertheless stepwise elution is most commonly used in large-scale chromatography due to the simpler apparatus requirements compared to gradient methods. This elution method is also ideally suited to batch adsorption methods where the adsorbent is simply given a series of batch elution washes. Stepwise elution uses conditions which increasingly favour protein desorption from the adsorbent.

Gradient elution is the most widely used method of protein desorption in laboratory-scale chromatography. The changing elution conditions provide improved resolution over stepwise methods since peak tailing is not encouraged. On a laboratory scale a simple gradient elution apparatus can be constructed using two beakers joined at their

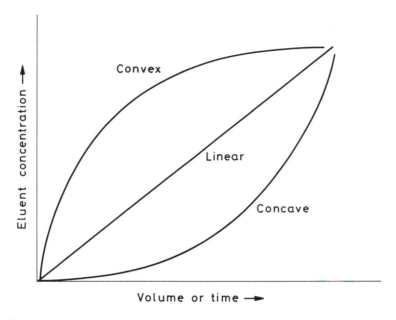

Figure 16. Alternative gradient shapes commonly used in protein elution.

Table 8. The three basic types of gradients used in column chromatography.

Form of gradient	Requirement	Result
Linear	$R_2 = 2R_1$	
Concave	$R_2 > 2R_1$	Better resolution initially
Convex	$R_2 = R_1$	Better resolution at end
	or $R_2 < 2R_1$	

R1, flow rate from limit buffer to mixing chamber.
R2, flow rate from mixing chamber to column.

base or connected via a peristaltic pump (*Figure 15*). One beaker acts as a mixing chamber while the other determines the upper limits of the gradient and is termed the 'limit' buffer. It is usually advisable to use a linear gradient initially and then, if more resolution is required, alter the gradient shape either to a convex or concave shape (*Figure 16*). The steeper the gradient, the closer proteins will be eluted. The form of the gradient is governed by the rate of mixing the limit buffer with the mixing chamber. A linear gradient is produced by using two pump channels. Buffer is removed from the mixing chamber at the same time as limit buffer is pumped in. The gradient shapes are produced as shown in *Table 8*.

In gradient elution the volume of eluent should be about five bed volumes (16). Larger volumes may result in band spreading and dilution since proteins will be eluted over a wide window of solvent composition. More complex gradients may be produced if two limit buffers are used or if the strength of the limit buffer is altered during elution. Generally, however, the gradient shapes described above are sufficient for most separations.

4.2.6 *Flow rate selection*

The optimum flow rate for a separation is always a compromise between achieving maximum resolution and reducing process time. Prolonged elution times resulting from excessively low flow rates may, however, result in band spreading through diffusion. Conversely, at a high flow rate, separation times are faster but back mixing limits the resolution obtainable. While some packing materials may allow very fast flow rates with minimal back pressure, mass transfer restrictions may limit the equilibration between the pores and the interstitial space. Thus the optimum flow rate is considerably less than the maximum. The nature of the sample, particularly its viscosity, may also restrict the choice of flow rate. An estimate of the appropriate flow rate is usually found in the manufacturer's literature and may then be followed by laboratory-scale studies to determine the influence of increasing flow rate on resolution. Protein purifications are usually performed at a flow rate of $10-30$ ml cm^{-2} h^{-1}. Some matrices (e.g. Sepharose Fast Flow) can be used at higher flow rates; however, flow rates used during fractionation should be much lower (20% of maximum value) to obtain optimum resolution. Higher flow rates may however be used during sample loading, washing and buffer re-equilibration stages. This is particularly useful in concentration steps where large volumes are processed.

A detailed discussion of the principles involved in column scale-up is outside the scope of this chapter. However, prior to scale-up the optimum linear flow rate (expressed as flow rate per unit of column cross-sectional area) and bed height are determined on a small scale. On scale-up the bed diameter is then increased and the same linear flow rate used (18,19). In addition the volumes used in sample loading, washing and elution are increased in proportion to the increase in bed volume. The object of scale-up of a chromatographic separation is to obtain the same percentage yield and product quality in the same time (20).

4.2.7 *Column regeneration*

Following elution of the desired protein, strongly bound protein and non-protein material should always be removed from the adsorbent to prevent a slow build-up of contamination which would result in column fouling, loss of resolution, blockage and contamination of the purified sample.

In column chromatography regeneration is frequently achieved *in situ*. The conditions used depend on the chemical stability of the matrix in use, the nature of the contamination and the application. Hydrogel based ion exchangers, for example, are commonly regenerated in a high salt concentration (e.g. 1 M NaCl) to remove bound protein. This is followed by washing in one column volume of 0.5 M NaOH to remove the very strongly bound material. This alkali treatment also serves to sterilize the column packing and remove any lipids and pyrogens. The alkali must be washed out from the column using several column volumes of equilibration buffer so as to prevent matrix deterioration.

More strongly bound proteins may be removed using 6 M urea or detergents provided they are washed free from the column after use. Pharmacia-LKB recommend detergents are removed using an increasing ethanol gradient (25–95%) followed by butanol, ethanol and distilled water.

More chemically stable matrices such as the synthetic resins can withstand stronger regenerating conditions such as cycling in 1 M NaOH and acid. Manufacturers' literature should always be consulted.

4.2.8 *Storage*

Prolonged storage of the matrix requires the addition of agents to prevent microbial growth using the matrix as a substrate. Commonly used preservatives include 20% ethanol and, in non-therapeutic applications (9), 0.2% merthiolate, 0.02% hibitane, 0.02% sodium azide or 0.5% chloretone. Long term storage of matrices in alkaline conditions should be avoided to prevent degradation of the matrix structure. Additional details for regeneration of specific adsorbents are given in later sections.

4.3 **Medium and high pressure equipment**

Both MPLC and HPLC equipment have certain requirements not necessary for the successful use of LPLC.

(i)   The pump must provide constant, pulse-free flow at increased back pressure.
(ii)   The column must be capable of withstanding the increased pressure.
(iii) The detector should have a fast response time since protein peaks may pass through in a matter of seconds.

In MPLC, equipment has been designed to minimize protein denaturation and allow the use of halide-containing buffers (21). Consequently stainless steel components commonly used in HPLC have been replaced with borosilicate glass, titanium and teflon. The requirements of each component are discussed below.

4.3.1 *Pumps*

A variety of pumps exist for use in high performance separations (*Table 9*). The majority are designed on the common feature of one or two plunger pumps providing flow through operation of a ball-type valve controlling alternate solvent delivery and suction. Pumps may direct solvent from two plungers (e.g. Waters 501, LKB 2150) or by using a single plunger with two pump heads (e.g. Gilson 302, Beckman 110B). Where salt-containing

Table 9. List of some commonly used HPLC/MPLC pumps.

Manufacturer	Name	Description
Beckman	110B	Single piston
Bio-Rad	1350 Soft Start	Dual pistons
Cecil	CE 1100	Dual overlapping pistons
Gilson	302/303	Single piston
LKB	2150	Dual piston
Perkin Elmer	Series 100	Single piston
Pharmacia	P500	Biocompatible dual piston (FPLC)
	P3500	Biocompatible dual piston (HPLC)
Philips	PU4100	Dual piston
Shimadzu	LC-6A	Single piston
Waters	501	Dual piston

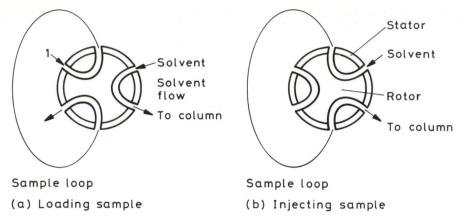

Figure 17. Loop injector design. The sample is first injected into a fixed volume loop (port 1) which is then incorporated into the solvent flow by rotating the inner cylinder.

buffers need to be used frequently stainless steel must be eliminated from all parts of the pump which come into contact with the fluid. In the Pharmacia P500 pump for example, a two plunger system is used consisting of borosilicate glass cylinders and fluoroplastic reinforced titanium plungers.

4.3.2 *Sample application*

In high performance separations a small volume of sample ($10-100$ μl) is usually injected into the solvent flow without interruption. This may be achieved using a Hamilton syringe directly to inject the sample through a hermetically sealing septum into the solvent flow. Loop injectors are however more common (e.g. Rheodyne) allowing the injection of a fixed volume of sample (e.g. 20 μl) into a sample loop which is then flushed through with solvent to inject the sample onto the column (*Figure 17*).

4.3.3 *Columns*

In HPLC the column and tubing must withstand the high operating pressure. Consequently stainless steel is used and sealing between the two is ensured by using metal ferrules compressed against steel nuts. Stainless steel columns are usually $15-30$ cm long with an internal diameter of $2-9.5$ mm, with 3.9 and 4.6 mm being most common. The trend towards the use of microbore systems for more rapid analysis using less solvent has, however, led to the use of even narrower and shorter column dimensions. Pre-packed columns are recommended for optimum resolution. In bio-compatible systems the elimination of stainless steel has been achieved through the use of borosilicate glass columns and, usually, teflon tubing.

Guard columns are recommended for high performance separations. These remove strongly adsorbed molecules so that fouling of the column head is prevented through use of these renewable inline pre-columns.

4.3.4 *Detectors*

The principle of detection used in high performance separation differs little from that

Table 10. List of some commonly used HPLC/MPLC UV detectors.

Manufacturer	Name	Description
Beckman	160	Fixed wavelength 214−546 nm
	163	Variable wavelength 195−350 nm
	164	Variable wavelength 190−700 nm
Cecil	CE 1200	Variable wavelength
		190−400 or 380−600 nm depending on lamp used
Gilson	111B	Fixed wavelength 254 or 280 nm
	116	Variable wavelength 190−380 nm
		Spectrum analysis
	HM/Holochrome	Variable wavelength 190−600 nm
LKB	2151	Variable wavelength 190−600 nm
Perkin Elmer	LC-90	Variable wavelength 195−390 nm
Pharmacia	UVI	Fixed wavelength for FPLC
	UVM	Fixed wavelength for FPLC
Philips	PU 4100	Variable wavelength
		190−700 nm. Scanning accessory.

used in low pressure chromatography. The rapid separation time means, however, that the speed of response of high performance detectors should be faster. UV detectors are most commonly used in protein separations and although a fixed wavelength (e.g. 280 nm) is usually needed, many detectors are fitted with a variable wavelength adjustment. Most HPLC detectors (*Table 10*) are high cost and consequently fitted with a deuterium lamp in preference to the mercury lamps commonly used in low pressure detectors. Some detectors allow the analysis of two wavelengths concurrently or have a spectrum analysis facility allowing the spectral analysis of any chromatogram peak (e.g. the Philips PU4100). Hence many detectors are suitable in a wide range of HPLC applications and have expensive facilities not necessarily required for routine protein purification. The choice will therefore depend on the dedication of the system in use.

4.4 Basic procedures in HPLC/MPLC

The basic procedure for HPLC separations is given in *Method Table 3*. The following points should, in particular, be noted for the optimal use of both HPLC and FPLC in protein purification.

(i)   In order to prevent gas bubbles forming and occluding the tubes the mobile phase should be de-gassed and filtered using 0.45 μm filters (e.g. Millipore, Whatman) and a vacuum filter apparatus.

(ii)   In order to prevent back pressure increase due to particulate fouling, the sample should be freed of precipitate and suspended solids using centrifugation or filtration. It should not form a precipitate when added to the mobile phase.

(iii)   A low volume pre-column should be used to remove particulates and strongly adsorbing components from the sample, thereby prolonging column life. The

Method Table 3. General procedure for protein purification using high performance techniques.

---

1. Degas and filter the mobile phase into a suitable bottle using a 0.45-μm filter and vacuum pump.
2. Put the pump inlet line into the mobile phase reservoir and draw the liquid along the tubing line using a syringe.
3. Ensure no air is trapped in the pump heads or pre-column tubing by disconnecting the column inlet and running the pump for a few minutes. This is termed 'Priming'. Stop the pump.
4. Connect up the column inlet and start the pump. A flow rate of 1 ml min^{-1} is common on an analytical scale. Run the pump for several minutes until the column is equilibrated. This is usually evident in a stable detector output.
5. Inject the sample. Sample volumes of up to 200 μl are usual on an analytical scale for HPLC but may be 2−5 ml using a 1 ml FPLC column.
6. Separation is commonly achieved using isocratic or gradient elution, typical separations last up to 1 h.
7. Optimize the degree of resolution further if necessary by adjusting the mobile phase composition or the gradient shape.
8. Prior to injecting another sample re-equilibrate the column to the initial running conditions.

---

pre-column or, if refillable, its packing, should be replaced as required by the application. Crude samples will require frequent pre-filter replacement. Deterioration of peak shape and resolution and increased pressure mean the pre-column should be replaced.

(iv) Solvents used in purification should be high quality (HPLC Grade). Manufacturers include Waters and Fisons. They should not interfere with protein detection and the use of low-UV grade solvents is often practised when detecting at the lower wavelengths. Many are toxic (e.g. acetonitrile and tetrahydrofuran) and care should be exercised in their handling and subsequent disposal.

(v) Caution should be exercised in the use of halide containing mobile phases in direct contact with metal equipment components. The resulting corrosion and abrasive effect of salt crystals may limit equipment life. The equipment should be thoroughly washed out with a non-halide-containing mobile phase after their use.

(vi) Silica based packings should not be used routinely above pH 8. The use of chelating agents with silicas should also be avoided and in the choice of column packings, 'end-capped' silicas are preferred. End-capping minimizes the presence of unreacted silanol groups which may cause unwanted interference effects during a separation.

5. SEPARATION ON THE BASIS OF CHARGE

Ion-exchange is the most commonly practised chromatographic method of protein purification. This stems, in part, from its ease of use and scale-up, wide applicability

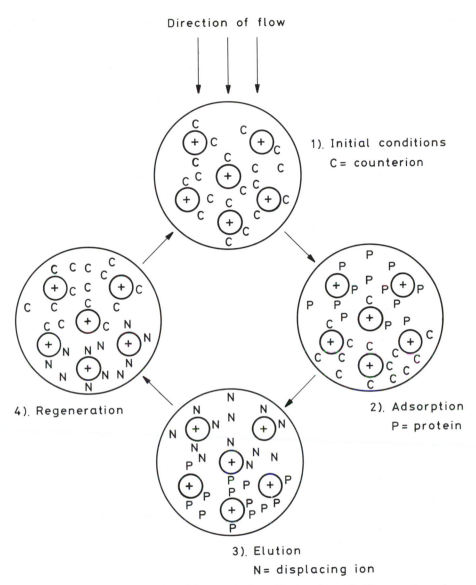

Figure 18. Diagramatic representation of the stages in the purification of a protein (**P**) using anion exchange. Desorption and regeneration may be accomplished in one step using the same counter-ion.

and low cost in comparison with other separation methods. Ion-exchange of proteins involves their adsorption to the charged groups of a solid support followed by their elution with fractionation and/or concentration in an aqueous buffer of higher ionic strength. The basic steps are illustrated in *Figure 18*. A brief introduction to ion-exchange theory will be presented in the next section followed by guidance on the use of this technique. For those seeking a more in-depth treatment of ion-exchange theory, the works of Vermeulen and co-workers (22), Scopes (8) and Osterman (3) are

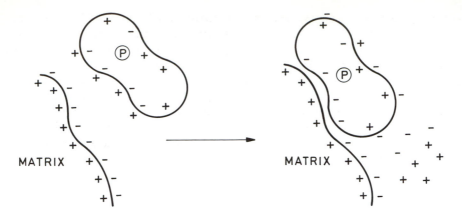

Figure 19. The dissociation of counter-ions occurs on adsorption of a protein (**P**). In this case multi-point binding has occurred, the adsorption is strong and a high counter-ion concentration may be required to promote elution.

recommended. Janson and Low (23), Scawen (10) and Chase (11) cover aspects of the large-scale use of ion-exchangers.

5.1 Theory of ion-exchange

Proteins carry both positive and negative charged groups on their surface, due largely to the side chains of acidic and basic amino acids. Positive charges are contributed by histidine, lysine, arginine and, to a lesser extent, N-terminal amines. Negative groups are due to aspartic and glutamic acids, C-terminal carboxyl groups and, to a lesser extent, cysteine residues. The net charge on a protein depends on the relative numbers of positive and negative charged groups; this varies with pH. The pH where a protein has an equal number of positive and negative charged groups is termed its isoelectric point (pI). Most proteins have a pI between pH 5 and 9. Above their pI proteins have a net negative charge while below it their overall charge is positive.

Ion-exchange is the separation of proteins on the basis of their charge and can be used to resolve proteins which differ only marginally in their charged groups. Separation of proteins is achieved by their difference in equilibrium distribution between a buffered mobile phase and a stationary phase consisting of a matrix to which charged inorganic groups are attached. For the effective use of ion-exchange in protein purification the stationary phase must therefore be capable of binding either positively-charged or negatively-charged proteins. To this end ion-exchange matrices are derivatized with positively-charged groups for the adsorption of anionic proteins (termed anion exchangers) or negatively-charged groups for the adsorption of cationic proteins (cation exchangers).

Associated with both stationary phase and protein charged groups are counter-ions which are simple, low molecular weight ions. In order for the protein to bind to the stationary phase, therefore, the counter-ions of both groups must become electrolytically dissociated (*Figure 19*).

Counter-ions 'screen' the exchanger groups, preventing their binding with a protein. Na^+ and H^+ are the common counter-ions for cation exchangers while Cl^- and OH^-

are usually used with cation exchangers. Counter-ions can be arranged in an 'activity' series (*Figure 20*) according to their strength of interaction with their respective ionogenic groups at equal concentration. Consequently chloride would replace hydroxide ions at equal strength as the counter-ion for an anion exchanger. Counter-ions do not permanently bind to an ionogenic group but stay in a state of equilibrium, continually shifting between the bulk solution and the exchanger groups. It follows that the ionogenic groups can become uncovered to allow the binding of a protein. The higher the counter-ion concentration, the less frequently do the ionogenic groups become uncovered. Prior to the use of an ion exchanger the counter-ion may require replacement with a different ion more suitable to the particular application. Matrix pre-treatment and conditioning is discussed in more detail in Section 5.3.

5.2 Selection of conditions for ion-exchange purifications

A wide variety of matrices, functional groups, and adsorption/desorption conditions are available. The factors involved in making each choice are described below so as to guide the practitioner in their selection.

Cations: $Ag^+ > Cs^+ > Rb^+ > K^+ \geq NH_4^+ > Na^+ > H^+ > Li^+$

Anions: $I^- > NO_3^- > PO_4^- > CN^- > HSO_3^- > Cl^- > HCO_3^- > HCOO^- > CH_3COO^- > OH^- > F^-$

Figure 20. The activity series of anionic and cationic counter-ions. The counter-ions at the beginning of each series have a stronger attraction for the ionogenic groups of the ion exchangers.

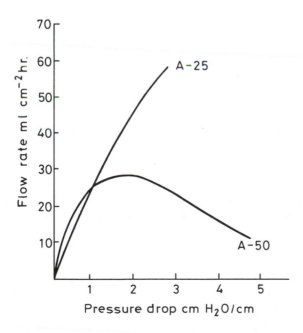

Figure 21. Variation in flow rate with pressure drop across a bed of DEAE−Sephadex. Bed dimensions = 2.5 × 45 cm, pH = 7, ionic strength = 0.1 M. Reproduced with kind permission of Pharmacia-LKB.

Table 11. List of commonly used ion exchangers.

Matrix	Manufacturer/ Trade name	Particle size (μm)	pH range	Comments
A. Low and medium pressure packings				
Cellulose				
Fibrous	Whatman			High flow rates (100 ml cm^{-2} h^{-1}) but limited resolution—ideal for initial fractionation-capacities up to 100 mg g^{-1}. Require pre-cycling.
	Bio-Rad Cellex	80 × 18		
	Serva	50–200		
Microgranular	Serva	100–200		Flow rates 100 ml cm^{-2} h^{-1}
	Whatman			Improved resolution over fibrous.
Beaded	Pharmacia	40–160	2–9	Flow rates 100 ml cm^{-2} h^{-1}
	Serva	50–160		Improved resolution over fibrous. Exclusion limit 10^6.
Agarose	Pharmacia Sepharose	45–165	2–14	M^{ex} 4 × 10^6
	Sepharose Fast Flow	45–165		Usable at flow rate >300 ml cm^{-2} h^{-1}
Dextran	Pharmacia A/C 25	40–125	2–13	A/C usable only up to 30 ml cm^{-2} h^{-1}. Matrices swell and shrink depending on ionic strength.
Polyacrylate	IBF-Trisacryl	40–80	< –13	M^{ex} >10^7. Chemically stable.
	Bio-Rad Biorex 70	Up to 1180		M^{ex} 7 × 10^4
	Anachem-Separon	50–80		35 nm pores—suitable for HPLC.
Polyvinyl	Merck Fractogel-5	25–50	1–14	Suitable for MPLC—resistant to 7 bar.
Composites (Kieselguhr -Agarose)	Sterling Organics	160–1000	4–10	M^{ex} >10^6. Highly porous and good hydrodynamic properties.

Polystyrene-dvb	Bio-Rad BDH Serva	Up to 1180 Up to 1000	1–14	Non-specific adsorption from hydrophobic matrix.
Silica	Serva	40–200	2–9	$M^{ex} > 10^6$. Suitable for MPLC. Good hydrodynamic properties. Not usable above pH 9.
Polyether	Pharmacia Mono Q Mono S	10 10	2–12 2–12	Used in HPLC (FPLC) system, Max. pressure 750 p.s.i.
B. High pressure packings				
Silica	Alltech Synchropak 300	6.5	<9	300 nm pores. Polymeric amine coated.
	Anachem Anagel TSK/IEC			Silica or resin based.
	Anachem Dynamax Hydropore	12		Polyethyleneimine coated.
	Bio-Rad Bio-5 300	5, 10	<9	Pores size up to 400 nm.
	Brownlee Aquapore	10		Polymeric amine coated.
Resins	Anachem Anagel TSK/IEC			
	Beckman Spherogel TSK			
	Bio-Rad TSK IEX	10–20	2.5–12	Hydroxylated polyether.
	Waters Protein-PAK	10		100 nm pores, hydrophilic polymer coated resin.

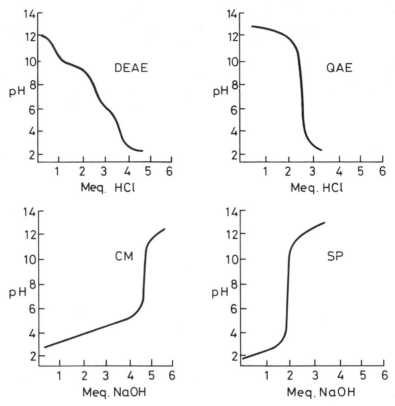

Figure 22. Typical titration curves for strong and weak anion and cation exchangers. The strong ion exchangers (QAE and SP groups) are fully ionized over the entire pH range used to purify proteins (3–11) while the weak ion exchangers (DEAE and CM) are ionized over a narrow pH range. Reproduced with kind permission from Pharmacia–LKB.

5.2.1 *Choice of matrix*

Section 3 should be consulted for guidance on the choice of support matrices. The characteristics of some of the more commonly used ion-exchange matrices (24,25) are summarized in *Table 11*. Related matrices can exhibit marked differences (see *Figure 21*), and manufacturers' literature should always be consulted.

5.2.2 *Selection of functional groups*

Ionogenic groups used in ion exchange are, as mentioned earlier, either positively-charged (anion exchangers), or negatively-charged (cation exchangers). Both types can be further divided into strong and weak groups. Strong ionogenic groups remain ionized over the whole operating pH normally used in protein purification while weak groups have a narrower effective pH range and are largely only partly ionized. Typical titration curves for both strong and weak ion exchangers are shown in *Figure 22*. The ionogenic groups used in the ion exchange of proteins are listed in *Table 12*, together with their structural formulae.

The most commonly used functionalities are the weak ionogenic groups. The

Table 12. Ion-exchange groups used in the purification of proteins.

Formula	Name	Abbreviation
Strong anion		
$-CH_2N^+(CH_3)_3$	Triethylaminoethyl	TAM$-$
$-C_2H_4N^+(C_2H_5)_3$	Triethylaminoethyl	TEAE$-$
$-C_2H_4N^+(C_2H_5)_2CH_2CH(OH)CH_3$	Diethyl-2-hydroxypropylaminoethyl	QAE$-$
Weak anion		
$-C_2H_4N^+H_3$	Aminoethyl	AE$-$
$-C_2H_4NH(C_2H_5)_2$	Diethylaminoethyl	DEAE$-$
Strong cation		
$-SO_3-$	Sulpho	S$-$
$-CH_2SO_3-$	Sulphomethyl	SM$-$
$-C_3H_6SO_3-$	Sulphopropyl	SP$-$
Weak cation		
$-COO-$	Carboxy	C$-$
$-CH_2COO-$	Carboxymethyl	CM$-$

diethylaminoethyl (DEAE) group is usually used in anion exchange to purify negatively-charged proteins while the carboxymethyl (CM) group is frequently used in cation exchange for the recovery of positively-charged groups. Strong ion exchangers are becoming more popular in protein purification, notably the Sepharose Fast Flow packings, based on sulphomethyl and triethylaminoethyl functionalities for strong cation and anion exchange, respectively.

In selecting the most suitable functional group for a purification the pH stability of the desired protein should be considered. Proteins are amphoteric; they may carry a net positive or negative charge depending on whether the buffer pH is below or above the isoelectric point, respectively. The choice must therefore be made between using an anion or cation exchanger. In practice the decision is sometimes restricted by the pH stability of the protein. If a protein is more stable above its pI then an anion exchanger should be chosen. Conversely if the protein is more stable below its pI then a cation exchanger should be chosen. In protein purification anion exchange functionalities (e.g. DEAE) are most frequently used since proteins of pI below 7 are more common.

The purification of α-amylase can be used as an example of exchanger selection. The pI of the enzyme is 5.2 so that weak ion exchangers can be used. However, the stability of the enzyme falls off rapidly below pH 5 (*Figure 23*). The effective use of cation exchangers below pH 5 is therefore prevented, restricting the choice to matrices bearing an anion exchange functionality. The pI of a protein can be determined using isoelectric focusing or by reference to lists of protein pI's to be found in the literature (26−28). According to Osterman (3), the average distance between ionogenic groups on an ion-exchange matrix can be as little as 1−3 nm. In highly porous matrices many ionogenic groups will of course have a much wider separation. It follows that globular proteins of molecular weight 60−100 kd, which have a diameter of 7−10 nm, have the opportunity to bind to several ionogenic groups at once, provided they are charged and become uncovered by counter-ions. With certain microporous resins the density

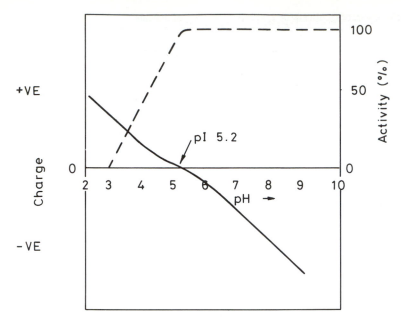

Figure 23. Simplified representation of pH stability (---) and net charge (—) of α-amylase. The isoelectric point of 5.2 prevents the effective use of cation exchangers.

of surface ionogenic groups is sufficiently high that multi-point protein adsorption can occur. This may result in the requirement for a high salt concentration (0.5 M NaCl) to promote elution; an 'all or nothing' adsorption elution pattern leading to minimal resolution, and even a possibility of protein denaturation. More porous supports have a less dense distribution of ionogenic groups so that the possibility of denaturation is minimal. With strong ion exchangers their ionogenic groups are all charged over a wide pH range (*Figure 22*). Consequently the chance for multi-point binding is higher than with weak ion exchangers. Where the aim of the ion exchange step is concentration this may be acceptable since a single, high concentration of salt is usually used for elution in a minimal volume and optimum resolution is not required. Weak ion exchangers however are normally chosen for the ion exchange of proteins between a pH of 6 and 9. They are recommended for the purification of labile proteins where mild eluting conditions are needed. Since weak cation exchangers lose their charge below pH 6 and weak anion exchangers above pH 9 (*Figure 22*), strong ion exchangers may be used for proteins of pI outside these pH values or for weakly-charged proteins which require an extreme pH to promote adsorption. Both strong and weak ion exchangers may therefore be used for protein adsorption. In the use of the former the increased potential for multi-point binding should be kept in mind with the result that stronger eluting conditions may be needed.

5.2.3 *Buffers used in ion exchange*

The mobile phase used in ion exchange is usually aqueous since the electrolyte properties of water contribute to the dissociation of the ionogenic groups and matrix swelling (8); both effects increase the rate of ion exchange.

Table 13. Buffers commonly used in the ion exchange of proteins[a].

Ion exchanger	Buffer	pK	Buffering range
Cation	Acetic	4.76	4.8−5.2
	Citric	4.76	4.2−5.2
	Mes	6.15	5.5−6.7
	Phosphate	7.20	6.7−7.6
	Hepes	7.55	7.6−8.2
Anion	L-Histidine	6.15	5.5−6.0
	Imidazole	7.0	6.6−7.1
	Triethanolamine	7.77	7.3−7.7
	Tris	8.16	7.5−8.0
	Diethanolamine	8.8	8.4−8.8

[a]See also Chapter 1.

In an aqueous solution the surface charge groups of a protein are associated with counter-ions in a similar way to the ionogenic groups on the exchanger. Following the exchange of protein with the exchanger-bound counter-ions an increase in solution ionic strength may occur due to the release of both protein and exchanger counter-ions. The solution pH may also change (to a lower pH in anion-exchange and to a higher pH in cation-exchange) causing protein denaturation. Consequently the mobile phase in ion exchange is always buffered to minimize pH fluctuations, which will also influence the charge on a protein and its equilibrium between stationary and mobile phases. The minimum buffering strength recommended for ion exchange is 10 mM within 0.3 pH units of its pK value (8). The choice of buffer strength is always a compromise between minimizing pH fluctuation and maximizing adsorbent capacity.

Commonly used buffers in the ion exchange of proteins are shown in *Table 13*. The most suitable buffer for a purification will depend on the choice of ion exchanger and the adsorption pH (see also Chapter 1).

Another reason for the use of buffers in ion exchange is to minimize the localized pH differences which can result from the Donnan effect. Since the counter-ions for ion exchangers are frequently H^+ and OH^-, the use of unbuffered solutions would cause localization of OH^- ions at the surface of anion exchangers and H^+ ions at the surface of cation exchangers, causing the Donnan effect which is a localized pH effect (*Figure 24*). This may result in an unpredictable ion-exchange process and even worse, protein denaturation. Consequently solutions are always buffered.

Lastly, two important rules should be remembered.

(i)   The charged form of the buffer should not interfere with the ion-exchange process (i.e. use negatively-charged buffers such as acetate in cation exchange).

(ii)   The temperature will affect the pK_a of the buffer such that its buffering capacity will be different in purifications carried out in the cold room.

5.2.4 *Selection of adsorption and elution pH*

The degree of ionic interaction between a protein and an ion exchanger can be controlled by adjusting the buffer pH to regulate the degree of ionization of both matrix and protein ionogenic groups. The pH of the buffer can therefore be adjusted in order to favour

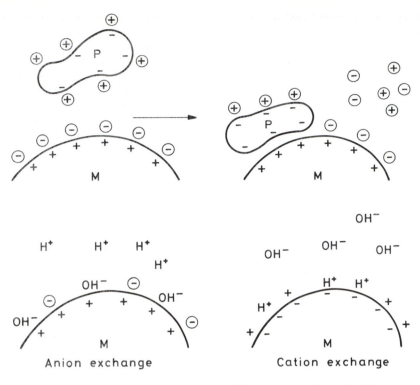

Figure 24. Donnan effect in anion- and cation-exchange which may cause localized pH differences in poorly buffered systems.

adsorption of protein to the ion exchanger or encourage protein – matrix dissociation and sample elution.

For protein adsorption a pH about 1 unit above or below the pI of the target protein is used. A larger difference in pH would lead to a greater net charge on the protein, multi-point adsorption and a requirement for stronger elution conditions. Normally a pH is chosen to be just sufficient to promote adsorption. This can be simply determined as shown in *Method Table 4* . If pH change is to be used as a means of protein elution, the same test outlined above can also be used to determine the most suitable elution pH (i.e. that at which protein adsorption is just prevented). A protein usually begins to dissociate from an exchanger at 0.5 of a pH unit from its pI at an ionic strength at 0.1 M NaCl (9). These simple tests give some idea of the starting and elution pH. In the resolution of multi-component mixtures some degree of trial and error is necessary since the titration curve (i.e. the net charge as a function of pH) of all the proteins present is not known. One important point should however be kept in mind. Although a protein has a net negative charge above its pI and a net positive charge below, the localization of charge clusters on a protein surface may occasionally allow its adsorption to an ion exchanger at a pH on the 'wrong' side of its pI (8). A protein with a localized area of positive charges but an overall negative charge could, therefore, by correct orientation on the ion-exchange surface, bind to a cation exchanger, and *vice versa*.

Method Table 4. Rapid method for the determination of the correct initial pH for ion exchange.

---

1.   Add 0.5 ml wet volume of ion exchanger to each of 10 test tubes.
2.   Equilibrate the matrix by washing 10 times in 0.5 M buffer, with each tube set up at a different pH. Use pH 4−8 for cation exchangers and 5−9 for anion exchangers. Use a pH interval of 0.5 units between each tube.
3.   Wash the matrix in a low ionic strength buffer (∼20 mM) five times, in each tube at the pre-selected pH used in (2).
4.   Add a known amount of the sample protein to each tube, and mix for 10 min[a].
5.   Allow to settle and assay the supernatant for the protein using A_{280} if sufficiently pure or a specific assay if the sample contains other proteins.
6.   The starting pH is that which just allows the protein to adsorb. The eluting pH is that which just prevents adsorption.

---

[a]An incubation time of 10 min is based on the use of hydrogel matrices. Other less porous supports may require longer.

5.2.5 *Selection of adsorption and elution ionic strength*

The ionic strength of the buffer used in ion exchange is used to control the degree of blocking of the ionogenic groups on both protein and stationary phase. The ionic strength is therefore critical in controlling the equilibrium distribution of a protein between the stationary and mobile phase. On the adsorption step the highest ionic strength which will allow protein binding is used while on elution the lowest ionic strength is recommended. The reasons for this are threefold.

(i)   If the ionic strength is too low on adsorption the protein will bind too tightly and effective elution will be made difficult.
(ii)   Keeping the ionic strength as high as possible on adsorption minimizes the binding of unwanted contaminants. Conversely keeping the ionic strength as low as possible on elution minimizes elution of bound contaminants.
(iii) The strategy outlined above simplifies the elution step.

The optimum ionic strength for adsorption and elution steps can be easily determined in a similar fashion to the measurement of optimum pH outlined above. A typical method is shown in *Method Table 5*. A salt concentration of 20−50 mM NaCl is usually used during adsorption and up to 0.5 M NaCl during elution.

5.3 **Procedures in ion-exchange separations**

Having established the optimum adsorption and elution conditions for the purification of a protein, certain other factors must be considered before starting a purification. These are the need for matrix pre-treatment, the adsorption mode and the elution mode.

5.3.1 *Matrix pre-treatment*

Preparation of the ion exchanger for use may involve removal of fines, swelling, washing and conversion to the correct counter-ion.

Method Table 5. Rapid method for the determination of optimum ionic strength for adsorption and elution.

---

1. Add 0.5 ml of ion exchanger to each of 10 tubes and equilibrate using ten 10 ml washes of 0.5 M buffer.
2. Equilibrate each tube to a different ionic strength using 10−450 mM NaCl in 10 ml of buffer at a constant pH. Use five washes per tube.
3. Add the sample to each tube and mix for 10 min[a].
4. Assay the supernatant for the protein of interest.
5. The optimum ionic strength for adsorption is that which just allows binding. The optimum ionic strength for elution is that which just prevents adsorption.

---

[a]An incubation time of 10 min is based on the use of hydrogel matrices. Other matrices which are less porous may require a longer time to allow for adsorption.

(i) *Fines removal.* Removal of fines has been described in Section 4.2.3. Many ion exchangers (e.g. agaroses, Sephacel, Trisacryl) are supplied relatively free of fines.

(ii) *Swelling.* Matrix swelling is necessary for supports supplied in a dry form (Sephadexes, resins, celluloses). The dry material is suspended in 10−15 bed volumes of water and left to swell for a period recommended by the manufacturer. Sephadexes require a prolonged swelling time of 2 days but this may be reduced to 2 h by incubating the slurry in a boiling water bath. Dry celluloses also require pre-cycling to disrupt hydrogen bonds and improve porosity, 0.5 M HCl and 0.5 M NaOH for anion exchangers and the reverse for cation exchangers, with each wash lasting 30 min. Precise details should be obtained from the manufacturers.

(iii) *Washing.* Certain matrices (e.g. resins) may also require washing to remove contaminants. Resins should be washed in a solvent such as methanol or acetone for 1 h to remove trapped air. They are then treated in 2 M alkali and washed in water. Cation exchangers are given an additional 2 M acid treatment and then washed again.

(iv) *Counter-ion conversion.* Ion exchangers as supplied by the manufacturer usually have a specified counter-ion associated with the ionogenic groups on the matrix. This is usually a Cl^- or OH^- counter-ion for anion exchangers and H^+ or Na^+ for cation exchangers. As mentioned earlier, counter-ions differ in their strength of attraction to ionogenic groups; if the eluent to be used contains a different counter-ion, the exchanger should be pre-treated to convert it to this form. Conditions for counter-ion conversion are shown in *Table 14*. Note that for conversion to a weaker counter-ion much larger volumes are required.

5.3.2 Adsorption method

Having determined the capacity and selected the optimum pH and ionic strength for both adsorption and elution steps, a choice should be made between using ion exchange in a batch or column mode.

(i) *Batch adsorption.* Batch protein fractionation is carried out in free solution and although inferior to column separations in efficiency, it is ideally suited to the initial treatment of large volumes of sample. Furthermore it does not suffer from the problems

Table 14. Conditions required for counter-ion conversion.

Original counter-ion	Required counter-ion	Procedure
H^+	Na^+	2 vols 0.1–1 M NaOH or 2 vols 3 M NaCl
OH^+	Cl^-	2 vols 0.1–1 M HCl or 2 vols 3 M NaCl
Na^+	H^+	30 vols 0.1–1 M HCl or 30 vols 3 M NaCl
Cl^-	OH^-	30 vols 0.1–1 mM NaOH or 30 vols 3 M NaCl

of bed swelling and shrinkage which are sometimes encountered in column separations, particularly those using dextran-based matrices. Elution following batch adsorption may be carried out in batch mode or the ion exchanger slurry may be packed into a column and then eluted. The entire capacity of the ion exchanger should be used during the adsorption step.

(ii) *Column adsorption.* As with batch adsorption packed bed ion exchangers can be used to purify a protein by adsorption of contaminants. This allows the desired protein to pass through the column without binding. While this is an acceptable means of purification, no concentration of the protein results. Usually, however, the required protein is adsorbed onto the support in preference to contaminants and then eluted with concentration and/or fractionation.

5.3.3 *Elution method*

It is important to distinguish between two alternative modes of ion-exchange protein purification.

(i) *Static ion exchange.* Here the protein is initially fully adsorbed to the bed and then completely eluted by displacement into the mobile phase using a small volume of a strong eluent. This method is useful for the concentration of protein from a large volume of sample.

(ii) *Dynamic ion exchange.* Here the separation of proteins is achieved by their relative speeds of migration through the column. In contrast to the static method, therefore, all the sample components migrate, but separate depending on their relative equilibrium distributions between stationary and mobile phases. Three choices of elution conditions exist.

(1) Isocratic elution. Here the sample volume should be between 1% and 5% of the bed volume. This is because the sample is only loosely bound and not concentrated during the adsorption step. A long column is used (up to 100 cm) with a diameter to length ratio of around 1:20. The starting buffer is used throughout the separation which may give good resolution of similar proteins but results in large elution volumes.

(2) Step-wise elution. This is achieved using a sequential, discontinuous change in pH and/or salt concentration (*Figure 25*). The column volume used is determined by the exchanger capacity and the sample volume. The sample should initially

213

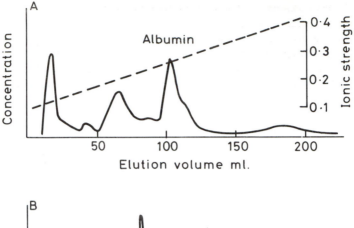

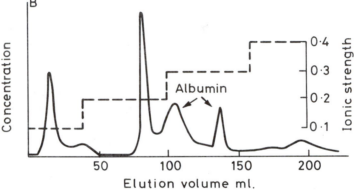

Figure 25. Continuous (**A**) and stepwise gradient elution of bovine serum on QAE-Sepharose A-50. In stepwise elution (**B**) two albumin peaks were found. This can often occur in stepwise elution when the buffer change is introduced too early. Bed dimensions, 1.5 × 26 cm; sample, 4 ml of 3% serum; eluent, 0.1 M Tris−HCl pH 6.5, 0.1−0.5 M NaCl; flow rate, 0.2 ml min^{-1}. Reproduced with permission from Pharmacia-LKB.

adsorb to 5−10% of the total bed capacity. The column length is usually shorter (20−40 cm).

(3)   Gradient elution. Here the composition of the eluent (pH and/or ionic strength) is changed continuously. As in stepwise elution sample protein content should be 5−10% of the bed capacity. The column length is usually 20−40 cm with a diameter to length ratio of not more than 1:5. The volume of buffer required for elution should be determined empirically. If the gradient is too steep resolution will be lost, while too shallow a gradient will result in unnecessary dilution and long separation times. The total volume of eluent should be about five times the bed volume (9).

An increasing pH gradient may be used for cation exchangers while a decreasing pH gradient is used for anion exchangers. Continuous pH gradients are rarely used because they have poor reproducibility (9). This is due to the titration of both protein and ion exchanger ionogenic groups as the pH is altered. A changing pH gradient is also difficult to produce at constant ionic strength. Consequently elution through pH change usually uses a stepwise method. Elution by a change in pH may also be combined with an increase in ionic strength.

Stepwise elution uses simpler apparatus and is usually used on a large scale. The resolution of eluted peaks may be poorer than in gradient elution since stepwise increases in elution conditions may cause the co-elution of several proteins. Furthermore, as seen in *Figure 25*, if a stepwise change in elution conditions is introduced too early false peaks may be produced.

5.3.4 *Regeneration and storage*

Regeneration of ion exchangers involves the removal of tightly bound contaminants and the conversion of the support to the required counter-ion form ready for equilibration and protein adsorption. Regeneration can be carried out in the column but for matrices such as Sephadex, where bed volume is dependent on ionic strength, the ion exchanger should be removed from the column and regenerated in free solution.

Removal of tightly bound protein is first achieved using 2 M NaCl and is then followed in some cases with an alkaline wash. The manufacturers' literature should be consulted for instructions on the precise regeneration conditions.

Ion exchangers stored in a wet state are susceptible to microbial degradation. This is particularly true for polysaccharide-based matrices. Recommended antimicrobial agents (9) for anion exchangers include 0.002% chlorohexidine (Hibitane), 0.001% phenyl mercuric salts in a mildly alkaline solution and 0.05% trichlorobutanol in a weakly acidic solution. The latter can also be used with cation exchangers, in addition to 0.005% merthiolate (ethyl mercuric thiosalicylate) in a mildly acidic solution. In applications where these preservatives cannot be used a high concentration of organic solvents (70% ethanol) is recommended.

5.4 **High performance ion-exchange chromatography**

A wide range of HPLC matrices are now available; some have been listed in *Table 11*. Ion-exchange HPLC of proteins is usually based on wide pore silica or resin matrices. For the purification of medium molecular weight proteins (30−100 kd) pore sizes should be at least 30 nm while for larger proteins (>100 kd) pore sizes should be 500 nm or even 1000 nm in diameter.

The principles of low pressure ion-exchange can be applied to high performance separations. Analytical scale columns (typically 4.6 mm × 7−10 cm) usually separate mg quantities of protein at a time using 200 μl injection volumes, flow rates of 0.3−1.0 ml min^{-1} and a pressure of 1000 p.s.i. Typical separation times are less than 1 h. The high cost of HPLC media usually restricts its use to small volumes of sample.

More recently developed medium pressure packings include Accell (Waters) and Monobeads (Pharmacia). Accell packing is a 37−55 μm silica-based material with large pores. It is used at flow rates of up to 200 ml cm^{-2} h^{-1} to provide protein separations in 30 min to 2 h either in pre-packed or laboratory-packed glass columns. Accell is available with anion (quaternary methylamine) and cation (carboxymethyl) functionalities; the typical capacity is 30−40 mg ml^{-1} (BSA). The monobead packings (Mono Q and Mono S) are based on a hydrophilic polymer resin of size 10 ± 0.5 μm. The matrix is highly porous (m^{ex} > 10^7), is typically used at flow rate of 150−200 ml

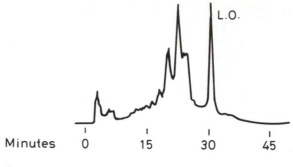

Column:	Bio-Gel TSK DEAE-5PW (Cl-form) 75 x 7.5mm
Sample:	Commercial preparation of lipoxidase, 1 mg in 0.1ml
Eluant:	pH 9.0 Solvent A: 0.02M ethanolamine HCl Solvent B: 0.5M NaCl in 0.02M ethanolamine-HCl
Linear Gradient:	0% to 100% B over 60 min.

Figure 26. Purification of commercial lipoxidase using ion exchange HPLC. Reproduced with permission from Bio-Rad.

$cm^{-2} h^{-1}$ and can withstand back pressures of up to 750 p.s.i. Separation times are usually between 20 min and 1 h. The capacity of the ion exchangers is quoted as 25 mg ml^{-1}.

For larger scale and lower cost purifications Pharmacia recommend using Sepharose Fast Flow which can only be used at back pressures of up to 25 p.s.i.; the separation times are therefore significantly longer but still an improvement on conventional low pressure packings.

5.5. Separation using chromatofocusing

Chromatofocusing, first described and experimentally verified by Sluyterman and co-workers (29,30) can be considered as an extension of isoelectric focusing and ion-exchange chromatography. In isoelectric focusing, proteins are separated by electrophoresis in a pH gradient in a matrix produced by a current. In ion exchange, proteins are bound to the column at an initial pH and may then be eluted by a changing pH gradient. This is produced by mixing a 'limit' buffer (of different pH) with the initial buffer in a mixing chamber and then pumping this through the column (*Figure 27*). The change in mobile phase pH partitions the protein into the mobile phase where it is eluted from the column. In chromatofocusing the pH gradient is produced inside the column by mixing an anion exchange matrix, pre-adjusted to one pH, with a buffer at a second lower pH (*Figure 27*). The protein sample is applied to the column in the

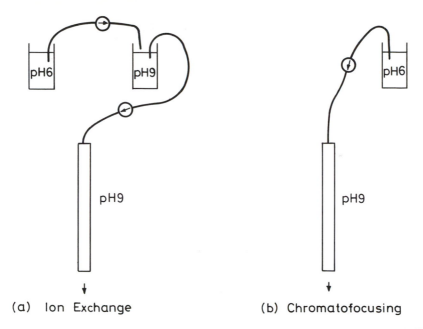

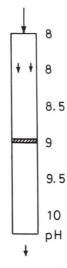

Figure 27. Chromatofocusing differs from ion exchange in that the anion exchange matrix has distinct buffering properties so that a pH gradient is produced on the column rather than outside the column.

Figure 28. A protein of pI 9 applied to a column pre-equilibrated to pH 10 will move rapidly down the column in the pH 8 elution buffer until it reaches its isoelectric pH where it will bind to the exchanger.

elution buffer or the starting buffer. A protein of pI 9 in a pH 8 buffer pumped onto a column pre-equilibrated at pH 10 will therefore move with the buffer down the column and experience rise in pH due to the buffering action of the column. When the protein reaches a pH just above its pI it will become negatively charged and bind to the positively-

217

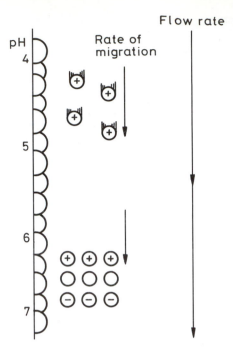

Figure 29. The focusing effect of chromatofocusing. Proteins migrate through the column slower than the flow rate of buffer. Proteins at a pH close to their pI migrate slower than proteins higher up the column at a lower pH, allowing them to catch the leading band up. Reproduced with kind permission from Pharmacia-LKB.

charged column matrix. (*Figure 28*). As elution buffer flows through the column the bound protein will experience a lowering of the pH until it becomes positively charged again and moves with the buffer down the column to reach a second point where the pH is again just above its pI and adsorption occurs. This sequence of events is repeated many times until the protein is eluted from the column, at its isoelectric pH. The protein is never subjected to a pH more extreme than its pI value; the separation is therefore mild.

A further important feature of chromatofocusing is the focusing effect which operates during the separation. A protein applied to the column will travel with the elution buffer to a point where it is negatively charged and then binds to the exchanger, until more elution buffer flows down the column, lowering the pH still further. Any additional protein applied in a second aliquot to the column can therefore catch up with the initial bound protein (*Figure 29*). This serves as a peak sharpening effect in chromatofocusing, allowing resolution of complex mixtures of proteins, providing they differ in their pI's.

Due to the focusing effect of chromatofocusing, proteins can be applied to the column in large volumes without impairing the resolution. The maximum volume in which a protein can be dissolved in order to emerge as a single band is termed the distribution volume. This is governed by the pI of the protein and the equilibration pH. If a protein's pI is close to the initial column pH the protein will be carried rapidly down the column in the elution buffer to be eluted in the void column; the distribution volume is then near zero. Conversely if the pI is close to the elution buffer pH then the protein distribution volume will be very large.

Table 15. Chromatofocusing media and buffers.

pH range	Exchanger	Elution buffer
11−8	PBE 118	Pharmalyte 8−10.5
9−6	PBE 94	Polybuffer 96
7−4	PBE 94	Polybuffer 74
12−2	Mono P[a]	Any of the above

[a]Mono P is a 10-μm matrix for use in the Pharmacia-LKB FPLC system.

Table 16. Starting buffers for use in chromatofocusing.

Upper pH unit	Buffer	Required pH
9	Ethanolamine−HCl	9.4
8	Tris−HCl	8.3
7	Imidazole−HCl	7.4
6	Histidine−HCl	6.2
5	Piperazine−HCl	5.5

5.5.1 *Experimental conditions and chromatofocusing*

The initial studies on the feasibility of chromatofocusing by Sluyterman and co-workers (30,31) used DEAE−Sepharose and acetate buffer. The conclusions from this work are important in the selection of both column media and suitable buffers.

(i) The column matrix must allow rapid pH equilibration between ion exchange groups and the buffer. Polyethyleneimine attached to epoxide-activated Sepharose was subsequently chosen in preference to DEAE−Sepharose.

(ii) The ionic strength of the buffer should be kept relatively constant during elution. The buffering capacity should be constant over the entire pH range so as to provide the near-linear pH gradient necessary for obtaining optimum resolution. This was best achieved using ampholyte buffers during elution although normal cationic buffers (e.g. triethylamine) are used for pre-equilibration.

Although chromatofocusing using conventional ion exchange materials is possible (31,32), the resolution of complex protein mixtures is best achieved using matrices designed to combine a high capacity with rapid equilibration and buffers providing linear, reproducible pH gradients. To this end Pharmacia-LKB market a range of matrices (*Table 15*) and ampholyte buffers (*Table 16*) for use in chromatofocusing (33), and a large number of protein separations have been achieved using these materials (34−41) using both low pressure and FPLC.

The choice of exchanger and operating buffers depends on the range of the pH gradient used in elution. This in turn is determined by the pI of the protein to be purified: the pI of the protein should be approximately midway between the pH extremes of the gradient (33).

A procedure for the use of chromatofocusing is described in *Method Table 6*. In addition, the following points should be noted.

(i) If the pI of the protein is not known a pH gradient from 7 to 4 is used initially. If the protein pI is above 7 it will not bind to the exchanger and a pH gradient from 9 to 6 should be used.

Method Table 6. Procedure for the separation of proteins using chromatofocusing.

1. Determine the protein pI and select the appropriate exchanger, starting and elution buffer from *Tables 15* and *16*.
2. Mix the exchanger with half its bed volume of starting buffer, pre-adjusted to ~0.3 pH units above the upper limit of the gradient. Degas and pour into the column. 10 ml of exchanger per 100 mg of protein is normally adequate. Use a long narrow column e.g. 1 × 40 cm.
3. Equilibrate the column with 10−15 bed volumes of starting buffer at a flow rate of 100 ml cm^{-2} h^{-1}.
4. Dilute the elution buffer 1−10 with water; adjust to the lower pH limit of the gradient. Equilibrate the sample with starting or elution buffer. Its volume may be large but is usually not over half the column void volume. Its ionic strength should be <0.05 M.
5. Apply the sample to the column at 40 ml cm^{-2} h^{-1} and then begin elution. A gradient volume of 10 column volumes is normally sufficient.

(ii) The amount of exchanger required for a separation will depend on the purity of the sample, its volume and its protein content. For samples containing up to 200 mg of protein per pH unit of the gradient, bed volumes of 20−30 ml are recommended. The volume of sample is not critical if loading onto the column is finished prior to elution of the protein of interest.

(iii) The pH of the start buffer is usually adjusted to 0.4 of a pH unit above the required pH to compensate for fluctuations in pH on application of the elution buffer. This is due to a release of protons from the exchanger during the early stages of elution.

(iv) The stronger the elution buffer the shorter the pH gradient and the lower the resolution. A pH gradient of over 3 units is not recommended. For the highest resolution, a narrow pH gradient should be used.

(v) Proteins should be monitored at the outlet using UV absorption at 280 nm, since elution buffers absorb at 240 nm and below.

Chromatofocusing can provide very good resolution of complex mixtures of proteins (*Figure 30*) provided they do not form a continuous distribution of isoelectric points, in which case adjacent protein peaks will overlap with loss of resolution. The presence of non-protein charged material (i.e. nucleic and fatty acids) may also reduce the efficiency of separation. Proteins can be eluted within a pH window of 0.04−0.05 of a unit giving sharp well resolved peaks provided the gradient is not run too fast, in which case loss of resolution will occur (33). Chromatofocusing has also been used with samples containing 7 M urea, 1% (v/v) Triton X-100, 1% (v/v) Tween 80, 5% (v/v) DMSO and 50% (v/v) ethylene glycol, but additives causing an increase in ionic strength should be avoided. The feasibility of using chromatofocusing for protein purification depends on the acceptability of both the cost of the ampholyte buffers required for elution and their presence in the protein sample after purification.

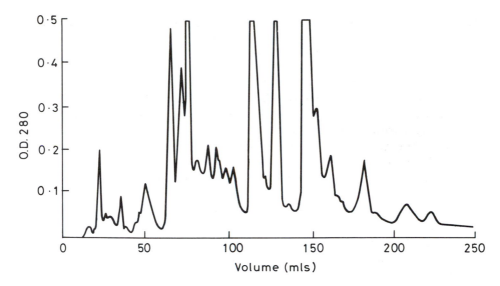

Figure 30. Fractionation of elk muscle proteins by chromatofocusing. Column 10 mm × 40 cm. Sample 5 ml of muscle extract. Start buffer 0.025 M ethanolamine−HCl, pH 9.4. Elution polybuffer 96, pH 6. Flow rate 20 cm^2 h^{-1}. Reproduced with kind permission of Pharmacia-LKB.

5.5.2 *Exchanger regeneration and storage*

Exchangers can be regenerated without removal from the column using 1 M NaCl. Strongly bound proteins can be removed using 0.1 M HCl provided the exchanger is re-equilibrated to a more neutral pH as soon as possible.

Exchangers should be stored in 24% ethanol, while elution buffers are supplied sterile (Pharmacia-LKB) but should be stored cold in the dark.

6. PURIFICATION BASED ON HYDROPHOBICITY

Reverse-phase chromatography (RPC) is the separation of solutes on the basis of their distribution between a polar mobile phase and an organic phase which is fixed to a matrix. Along with the contrasting technique of normal phase chromatography (using a polar stationary phase and organic mobile phase) these techniques were originally termed partition chromatography, since the two separation phases were essentially liquid, one being immobilized to a stationary support. The slow leakage of the immobilized liquid phase meant, however, that steps were subsequently taken chemically to modify the support, thereby fixing the surface film in place. RPC commonly uses aliphatic chains of up to 18 carbon atoms (C$_{18}$) chemically bonded to silica. The mobile phase is polar, consisting mostly of water with the addition of polar solvents such as methanol, propanol, ethanol and acetonitrile to promote displacement of the solute from the non-ionic stationary phase into the mobile phase. The combination of a highly hydrophobic stationary phase and the solvents used to promote elution may cause protein denaturation. This is largely due to the disruption of protein tertiary structure caused by interaction with the high density of alkyl groups attached to the silica packing (3). In order to minimize the risk of denaturation, less densely clustered alkyl groups of a milder hydrophobic nature (e.g. C_8) are required. Under these conditions, proteins can usually be eluted using a decreasing salt gradient. This has been termed hydrophobic interaction chromatography (HIC). The hydrophobicity of a support therefore determines whether it is used in reverse-phase or hydrophobic mode.

The occurrence of non-ionic interactions in protein separation was originally reported as an interference effect resulting from the aliphatic spacer arms used in affinity chromatography (42). Subsequent work on the coupling of alkyl and aryl-amines to agarose gels (43−46) using the cyanogen bromide technique (47) provided amphiphilic *N*-substituted isoureas with both ionic and hydrophobic groups. The combination of both effects led to a complex mode of adsorption involving both non-ionic and electrostatic interactions (45,46). Electrically neutral adsorbents based on alkyl chains attached to agarose were consequently synthesized (48) allowing hydrophobic interaction without electrostatic interaction.

6.1 Theory of hydrophobic interaction chromatography

The simplistic model of protein tertiary structure envisages an essentially hydrophilic outer shell surrounding a hydrophobic core. However, surface hydrophobicity does occur due to the presence at the surface of the side chains of non-polar amino acids such as alanine, methionine, tryptophan and phenylalanine (49). It is likely that surface hydrophobicity not only helps to stabilize protein conformation but forms the basis of specific interactions connected with the biological function of the protein. These may include antigen−antibody, hormone−receptor and enzyme−substrate type interactions and are therefore of significant biological importance (50). The surface hydrophobic amino acids are usually arranged in patches, interspersed with more hydrophilic domains. The number, size and distribution of these non-ionic regions is a characteristic of each protein and can therefore be used as a basis for their separation (51).

A protein molecule in solution holds a film of water in an ordered structure at its surface which must be removed from non-ionic domains before hydrophobic interaction can occur as outlined in *Figure 31*. Removal of bound water molecules from the protein surface into the less ordered bulk solution results in an increase in entropy ($\Delta S > 0$). The overall free energy change (ΔG) for the interaction of two non-ionic groups in the adsorption step is related to entropy and enthalpy as follows:

$$\Delta G = \Delta H - T\Delta S \qquad (1)$$

Since the enthalpy change (ΔH) in the interaction is small, the free energy change of the process is negative and therefore proceeds spontaneously (50−52). The likelihood of a non-ionic interaction is further increased if the excluded water molecules are trapped in the bulk solution by the hydration of salt ions. Consequently it is convenient to use HIC following ammonium sulphate precipitation since the ions which are most effective at salting proteins out of solution are generally those which create the most structuring in water. The hydrophobic interaction between protein surface non-ionic groups at high salt concentration forms the basis of protein precipitation using ammonium sulphate (see Chapter 3). In HIC, however, the interacting non-ionic group (e.g. octyl/phenyl) is provided by a hydrophobic functionality attached to an inert matrix such as agarose.

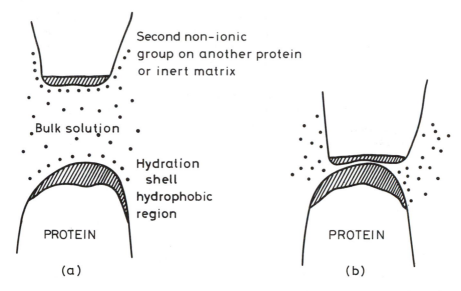

Figure 31. Model of non-ionic exchange interaction. (a) Protein in solution with a shell of water around its hydrophobic region. (b) Exclusion of water into bulk solution upon approach of another non-ionic group excludes the bound water in a spontaneous process driven by the gain in entropy.

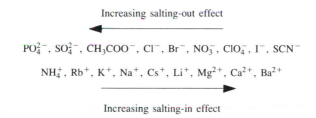

Figure 32. The influence of ions on hydrophobic interactions. (The Hofmeister series.)

Nevertheless, the adsorption step often requires the presence of salting-out ions such as sodium chloride or ammonium sulphate. The influence of certain ions on protein structure was first reported by Hofmeister (53). Salting-out ions decrease the availability of water molecules in solution, increase the surface tension and enhance hydrophobic interaction. In contrast, salting-in or chaotropic ions such as sodium thiocyanate prevent non-ionic interaction by ordering the structure of water (15). Their use in elution from hydrophobic matrices is, however, usually avoided since they may denature proteins. Chaotropic and salting-out ions are shown in *Figure 32*.

Temperature also influences the level of hydrophobic interaction. As equation (1) states, the free energy of the process becomes more negative with increasing temperature (up to ~60°C where the additional stability provided by electrostatic and Van der Waal's forces disappears). The reduced strength of HIC interaction at lower temperatures is not, however, usually significant enough to be used as a means of protein elution.

Protein and adsorbent charge may also have an important effect on the degree of hydrophobic adsorption. A protein with no net charge (i.e. at its pI) will have maximum

Table 17. List of some commonly used hydrophobic adsorbents.

Manufacturer/ Supplier	Packing	Comments
Low/medium pressure packings		
Miles	Butyl, hexyl, octyl agarose	Capacity—10 mg protein ml^{-1}
Pharmacia	Octyl, Phenyl Sepharose	Capacity—20 mg ml^{-1}
	Phenyl Superose	Capacity—10 mg ml^{-1}
	Alkyl Superose	Particle size 10 μm
		For use in FPLC
Serva	Daltosil Octyl	Silica based
	Phenyl, Aminohexyl	Pore size 50 nm
	Aminopropyl	Particle size 0.1−0.2 mm
	Aminophenyl	
Waters	Protein-Pak HIC	10 μm particle size
	Phenyl 5PW	Used in glass columns
HPLC packings		
Alltech	Synchropack Propyl	6.5 μm silica
Anachem	Anagel TSK/HIC	
Beckman	Spherogel TSK-Phenyl	
	Spherogel CAA-HIC	5 μm particles 30 nm pores
	[a]Ultrapore C3	5 μm silica
Bio-Rad	Bio-Gel Tsk Phenyl	10 μm particles, 100 nm pore
	[a]Hi-Pore RP	Silica C4
Brownlee	Aquapore RP300	Butyl or Phenyl on 7 μm silica
Perkin Elmer	TSK-Phenyl	
[a]Phase-Sep	Spherisorb C_1, C_3	30 nm pores 5, 10 μm particles
	Octyl, Phenyl	
[a]Shandon	Hypersil—Butyl, Octyl	5, 10 μm silica 30 nm pores
[a]Vydac	C_4, diphenyl	
Waters	μBondapak-Phenyl	

[a]Normally used in reverse-phase chromatography.

hydrophobicity. At a pH where the protein and adsorbent have similar charges repulsion may occur, resulting in reduced adsorption.

Other conditions which reduce adsorption include polarity lowering agents such as water miscible solvents (e.g. ethanol and ethylene glycol) and detergents. Their use in elution is usually considered a last resort when other milder conditions such as reduced conductivity do not promote protein recovery.

6.2 Experimental conditions for low pressure HIC

6.2.1 *Adsorption*

The most commonly used non-ionic adsorbents are based on agarose using octyl or phenyl functionalities. Many others exist and some are listed, along with HPLC packings in *Table 17*. The stationary phase should be charge-free or subsequent elution using a decrease in ionic strength will encourage electrostatic interaction and prevent effective desorption. The choice of the hydrophobic ligand is of considerable importance in determining the ease of subsequent protein elution.

Phenyl Sepharose, for example, is less hydrophobic than Octyl Sepharose (54). Extremely hydrophobic proteins (e.g. membrane proteins) may adsorb too strongly to Octyl Sepharose and require very strong eluting conditions. Phenyl Sepharose should therefore be used. In contrast, mildly hydrophobic proteins may not adsorb to Phenyl Sepharose and will therefore require use of more hydrophobic matrices.

The solution ionic strength will also control the protein hydrophobicity and therefore its degree of adsorption. Mildly hydrophobic proteins may not be adsorbed until the ionic strength is increased to just below that required for precipitation. The more hydrophobic proteins such as globulins will associate with other non-ionic groups even at low salt concentration (20−40% w/v ammonium sulphate). Prior to HIC, the sample ionic strength should be adjusted with salt and buffered preferably near the pI of the required protein to enhance non-ionic adsorption. The packed column should be similarly equilibrated. The sample is then applied to the matrix and washed with the same buffer. The capacities of hydrophobic matrices are usually as high as for ion exchangers (i.e. $10-100$ mg ml^{-1} of adsorbent) (8).

6.2.2 *Elution conditions*

A variety of conditions can be used for elution from hydrophobic matrices, providing a potentially powerful method for the resolution of complex mixtures of proteins. These include the following.

(i)   Reducing the ionic strength using either isocratic or gradient elution.

(ii)   Increasing the pH: most proteins gain a net negative charge and become more hydrophilic under mildly alkaline conditions.

(iii)   Reducing the temperature should theoretically promote elution as described earlier. Generally however, the temperature effect is too small for it to be used on its own as an effective means of elution.

(iv)   Displacement methods: addition of a component which has a stronger attraction for the ligand or makes the protein more hydrophilic.

        (1)   *Aliphatic alcohols.* (e.g. propanol, butanol and ethylene glycol). These reduce the polarity of the solution and disrupt hydrophobic interaction.

        (2)   *Aliphatic amines.* (e.g. butylamine). These have the effect of reducing the solution polarity causing desorption. The amines may bind to the protein or matrix hydrophobic groups.

        (3)   *Detergents.* Non-ionic detergents (e.g. Tween 20, Triton X-100) probably displace bound proteins by binding to both protein and column packing.

Ionic detergents (e.g. sodium dodecyl sulphate) are more easily removed from the

Method Table 7. Procedure for protein isolation by HIC using sequential elution with reduced ionic strength and a displacer.

1.	Suspend the matrix in half its bed volume of 0.8 M ammonium sulphate in 0.025 M potassium phosphate buffer pH 7, degas and pour as a slurry into the column.
2.	Wash the column with three column volumes of the starting buffer.
3.	Apply the sample in the same buffer.
4.	Elute with 3.5 bed volumes of a decreasing linear gradient of ammonium sulphate (0.8−0 M) in 0.025 M potassium phosphate buffer pH 7.0.
5.	Elute with four bed volumes of a linear increasing gradient of 0−80% (v/v) ethylene glycol.

Method Table 8. Hydrophobic interaction chromatography involving the use of a displacer during adsorption.

Matrix : Phenyl Sepharose CL-4B (50 ml)
Column : K-26/40 (2.9 × 9.4 cm)
Application : Purification of natural human fibroblast interferon (IFN).

1.	Remove fines and degas matrix, pack into column.
2.	Rinse with 0.5 litres of 40% (v/v) ethylene glycol in 0.02 M sodium phosphate buffer pH 7 + 0.15 M NaCl (PBS).
3.	Apply 1 litre of preparation containing 40% ethylene glycol in PBS at a flow rate of 10.5 ml cm^{-2} h^{-1}.
4.	Wash with 0.3 litres of 40% (v/v) ethylene glycol in PBS.
5.	Recover IFN with 0.3 litre of 75% (v/v) ethylene glycol in PBS. Dilute immediately with 0.25 litre PBS.

column after protein desorption. They tend, however, to cause protein denaturation. Non-ionic detergents are milder and can be used at levels of 1−3% without causing loss of activity (8).

Frequently, a gradient of decreasing ionic strength is used followed by a gradient of an increasing concentration of a displacer to ensure desorption as given in *Method Table 7*. If the desired protein is very hydrophobic, displacers (55) may be used in the initial adsorption conditions to prevent adsorption of other less hydrophobic contaminants (*Method Table 8*). Protein desorption may also be achieved using one-step elution (56) at reduced ionic strength. An example is given in *Method Table 9*. Both reducing ionic strength and increasing displacer gradients (57) may however be applied at the same time (*Method Table 10*).

Detergents may also be used in tandem in applications where strong hydrophobic adsorption has occurred. In the example shown in *Method Table 11*, 1.5% Tween 80 is used to remove cholate from the column during cytochrome *c* oxidase purification (58). The enzyme itself is then eluted using 1% (w/v) Triton X-100 which acts as a stronger non-ionic detergent.

Method Table 9. One-step elution using reduced ionic strength.

Matrix : Phenyl Sepharose CL-4B
Column : 2.6 × 26 cm
Application : Human interleukin 1 (IL1) purification

1. Remove fines, degas and pack column.
2. Equilibrate with 1 M ammonium sulphate in 10 mM potassium phosphate buffer pH 6.8.
3. Apply sample.
4. Wash with 450 ml of the same buffer.
5. Elute with 800 ml of 0.2 M ammonium sulphate and 10 mM phosphate buffer pH 6.8.

NB In this example it was necessary to avoid the use of ethylene glycol since it denatures human IL1.

Method Table 10. Gradient elution using reduced ionic strength and increasing displacer.

Matrix : Octyl Sepharose CL-4B
Column : K16/20 (bed volume 30 ml)
Application: β-amylase purification

1. Remove fines and degas matrix, pack into column.
2. Wash with 0.01 M sodium phosphate buffer pH 6.8 + 25% (w/v) ammonium sulphate.
3. Apply sample (40 ml in the same buffer). Flow rate = 25 ml h^{-1}.
4. Wash with 85 ml of the same buffer.
5. Elute with a gradient of decreasing ammonium sulphate concentration (25−0%) and increasing ethylene glycol (0−5%) simultaneously.

The elution conditions necessary for optimum protein separation must be determined empirically. A decreasing ionic strength gradient (down to just water), should, however, always be used initially. Failure to elute the required protein can then be overcome by using stronger elution conditions (e.g. ethylene glycol) or by using a less hydrophobic matrix.

6.2.3 *Matrix regeneration and storage*

Prior to re-use, any tightly bound protein should be removed from the matrix with 6 M urea, and the support washed in starting buffer. Any detergents used during protein elution should be removed with an increasing ethanol gradient (up to 95%) followed by two bed volumes of butan-1-ol, one bed volume of ethanol and one bed volume of distilled water (54). The bed is then washed with starting buffer. For prolonged storage the gels are suspended in 0.02% (v/v) merthiolate or 20−25% (v/v) ethanol and stored at 2−8°C.

Method Table 11. Use of two detergents in hydrophobic interaction chromatography.

Matrix : Octyl Sepharose CL-4B
Application : Cytochrome *c* oxidase purification

1.   Remove fines, degas matrix and pack into column.
2.   Equilibrate with Tris−cholate buffer pH 8 + 15% (w/v) ammonium sulphate, 1 mM EDTA.
3.   Apply sample to column in same buffer at 4 ml min^{-1}.
4.   Wash with 10% (w/v) cholate/50 mM Tris−sulphate (pH 8) + 1 mM EDTA.
5.   Elute cholate with 1.5% (v/v) Tween-80 in 50 mM Tris−sulphate.
6.   Elute cytochrome *c* oxidase with 1% (v/v) Triton X-100 in 50 mM Tris−sulphate.

6.3 High performance HIC

The influence of support hydrophobicity on the risk of protein denaturation has already been mentioned and should be kept in mind in the selection of HPLC packings. Certain matrices are now recommended for HIC and these have been listed in *Table 17*.

In HPHIC, separations can be achieved in 30−60 min using flow rates of 0.5−1 ml min^{-1} (*Figure 33*). Slower flow rates (60−90 ml cm^{-2} h^{-1}; <0.4 ml min^{-1}) are however preferred. In the selection of column packings wide pore supports (>30 nm) are preferable for improved resolution; with pore sizes of 6−10 nm only surface adsorption of protein will occur (3).

Finally, care should be taken to wash the HPLC system out with water after use to remove all traces of salts and minimize corrosion of metal pump components.

Pharmacia-LKB also produce hydrophobic matrices for use in their FPLC system. The packings are based on 10 μm agarose particles (Superose) for use at flow rates of 150 ml cm^{-2} h^{-1} at a pressure of up to 400 p.s.i. The matrix has a capacity of 10 mg ml^{-1} allowing purification of up to 80 mg of protein at a time.

6.4 Advantages and disadvantages of HIC

Hydrophobic interaction chromatography provides a powerful additional means of separation which is applicable to the purification of most proteins. It is ideal for use immediately after salt precipitation where the ionic strength of the sample will enhance hydrophobic interaction. In purification where the required protein is eluted in a gradient of decreasing ionic strength, it can be followed by ion exchange with little need for buffer change.

The diversity of potential eluting conditions can enable the resolution of even complex mixtures of proteins which would be difficult to separate by other chromatographic techniques. Predicting the best conditions for separation is, however, difficult and an element of trial and error is involved in process optimization. The effectiveness of HIC is generally reduced by the presence of hydrophobic contaminants in the feed. An additional problem encountered in the use of HIC is the strong conditions which are frequently required to elute proteins from hydrophobic matrices. This may involve the

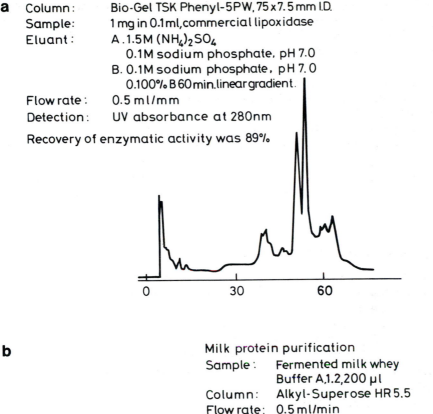

a Column :   Bio-Gel TSK Phenyl-5PW, 75 x 7.5 mm I.D.
     Sample: 1 mg in 0.1 ml, commercial lipoxidase
     Eluant :   A . 1.5M $(NH_4)_2SO_4$
                   0.1M sodium phosphate, pH 7.0
                   B. 0.1M sodium phosphate, pH 7.0
                   0.100% B 60 min. linear gradient.
     Flow rate :   0.5 ml/mm
     Detection :   UV absorbance at 280nm

Recovery of enzymatic activity was 89%

b

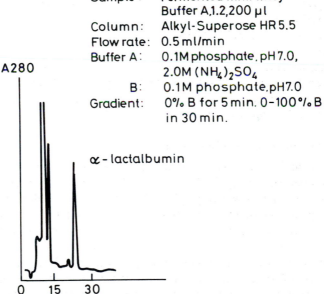

Milk protein purification
Sample :   Fermented milk whey
           Buffer A,1,2,200 µl
Column: Alkyl-Superose HR 5.5
Flow rate: 0.5 ml/min
Buffer A: 0.1M phosphate, pH 7.0,
          2.0M $(NH_4)_2SO_4$
B: 0.1M phosphate, pH 7.0
Gradient: 0% B for 5 min. 0–100% B
          in 30 min.

A280

α - lactalbumin

Figure 33. (a) HPLC chromatogram for the purification of lipoxidase using hydrophobic interaction. Reproduced with the permission of Bio-Rad. (b) FPLC purification of milk proteins. Reproduced with the permission of Pharmacia-LKB.

229

Table 18. List of some commonly used reverse-phase packings.

Manufacturer	Packing name	Comments
Alltech	Synchrom RP	6.5 μm silica, 30 nm pores C_1, C_4, C_8, C_{18}
Amicon	Matrex	$10-20$ μm silica C_8, C_{18}
Anachem	Dynamax	12 μm silica C_8
Anachem	Buio-Series PEP RP1	5 μm silica C_8
Beckman	Ultrapore C_3	5 μm silica 30 nm pores
Bio-Rad	Hi-Pore RP	Silica C_4 or C_{18}
DuPont	Zorbax PolyF	20 μm
DuPont	Zorbax Bio-Series	4 μm silica C_8
Perkin Elmer	HCODs C_{18}	5, 10 μm silica
Pharmacia-LKB	Pro RPC	5, 15 μm silica
Phase-Sep	Spherisorb	C_1-C_{18} Octyl, Phenyl
Shandon	Hypersil WP300	Butyl, Octyl, 5, 10 μm silica
Shandon	ODS, MOS Hypersil	C_8, C_{18} silica
Vydac	Vydac C_4, C_{18}	5 and 10 μm silica
Waters	μ Bondapak	C_{18}, CN, Phenyl
	Delta-Pak	15 μm silica C_4, C_{18}
Whatman	Protesil 300	Octyl silica 30 nm pores

Table 19. Eleutropic series of solvents for use in reverse-phase chromatography: stronger solvents have a greater eluting power for proteins on reverse-phase columns.

Increasing polarity	↑	Water Methanol Acetonitrile Ethanol Tetrahydrofuran N-Propanol	↓	Decreasing polarity

use of non-ionic detergents or ethylene glycol which can increase the likelihood of protein denaturation. The use of strong elution conditions may also be required when the reduction in ionic strength during elution encourages electrostatic interactions from charged groups on the adsorbent matrix. The use of HPLC for HIC must involve the regular washing out of salts so as to prevent corrosion.

6.5 Reverse-phase chromatography (RPC)

Reverse-phase techniques have traditionally been applied to the analysis of low molecular weight compounds using HPLC. It is characterized by the use of silica derivatized with alkyl functionalities (typically C_2-C_{18} alkyl chains) and an aqueous mobile phase containing an organic solvent such as methanol, propanol, acetonitrile and ethanol. Some commercially available reverse-phase packings are listed in *Table 18*.

In addition to ligand density the length of the alkyl chain is proportional to the packing hydrophobicity. Consequently C_{18} packings bind proteins more tightly and are more likely to cause denaturation than C_8 or C_4 packings. Packings should always be chosen in which unreacted silanol groups are 'end-capped' using blocking agents such as trimethyl-chlorosilane. This minimizes interference adsorption of protein onto the

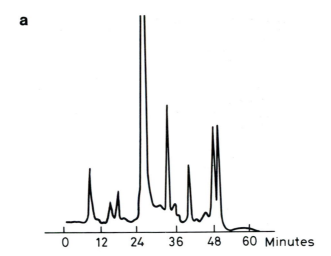

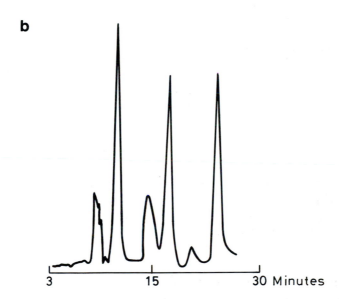

Figure 34. HPLC separations of proteins using reverse-phase chromatography. Reproduced with permission from Whatman. (**a**) Separation of human serum proteins using Protesil 300 Diphenyl. Flow rate: 0.5 ml min^{-1}, gradient 0−100% 1-propanol. (**b**) Separation of protein standards using Protesil 300. Flow rate 0.5 ml min^{-1}.

unreacted silanol groups which may cause loss of resolution and poor reproducibility. Furthermore, for protein purification, wide pore matrices (∼30 nm) should be used in order to maximize capacity and resolution. Preservation of biological activity, when essential, should be checked before proceeding further with any selected column packing.

The mobile phase may be buffered (1 M strength) using sodium acetate or pyridinium formate at pH 4−6 to minimize protein adsorption to unreacted silanol groups; this can be visualized as a narrowing of peak width. Protein retention is controlled using alteration of the polarity of the mobile phase by the addition of organic solvents. The most commonly used solvents may be arranged in an eleutropic series (*Table 19*). Inclusion of 'stronger' solvents such as acetonitrile and tetrahydrofuran in the mobile phase will give a greater reduction in retention time than the 'weaker' solvents such as ethanol and methanol. The optimum solvent composition depends on the packing used and the sample composition and is therefore a matter of trial and error. HPLC grade low-UV absorbance solvents should always be used and caution should be exercised in their handling (particularly tetrahydrofuran and acetonitrile). Examples of reverse-phase HPLC separations are shown in *Figure 34*.

Elution is achieved using gradient or isocratic conditions. Gradient conditions may be preferred to reduce the separation time or improve the resolution of certain areas of chromatogram. Usually the strength of the solvent is increased during gradient elution, to 40−80% (v/v) propanol or acetonitrile in water. Recently the use of the RPC has been extended to the Pharmacia FPLC system. The matrices are based on highly porous silica (5 or 1.5 μm) for use in glass columns at pressures up to 1500 p.s.i. The matrix recommended for protein purification is Pro RPC; like all silica matrices, its pH stability is restricted to between 2 and 8. Pro RPC is a wide pore matrix (300 nm) with a mixture of functional groups (C_1 or C_2 mixed with C_8). Separations typically take 30−90 min at a flow rate of 60 ml cm^{-2} h^{-1}.

Protein retention may also be controlled by the addition of counter-ions. These are usually aliphatic molecules with a charged group which associate with the charged groups of a protein, increasing its hydrophobicity. Commonly used counter-ions (ion-pairing agents) include heptane sulphonate, tetra-*n*-butylammonium hydroxide, trifluoroacetic acid (TFA) or triethylammonium phosphate/acetate. TFA is volatile and therefore easy to remove after purification.

Another method for altering the retention of proteins in RPHPLC is to change the operating pH such that relative hydrophobicity of the protein components is altered.

6.6 Advantages and disadvantages of reverse-phase chromatography

In the purification of peptides or proteins where retention of the native state is not essential, RPC provides a high resolution technique which, using HPLC, can often separate proteins not resolvable using other chromatographic methods. In the purification of peptide hormones for example, it therefore represents a powerful technique. Its drawbacks are in the frequent use of toxic solvents such as acetonitrile and in its limited application to protein purification where denaturation is unacceptable. Note that silica based matrices will dissolve above pH 7.5.

7. METAL CHELATE CHROMATOGRAPHY

Metal chelate chromatography (59), developed by Porath and co-workers in 1975, is a refinement of ligand exchange chromatography (60) for the purification of high molecular weight and conformationally unstable biopolymers. In ligand exchange chromatography sorption occurs between the sorbate and matrix via coordination bonds of a complex-forming ion. This has traditionally been accomplished using sorbate complexation to metal cations retained on chelating resins and cation exchange resins. In metal chelate chromatography of proteins sorption occurs by the coordination of surface histidine, cysteine and tryptophan residues to transition metal ions such as Cu^{2+} and Zn^{2+}.

The metal ion must be held in an accessible position away from the matrix so as to ensure efficient adsorption. This was achieved using iminodiacetate groups coupled to epichlorohydrin-activated agarose to form a bis-carboxymethylamino agarose matrix which would then complex with metal cations. The coupling of iminodiacetate to oxirane-activated agarose may be achieved in the laboratory (59,61) (*Method Table 12*). However, iminodiacetate-substituted agarose is also available from Pharmacia-LKB. Iminodiacetate has also been coupled to other supports including Trisacryl GF 2000, a macroporous polyhydroxymethyl polymer (62) and epoxysilylated silica for use in high performance metal chelate chromatography (63). Chelating Sepharose-6B (Pharmacia-LKB) has been used to purify a large number of proteins, including trypsin isoinhibitors (64), mammalian interferons (65), fibronectin (66), macroglobulin (76) and albumin (68).

7.1 Theory of metal chelate chromatography

The complexation of the transition metals zinc and copper with histidine imidazole and cysteine thiol groups (59) is pH-dependent with neutral pH conditions providing the strongest level of adsorption. Under alkaline conditions coordination with amino groups provides a stronger but less selective adsorption. Other metal ions which can coordinate with histidine and cysteine residues include cobalt, nickel, manganese, cadmium, magnesium and calcium, although the affinity of proteins for these cations is generally less than for copper and zinc (69). The complexation of tryptophan residues with these metal cations is also suspected (65).

Metal chelate chromatography therefore differs from conventional purification methods such as ion exchange and sorption since the interaction between protein and matrix does not take place directly but occurs via the complex-forming metal ion (60). The complex once formed must be sufficiently stable so that protein binding can occur irrespective of salt and non-electrolyte concentration. The technique thus separates proteins according to the surface density and number of imidazole and thiol residues and therefore involves a highly selective sorption process.

Method Table 12. Method for the preparation of iminodiacetate-substituted agarose.

---

1. Add 125 g of epoxy-activated agarose to 100 ml of 2 M sodium carbonate containing 20 g of the disodium salt of iminodiacetic acid.
2. Shake gently for 24 h at $60-65°C$.
3. Cool and wash the gel on a glass filter funnel using:
    - (i)   0.1 M sodium carbonate;
    - (ii) 0.01 M sodium acetate;
    - (iii) water.

Method Table 13. Procedure for protein purification using metal chelate chromatography.

---

1. Degas matrix and pack into column.
2. Charge with metal ion in water (e.g. $ZnCl_2$, or $CuSO_4 \cdot 5H_2O$, 1 mg ml^{-1}). The most appropriate metal ion should be determined empirically beforehand.
3. Equilibrate the column with a neutral pH buffer containing salt to minimize electrostatic effects (e.g. 0.02 M phosphate, pH 7.5 + 0.5 M NaCl).
4. Apply the sample pre-equilibrated in the same buffer.
5. Wash the column with equilibration buffer.
6. Elute protein using a reduced pH buffer e.g. 0.1 M sodium acetate, pH 4−6 + 0.5 M NaCl.
7. Elute strongly bound proteins using a chelating agent (e.g. 50 mM EDTA in 50 mM sodium phosphate buffer pH 7 + 0.5 M NaCl).

Subsequent elution is commonly achieved using a lower pH which destabilizes the protein−metal chelate complex (68). Although chelating agents (e.g. EDTA) provide little selective elution, they are frequently used to remove more strongly bound proteins (59,64,66). The most appropriate metal ion for a particular application must generally be determined empirically (69). Porath and co-workers (59) found that for serum protein purification copper and zinc ions were most effective and these are the metal ions of choice in most applications.

While much data exists on the binding of metal ions to proteins in free solution it is not always applicable to sorption onto immobilized metal chelates (65). Human serum albumin binds to zinc via almost all of its 16 imidazole groups in free solution, but it is not retained on a zinc chelate matrix.

7.2 Experimental conditions

The procedure for protein adsorption using metal chelate adsorbents is shown in *Method Table 13*.

A similar quantity of agarose gel to that used in ion exchange is recommended (69), while a flow rate of 20−25 ml cm^{-2} h^{-1} should provide optimum resolution (68). A range of metal ions should be tested initially on a protein sample to determine the most suitable conditions. This is best achieved using a series of small test columns each pre-equilibrated with a different metal ion (e.g. $CaCl_2 \cdot 2H_2O$, $MgCl_2 \cdot 6H_2O$, $MnCl_2$, $CuSO_4$: each at 1−5 mg ml^{-1}). As mentioned previously, protein adsorption is best achieved under neutral conditions in the absence of chelating agents. If the metal ion−protein binding is strong, the chelating properties of Tris−HCl buffer can be used to reduce the strength of binding. Salt (0.5−1.0 M NaCl) should be added in all buffers to minimize electrostatic interaction (68).

Elution is typically achieved using 0.05−0.1 M sodium acetate buffer at pH 4−6: the lower the pH the more effective the elution.

Displacement elution conditions using imidazole, histidine, glycine or ammonium chloride can also be used, while 0.05−0.1 M Tris−HCl, pH 8 or Tris−acetate pH 3 may also prove successful (67). All elution buffers should contain salt.

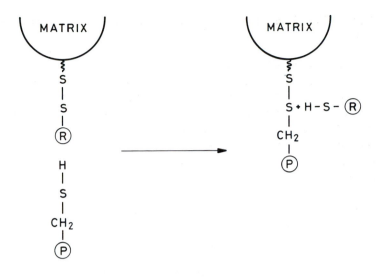

Figure 35. Principle of covalent chromatography. A protein (**P**) with a surface thiol provided by a cysteine residue reduces the ligand disulphide with the release of the end-capping functionality (**R**). A new disulphide is formed between the protein surface cysteine and the matrix thiol.

EDTA (0.05 M) should be used at the end of every purification to remove strongly bound protein and regenerate the column by stripping off bound metal ions.

8. COVALENT CHROMATOGRAPHY

Covalent chromatography was originally developed by Eldjam and Jellum (70) for the isolation of thiol-containing proteins using organomercurial-polysaccharide and involves a chemical reaction between the chromatographic material and a component in solution (71). Covalent chromatography is now generally used for the isolation of thiol-containing proteins by thiol−disulphide interaction, a method introduced by Bocklehurst and co-workers in 1973 (72,73). Protein tertiary and quaternary structure is frequently stabilized by intra- and inter-chain disulphide bridges formed by the oxidation of two cysteine residues. However, many proteins have non-oxidized surface cysteine side chains which can be oxidized to form a mixed disulphide with a secondary thiol attached to a stationary phase (*Figure 35*).

The isolation of proteins by thiol−disulphide interchange involves the reaction of a matrix-attached disulphide 2'-pyridyl group with a thiol-containing protein (*Figure 36*).

The chromophoric pyridine-2-thione released can be monitored spectrophotometrically at 343 nm, allowing the adsorption of proteins to be simply assessed. After thiol-containing proteins are bound, loosely adsorbed contaminants are washed away and elution promoted by reducing agents such as cysteine, glutathione and dithiothreitol (74). After use the matrix is re-activated with 2,2'-dipyridyl disulphide. While cellulose, cross-linked dextran and polyacrylamide have all been used as matrix supports in covalent chromatography, agarose has been adopted as the matrix of choice in the largest number of separations (75).

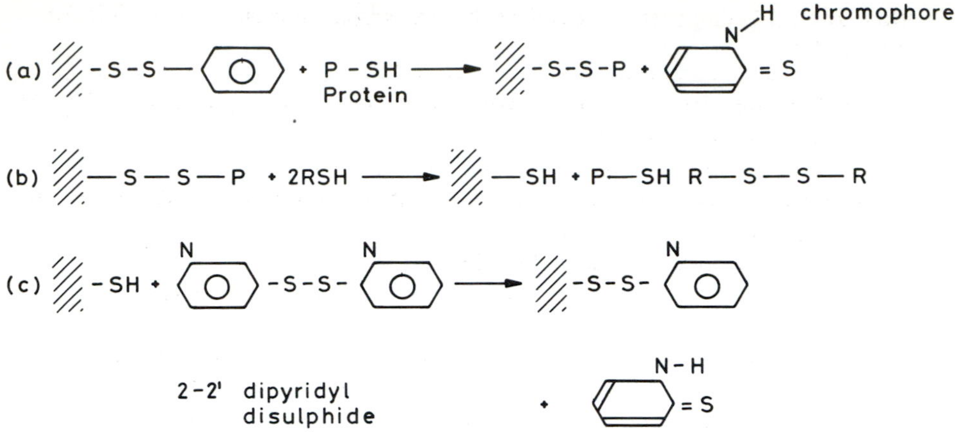

Figure 36. Adsorption, elution and regeneration steps in covalent chromatography by thiol−disulphide interchange. Reproduced with kind permission of Pharmacia-LKB.

8.1 Experimental techniques

8.1.1 *Adsorption*

This separation technique is usually applied to partially purified products with minimal non-protein contamination. It is therefore used following initial sample clean-up using salt-precipitation and ion-exchange chromatography (74). There are cases, however, where it has been used for protein isolation from relatively crude preparations (76).

An example of a procedure for covalent chromatography using Thiopropyl Sepharose is shown in *Method Table 14*. The pyridyl disulphide is attached to the cross-linked agarose matrix by a hydroxypropyl spacer arm. For the separation of larger proteins, activated Thiol Sepharose-4B is preferred. The lower degree of cross-linking and the longer spacer arm used in Thiol Sepharose (glutathione) allow adsorption of larger proteins. The gel capacity is, however, rather lower than for Thiopropyl Sepharose-6B. Adsorption can be carried out using column or batch techniques. Both initial gel pre-treatment and sample application use mild conditions with buffering at neutral or mildly acidic pH (74). At pH 8 most thiol-containing proteins will react readily while at an acidic pH ($\sim$4) the selectivity of protein binding can be increased by the adsorption of proteins having an abnormally low pK_a value. This includes many cysteine proteinases such as papain which was selectively adsorbed at a low pH by Brocklehurst and co-workers (73). Papain has an active site cysteine residue which has an unusually low pK_a value due to intramolecular hydrogen bonding.

8.1.2 *Elution*

Elution of bound proteins is achieved using low molecular weight thiols at a neutral pH. Removal of unreacted thiopyridyl groups prior to protein elution can, however, be accomplished using 4 mM dithiothreitol or 2-mercaptoethanol in 0.1 M sodium acetate, pH 4 (74). Protein elution generally uses 20−50 mM dithiothreitol or 2-mercaptoethanol in a neutral pH buffer.

Method Table 14. Procedure for covalent chromatography of proteins using Thiopropyl Sepharose-6B.

---

1. Swell gel for 15 min at room temperature in a buffer at neutral pH (e.g. 20 mM sodium phosphate, pH 7).
2. Pack into column and wash through with the same buffer.
3. Apply sample and wash off contaminants with the starter buffer. The adsorption step can be monitored at 343 nm.
4. Elute bound protein sequentially with:
    (i)   5–25 mM L-cysteine + 1 mM EDTA in a neutral buffer, pH 7–8.
    (ii) 50 mM reduced glutathione + 1 mM EDTA in elution buffer.
    (iii) 20–50 mM 2-mercaptoethanol + 1 mM EDTA in elution buffer.
    (iv) 20–50 mM dithiothreitol + 1 mM EDTA in elution buffer.

Hillson (77) developed a sequential elution method allowing increased selectivity during the desorption step. By using a series of thiols of increasing reducing strength or increasing concentrations of the same thiol, the degree of resolution of thiol-containing proteins can be improved (*Method Table 14*).

8.1.3 *Regeneration and storage*

Following protein elution, reducing agents must be displaced from the gel using 30–40 mg ml^{-1} 2,2'-dipyridyl disulphide pH 8 (activated Thiol Sepharose). The recommended procedure for Thiopropyl Sepharose is as follows.

(i)   Mix one volume of 2,2'-dipyridyl disulphide (30–40 mg ml^{-1}) in ethanol or isopropanol with four volumes of gel in 0.1 M borate buffer pH 8 + 1 mM EDTA.
(ii) Reflux at 80°C for 3 h.
(iii) Wash gel in ethanol and re-equilibrate in starting buffer.

After use agarose gels should be stored refrigerated at pH 4 in a sealed bottle containing 0.02% sodium azide. Do not allow the pH to drop below 4. Thiol-containing preservatives should be avoided.

8.2 **Advantages and disadvantages**

Covalent chromatography is a useful addition to the range of protein separation techniques currently available. Its selectivity for removing thiol-containing proteins has been used in a large number of applications. The specificity of the technique can be further increased by adjustment of the adsorption pH and by using the sequential elution method. The high degree of specificity during the adsorption stage may allow its use in relatively crude applications (74) but at present the large-scale use of covalent chromatography is limited by its cost (71). Although this is offset by the re-usability of gel matrix, the lengthy regeneration procedure involving a 3 h reflux step, is one drawback in its use.

9. OTHER ADSORPTION TECHNIQUES

Apart from the adsorption phenomena already described, other mediators of protein adsorption include the following.

(i)   Hydrogen bonds formed by the interaction of a hydrogen atom covalently bonded to an electronegative atom with another electronegative atom having a lone electron pair.

(ii)   Van der Waals forces caused by the perturbations in the electron clouds of molecules.

(iii)   Dipole interactions between molecules with partial charge separations.

Many different types of adsorbents thought to interact via these forces have been used in protein purification, for example, titania, alumina, Fullers Earth, Bentonite and Celite. However, since the precise nature of the adsorption mechanism is not clear an empirical approach is frequently used in the choice of adsorption and elution conditions. Most of these materials have been used in the form of particulates and have the advantage of being relatively inert, easily sterilized, cheap and of low toxicity (8). However, significant problems are associated with their use including irreversible adsorption, column clogging, variable capacities and slow adsorption/desorption kinetics. With the advent of superior matrices based on polysaccharides such as cellulose their use is therefore limited. Only one adsorbent, hydroxylapatite, continues to attract much attention and its use is described below.

9.1 **Hydroxylapatite**

Calcium phosphate gels have been used in protein purification for some time (8). However, it was not until the development of crystalline calcium phosphate or hydroxylapatite by Tiselius and co-workers (78,79) that the flow characteristics of this adsorbent were improved sufficiently to allow its successful use in column chromatography. Nevertheless, the widespread use of hydroxylapatite (HA) has in the past been limited, due in part to predictability, the availability of alternative adsorbents with superior chromatographic properties and the lengthy laboratory preparation necessary for its use (80).

Crystalline hydroxylapatite $[Ca_{10}(PO_4)_6(OH)_2]$ is prepared by slowly mixing calcium chloride and sodium phosphate to form a precipitate of brushite ($Ca_2 HPO_4 \cdot 2H_2O$). The brushite is then boiled with sodium hydroxide or ammonia to convert it to hydroxylapatite. The mechanism of protein adsorption onto HA is thought to involve both Ca^{2+} and PO_4^{3-} groups on the crystal surface (81,82). Since these charged groups are closely arranged on the crystal it is likely that dipole–dipole interactions exist between adsorbent and protein although purely electrostatic interactions cannot be ruled out. Bernadi (80,82) has suggested that acidic and neutral proteins bind to the hydroxylapatite calcium while basic proteins adsorb to surface phosphate groups. Exceptions to this rule include the phosphate binding capability of certain acidic proteins such as glycolytic and pentose phosphate pathway enzymes and aminoacyl tRNA synthetases.

Commercially available hydroxylapatites (83,84,85) and calcium phosphate-coated gels are now available from several manufacturers (*Table 20*), eliminating the need for laboratory preparation of HA. It should be noted, however, that commercial HA may be inferior to fresh laboratory-prepared material.

Table 20. Characteristics of some commercially available hydroxylapatite preparations.

Commercial name	Manufacturer	Matrix	Typical capacity (mg ml^{-1})	Additional data
HA-Ultrogel	IBF	Agarose	10 (Cyt c) 2 (BSA)	Bead diameter 60 – 180 μm Exclusion limited 5×10^6 (globular proteins)
Biogel HT/HTP	Bio-Rad	Crystalline HA	10 (BSA)	Prepared by the conventional method
Macrosorb C	Sterling Organics	Fabricated Macroporous HA	7 (BSA)	Pore size 1000 nm Bead diameter 250 – 500 μm
Hydroxylapatite-spheroidal	BDH	Spheroidal HA	2.5 (BSA)	Has greater mechanical stability than crystalline HA
Bio-Gel HPHT	Bio-Rad	Crystalline HA	5 mg ml^{-1} for optimum resolution	High performance material

Since hydroxylapatite adsorbs proteins by a different mechanism to other separation techniques, it is a useful additional chromatographic procedure which can often be used to provide resolution of protein mixtures not achievable by alternative methods such as ion exchange and hydrophobic interaction chromatography.

9.1.1 *Adsorption conditions*

The adsorption of proteins using HA can be carried out using batch or column methods. For large volume samples the former should be preferred. Adsorption is typically carried out in a low concentration of sodium or phosphate buffer (<20 mM) at a neutral pH (80,86). For batch conditions an adsorption time of 30 min has been used with gentle agitation. It is essential to avoid the presence of substances with a stronger affinity for calcium than phosphate (eg. EDTA and citrate), since this will decrease the hydroxyl-apatite capacity (80). In column chromatography the hydroxylapatite may be pre-washed in 0.5 M phosphate buffer pH 6.8 to ensure the removal of any adsorbed contaminants (86). The HA is then washed in a low concentration of phosphate buffer (pH 6.5−7) prior to sample application. While the adsorption of acidic proteins is little affected by the presence of sodium, potassium or calcium chloride, these salts may reduce the capacity of HA for basic proteins (80). Lysozyme, a basic protein, will not adsorb onto HA in the presence of 2 M KCl (80).

Typical flow rates used in column chromatography are $10-25$ ml cm^{-2} h^{-1} with crystalline hydroxylapatite but the BDH spherical porous material may be used at up to 100 ml cm^{-2} h^{-1}.

The capacity of HA for proteins is generally at its highest close to neutrality. Atkinson *et al.* (86), showed that the capacity of laboratory-prepared material for bovine plasma albumin was relatively constant at about 40 mg ml^{-1} of HA between pH 6.5 and 7, but fell off rapidly above pH 7 (86). Commercially available hydroxylapatite preparations have quoted capacities of nearer 10 mg ml^{-1} for protein (84,85). Following adsorption, the hydroxylapatite is usually washed with $1-2$ volumes of a low concentration of phosphate buffer (~ 20 mM).

9.1.2 *Elution conditions*

Protein elution from hydroxylapatite is usually achieved with a stepwise or continuous gradient of increasing phosphate concentration up to 500 mM (80,86). Potassium phosphate is preferred to sodium phosphate at higher concentration and at 4°C due to its superior solubility.

Basic proteins (e.g. lysozyme) absorb strongly to HA and are usually eluted at a phosphate concentration above 120 mM, pH 6.8. Acidic and neutral proteins may be eluted at lower concentrations ($30-120$ mM). Very basic proteins (e.g. lysine-rich histones) may require much stronger eluting conditions of up to 500 mM phosphate. An example of a procedure (87) for the fractionation of proteins using HA is given in *Method Table 15*. The stepwise elution of proteins using an increasing concentration of phosphate may lead to severe peak tailing when used in column chromatography (80) and is best used under batch conditions.

The pH of the eluent may have a significant effect on the concentration of phosphate required to desorb a protein. Atkinson and coworkers (86), found that although

Method Table 15. Procedure for the fractionation of proteins using hydroxylapatite.

1. Suspend the hydroxylapatite in 20 mM potassium phosphate buffer pH 6.8. 1 g of dry material should be used per 10 mg of protein.
2. Pour the hydroxylapatite slurry into the column (typical dimensions 2.5 × 25 cm) and wash the column through with five bed volumes of 20 mM potassium phosphate, pH 6.8, using a flow rate of ~20 ml cm^{-2} h^{-1}.
3. Apply the protein sample, preferably in the starting buffer.
4. Wash the bed with two bed volumes of starting buffer at 20 ml cm^{-2} h^{-1}.
5. Elute the protein with a linear gradient of an increasing phosphate concentration, e.g. 0.02−0.5 M potassium phosphate, pH 6.8. Up to 10 bed volumes should normally be used.
6. Remove tightly bound protein by continuing to elute with 0.5 M buffer at pH 6.8.

NB Purifications carried out in the cold should not use phosphate buffers over 0.5 M in concentration.

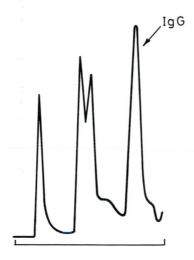

Column :	Bio−Gel HPHT system
Flow rate :	1 ml/min
Gradient :	0.01M Sodium phosphate 0.30 mM Calcium chloride pH7 to 0.01 mM Calcium chloride pH7
Sample :	0.5ml Ascites fluid.

Figure 37. HPLC purification of monoclonal antibodies using hydroxylapatite. Reproduced with permission from Bio-Rad.

tryptophan−tRNA synthetase, an acidic protein, requires a surprisingly high phosphate concentration (400 mM) to promote elution at pH 6.8, a concentration of less than 250 mM is adequate at pH 7.5. Optimum resolution is obtained using hydroxylapatite in columns with a length: diameter ratio of between 5:1 and 15:1. Hydroxylapatite has a tendency to adsorb carbon dioxide to form a hard crust as a layer on the column top. This may cause an increase in back pressure and reduce the flow rate. The top layer (~ 1 cm) of the column should therefore occasionally be removed. Bio-Rad also recommend that all water used in buffers should be boiled to remove carbon dioxide.

9.1.3 *Regeneration and storage*

Following elution the column should be washed thoroughly with 0.5 M phosphate buffer, pH 6.8 and then re-equilibrated in the starting buffer. Hydroxylapatite should be stored in a low concentration of phosphate buffer containing 0.03% (v/v) toluene or 0.03% (w/v) sodium azide.

9.2 **High performance hydroxylapatite**

Bio-Rad also market a hydroxylapatite material pre-packed in a stainless steel column for use in HPLC. The analytical scale column (4.8 ml bed volume) will purify 20 mg of protein per separation in less than 2 h using a flow rate of $0.5-1$ ml min^{-1} (83). Purification of larger amounts of protein (up to 100 mg per injection) necessitates the use of preparative-scale HPLC (5×2.5 cm column). A chromatogram for the purification of monoclonal antibodies using high performance hydroxylapatite is shown in *Figure 37*. The conditions for protein separation, using an increasing concentration of phosphate for elution, are similar to those used in low pressure chromatography.

10. REFERENCES

1. Cooney,C.L. (1985) *Comprehensive Biotechnology*. Moo-Young,M. (ed.), Pergamon Press, Oxford, Vol. 2, p. 233.
2. Rausch,C.W., Heckendorf,A.H. (1985) *Comprehensive Biotechnology*. Moo-Young,M. (ed.), Pergamon Press, Oxford, Vol. 2, p. 537.
3. Osterman,L.A. (1986) *Methods of Protein and Nucleic Acid Research, Part 3. Chromatography*. Springer-Verlag, New York.
4. Morris,C.J. and Morris,P. (1976) *Separation Methods in Biochemistry*. Pitman, London.
5. Deyl,Z. (ed.) (1984) *Separation Methods. New Comprehensive Biochemistry*. Elsevier, Amsterdam, Vol. 8.
6. Yang,C. and Tsaro,G.T. (1982) *Chromatography Advances in Biochemical Engineering*. Springer-Verlag, New York, Vol. 25.
7. Ruthven,D.M. (1984) *Principles of Adsorption and Adsorption Processes*. J. Wiley and Sons, Chichester.
8. Scopes,R. (1982) *Protein Purification − Principles and Practice*. Springer-Verlag, New York, Heidelberg, Berlin.
9. *Ion Exchange Chromatography—Principles and Methods*. Pharmacia-LKB.
10. Scawen,M. (1985) *Enzymes, Org. Synth; III Symp.*, **40**.
11. Chase,H.A. (1984) In *Ion Exchange Technology*. Nader,N. and Streat,M. (eds), Ellis Horwood Ltd, p. 400.
12. Asenjo,J.A. and Patrick,I. (1989) In *Protein Purification Applications: A Practical Approach*. Harris,E.L.V. and Angal,S. (eds), IRL Press, Oxford.
13. Osterman,L.A. (1984) *Methods of Protein and Nucleic Acid Research*. Springer-Verlag, New York, Vol. 1.
14. Sterling Organics, Manufacturers Literature.
15. Bernardi,G. (1971) In *Methods in Enzymology*, Jakoby,W.B. (ed.), Academic Press, New York, Vol. 22, p.325.
16. Haller,W. (1965) *Nature*, **206**, 693.

17. Rohm & Haas Technical Information.
18. Amicon. *A Practical Guide to Industrial Scale Protein Purification.*
19. Pharmacia Fine Chemicals A.B. (1984). *Scale-up to Process Chromatography—A Systematic Guide to the Design and Scale-up of Protein Purification for the Biotechnology Industry.*
20. Voser,W. and Walliser,H.P. (1985) *Discovery and Isolation of Microbial Products.* Verrall,M.S., (ed.), Ellis Horwood Ltd, Vol. 116.
21. Pharmacia Fine Chemicals. *The Pharmacia FPLC System.*
22. Vermeulen,T., Le Van,M.D., Heister,N.K. and Klein,G. (1984) *Perrys Chemical Engineers Handbook.* 6th edition. Perry,R.H., Green,D.W. and Maloney,J.O. (eds), McGraw Hill Book Company, Section 16, 16.1−16.48.
23. Janson,J.C. and Low,D. (1984) *World Biotech Report Vol. 1,* 305.
24. Pharmacia Fine Chemicals. *Sephadex based Ion Exchange Media.*
25. Pharmacia Fine Chemicals. *Sepharose based Ion Exchange media.*
26. Malamud,D. and Drysdale,J.V. (1978) *Anal. Biochem.,* **86**, 620.
27. Righetti,P.G. and Caravaggio,T. (1976) *J. Chromatogr.,* **127**, 1.
28. Righetti,P.G., Tudor,G. and Ek,K. (1981) *J. Chromatogr.,* **220**, 115.
29. Sluyterman,L.A. and Elgersma,O. (1978) *J. Chromatogr.,* **150**, 17.
30. Sluyterman,L.A. and Wijdenes,J. (1981) *J. Chromatogr.,* **206**, 441.
31. Sluyterman,L.A. and Wijdenes,J. (1981) *J. Chromatogr.,* **206**, 429.
32. Sluyterman,L.A. and Wijdenes,J. (1978) *J. Chromatogr.,* **150**, 31.
33. Pharmacia Fine Chemicals. *Chromatofocusing with Polybuffer and PBE.*
34. Burness,A.T. and Pardoe,I.E. (1983) *J. Chromatogr.,* **259**, 423.
35. Alexander,N.M. and Neeley,W.E. (1982) *J. Chromatogr.,* **230**, 137.
36. Bedi,G.S. and Back,N. (1984) *Prep. Biochem.,* **14**, 257.
37. Kovaco,K.L., Tigyi,G. and Alfonz,H. (1985) *Prep. Biochem.,* **15**, 321.
38. Gal, Toth-Martinez,B.L. (1983) *J. Chromatogr.,* **264**, 170.
39. Wette,W., Hudig,H., Wacker,T. and Kreutz,W. (1983) *J. Chromatogr.,* **259**, 341.
40. Addison,J., Lewis,W.G. and Harrison,P.M. (1984) *J. Chromatogr.,* **292**, 454.
41. Campeau,J.D., Marrs,R.P. and Di Zerega,G.S. (1983) *J. Chromatogr.,* **262**, 334.
42. O'Carra,P., Barry,S. and Griffin,T. (1974) *FEBS Lett.,* **43**, 169.
43. Er-El,Z., Zaindenzaig,Y. and Shaltiel,E. (1972) *Biochem. Biophys. Res. Commun.,* **49**, 383.
44. Yon,R.J. (1972) *Biochem. J.,* **126**, 765.
45. Hofstee,B.H.J. (1973) *Anal. Biochem.,* **52**, 430.
46. Hjerten,S. (1973) *J. Chromatogr.,* **87**, 325.
47. Axen,R., Porath,J. and Ernback,S. (1967) *Nature,* **214**, 1302.
48. Porath,J., Sundberg,L., Fornstedt,N. and Olsson,I. (1973) *Nature,* **245**, 465.
49. Hofstee,B.H.J. (1975) *Biochem. Biophys. Res. Commun.,* **63**, 618.
50. Hjerten,S. (1976) *Methods of Protein Separation.* Catsimpoolis,N. (cd.), Plenum, Vol. 2, p. 233.
51. Ochoa,J.L. (1978) *Biochemie,* **60**, 1.
52. Hjerten,S. (1976) *Protides. Biol. Fluids,* **23**, 667.
53. Hofmeister,F. (1888) *Arch. Exp. Pathol. Pharmakol.,* **24**, 247.
54. Pharmacia Fine Chemicals. *Media for Hydrophobic Interaction Chromatography.*
55. Dembinski,W.E. and Sulkowski,E. (1986) *Prep. Biochem.,* **16**, 175.
56. Krakauer,T. (1984−85) *Prep. Biochem.,* **14**, 449.
57. Pharmacia Fine Chemicals. *Octyl Phenyl Sepharose for Hydrophobic Interaction Chromatography.*
58. Rosen,S. (1978) *Biochem-Biophys. Acta,* **523**, 314.
59. Porath,J., Carlsson,J., Olsson,I. and Belfrage,G. (1975) *Nature,* **258**, 598.
60. Dakanov,V.A. and Semechkin,A.V. (1977) *J. Chromatogr.,* **141**, 313.
61. Hubert,P. and Porath,J. (1980) *J. Chromatogr.,* **198**, 247.
62. Moroux,Y. and Boschetti,E. (1983) In *Affinity Chromatography and Biological Recognition.* Proc. 5th Int. Symp., Chaiken,I.M. *et al.* (eds), p. 267.
63. Small,D., Atkinson,T., Lowe,C.R. (1983) In *Affinity Chromatography and Biological Recognition.* Proc. 5th Symp., Chaiken,I.M. *et al.* (eds), p. 267.
64. Salier,J.P., Martin,J.P., Lambin,P., McPhee,H. and Hochstrasser,K. (1980) *Anal. Biochem.,* **109**, 273.
65. Sulkowski,E., Vastola,K., Oleszek,D. and Von Muerichhausen,W. (1982) *Affinity Chromatography and Related Techniques.* Elsevier, Holland, p. 313.
66. Sinosich,M.J., Davey,M.W., Teisner,B. and Grudzinskas,J.G. (1983) *Biochem. Int.,* **7**, 33.
67. Harrison,H. and Kagedal,L. (1981) *J. Chromatogr.,* **333**,
68. Bollin,E. and Sulkowski,E. (1978) *Archiv. Virol.,* **58**, 149.
69. Pharmacia Fine Chemicals. *Chelating Sepharose 6B.*
70. Edljam,L. and Jellum,E. (1963) *Acta. Chem. Scan.,* **17**, 2610.

71. Brocklehurst,K., Carlsson,J. and Kierstan,M.P.J. (1985) *Topics in Enzyme and Fermentation Biotechnology*.
72. Brocklehurst,K., Carlsson,J., Kierstan,M.P.J. and Crook,E.M. (1973) *Biochem. J.*, **133**, 573.
73. Brocklehurst,K., Carlsson,J., Kierstan,M.P.J. and Crook,E.M. (1974) In *Methods in Enzymology*. Jakoby,W.B. and Wilchek,M. (eds), Academic Press, New York, vol. 34B, p. 531.
74. Pharmacia Fine Chemicals. *Media for Covalent Chromatography*.
75. Separation News (1987) Pharmacia Fine Chemicals, Vol. 13.6.
76. Carlsson,J. and Svenson,A. (1974) *FEBS Lett.*, **42**, 1183.
77. Hillson,D.A. (1981) *J. Biochem. Biophys. Methods.*, **4**, 101.
78. Tiselius,A., Hjerten,S. and Levin,O. (1956) *Arch. Biochem. Biophys.*, **65**, 132.
79. Hjerten,S. (1956) *Biochem. Biophys. Acta.*, **31**, 216.
80. Bernardi,G. (1971) In *Methods in Enzymology*, Jakoby,W.B. (ed.), Academic Press, New York, vol.22, p. 325.
81. Bernardi,G. and Kawasaki,T. (1968) *Biochem. Biophys. Acta.*, **160.**, 301.
82. Bernardi,G., Giro,M-G. and Gaillard,C. (1972) *Biochem. Biophys. Acta.*, **278**, 409.
83. Bio-Rad Laboratories Ltd Catalogue 1987.
84. HA Ultrogel, *Hydroxylapatite-agarose gel for adsorption chromatography*, IBF.
85. Sterling Organics Ltd. Macrosorb—*Adsorbents for industrial scale biochemical separations*.
86. Atkinson,A., Bradford,P. and Selmes,I.P. (1973) *J. Appl. Chem. Biotechnol.*, **23**, 517.
87. Kerry,J.A. and Kuok,F. (1986) *Prep. Biochem.*, **16**, 199.

CHAPTER 5

Purification by exploitation of activity

1. INTRODUCTION—S.Angal and P.D.G.Dean

Proteins carry out their biological functions through one or more binding activities and consequently contain binding sites for interaction with other biomolecules, called ligands. Ligands may be small molecules such as substrates for enzymes or larger molecules such as peptide hormones. The interaction of a binding site with a ligand is determined by the overall size and shape of the ligand as well as the number and distribution of complementary surfaces. These complementary surfaces may involve a combination of charged and hydrophobic moieties and exhibit other short range molecular interactions such as hydrogen bonds. This binding activity of a protein, which is stereoselective and often of a high affinity, can be exploited for the purification of the protein in a technique commonly known as affinity chromatography.

The operation of affinity chromatography involves the following steps.

(i)   Choice of an appropriate ligand.
(ii)   Immobilization of the ligand onto a support matrix.
(iii)   Contacting the protein mixture of interest with the matrix.
(iv)   Removal of non-specifically bound proteins.
(v)   Elution of the protein of interest in a purified form.

At best, affinity chromatography is the most powerful technique for protein purification since its high selectivity can, in principle, allow purification of a single protein of low abundance from a crude mixture of proteins at higher concentrations. Secondly, if the affinity of the ligand for the protein is sufficiently high, the technique offers simultaneous concentration from a large volume. In practice, such single-step purifications are not common and successful affinity chromatography requires careful consideration of a number of parameters involved. The remainder of this chapter attempts to guide the experimenter in the selection and use of affinity adsorbents for protein purification. For more extensive information on this technique the reader is advised to consult the many excellent texts on this subject (1−7) as well as proceedings of symposia (8−11).

2. DESIGN AND PREPARATION OF AFFINITY ADSORBENTS—S.Angal and P.D.G.Dean

The construction of an affinity adsorbent for the purification of a particular protein involves three major factors.

(i)   Choice of a suitable ligand.
(ii)   Selection of a support matrix and spacer.
(iii)   Attachment of the ligand to a support matrix.

The criteria for making these decisions are discussed in the following sections.

Table 1. Examples of ligands suitable for the purification of proteins by affinity chromatography.

Ligands[a]	For purification of:
Protein A/protein G	Immunoglobulins from various species
Monoclonal antibodies	Antigens, protein A fusions
Antigens	Specific monoclonal/polyclonal antibodies
Steroids	Steroid receptors, steroid binding proteins
Fatty acids	Fatty acid-binding proteins, albumin
Nucleotides	Nucleic acid-binding proteins
	Nucleotide-requiring enzymes
Protease inhibitors	Proteases
Lysine/arginine	Plasmin, plasminogen activators
Polymixin B	Bacterial endotoxin C, removal of endotoxin from protein
Phenylboronate	Glycated proteins, glycoproteins
Lectins	Glycoproteins
Sugars	Lectins, glycosidases
Phosphoric acid	Phosphatases
Biotin	Biotin-binding proteins, avidin, streptavidin
Avidin	Biotin-containing enzymes
Heparin	Coagulation factors, lipases, connective tissue proteases, DNA polymerases
Gelatin	Fibronectin
Calmodulin	Calmodulin-binding enzymes
Triazine dyes	Dehydrogenases, kinases, polymerases, interferons, restriction enzymes

[a] All the above ligands are available in the immobilized form from a variety of suppliers.

2.1 **Choice of ligand**

Suitable pairs of protein and ligand combinations for affinity chromatography are antigen−antibody, hormone−receptor, glycoprotein−lectin or enzyme−substrate/cofactor/effector. In practice, the natural biological ligand may be very expensive or difficult to obtain and analogues or mimics (pseudo-ligands) are preferentially employed. The factors to consider when selecting a ligand for protein purification are as follows.

(i) *Specificity.* Ideally, the ligand should recognize only the protein to be purified (e.g. monospecific antibody). The choice may be evident from the known biological properties of the protein to be purified. If this is not possible a ligand known to recognize a group of proteins should be selected. Examples of ligands are given in *Table 1*. Group-specific ligands, by definition, allow the purification of related proteins or protein families without the need to invest a large amount of time and labour in the preparation of monospecific adsorbents. Their wide application also makes them popular with manufacturers of affinity adsorbents and many are therefore available in ready-to-use format. In combination with other increasingly high resolution chromatographic techniques they offer the advantage of speed over tailor-made monospecific adsorbents.

(ii) *Reversibility.* The ligand should form a reversible complex with the protein to be purified such that the complex is resistant to the composition of the feedstream and washing buffers but is easily dissociable without requiring denaturing conditions for elution.

(iii) *Stability*. The ligand should be stable to the conditions to be used for immobilization (e.g. organic solvents in some cases) as well as the conditions of use. This includes resistance to proteolysis and to denaturation by eluents or cleaning agents.

(iv) *Size*. The ligand should be large enough such that it contains several groups able to interact with the protein resulting in sufficient stereoselectivity and affinity. In addition, the ligand should contain a functional group which can be used for immobilization without significantly affecting the protein binding characteristics. If the ligand is so small that the matrix backbone would interfere with access to protein molecules it is preferable to interpose a spacer molecule between the ligand and matrix. On the other hand, a ligand which is very large is likely to be more susceptible to denaturation or degradation and can cause increased non-specific binding through other parts of the molecule. The size of the ligand determines the amount which can be coupled to a matrix and its efficiency. This matter is discussed in more detail in Sections 3.3 and 3.4.

(v) *Affinity*. The interaction of a protein P and a ligand L can be described by the equation:

$$P + L \leftrightharpoons PL$$

where the equilibrium dissociation constant, K_L, is defined by:

$$\frac{[P]\ [L]}{[PL]}$$

A simple rearrangement gives:

$$\frac{[L]}{K_L} = \frac{[PL]}{[P]}$$

which implies that for substantial adsorption of the protein from solution (e.g. $[PL]{:}[P] = 95{:}5$) the value of K_L must be about two orders of magnitude less than the concentration of immobilized ligand.

In practice, for a protein ligand of M_r 10 000 immobilized at 10 mg ml^{-1} (i.e. 1 mM) a value for K_L less than 10^{-5} M would be needed for substantial adsorption of the desired protein from similar volumes of matrix and solution (operation in batch mode). For concentration from a 10-fold volume the K_L should be less than 10^{-6} M. Smaller ligands can be immobilized at $10-100$-fold higher concentrations and would, by analogy, be able to cope with a higher K_L. For higher ligand affinities ($K_L < 10^{-8}$ M) dissociation of the ligand$-$protein may prove difficult. As a general rule affinity techniques operate well between $K_L = 10^{-4}$ M and 10^{-8} M. Batch techniques and low ligand concentrations may be successful at the lower K_L whereas columns and higher ligand concentrations will be required for $K_L = 10^{-4}$ M. These numbers can only be used as guidelines since the K_L values determined for free ligand may be different to those for the immobilized ligand and the proportion of immobilized ligand which is functional varies with ligand size as will be discussed later (Section 3.4).

2.2 Selection of the matrix

A chosen affinity ligand is immobilized to a solid support or matrix via one or more covalent bonds. The effectiveness of the immobilized ligand in purifications can be markedly dependent on the structure of the matrix (12). It is therefore important to consider some criteria for the selection of matrices.

(i)   The matrix should have a high degree of porosity so that large proteins can have unhindered access to ligand immobilized on the interior portions of the lattice.
(ii) The matrix should be chemically stable under the conditions used for activation and coupling as well as those used during operation and regeneration.
(iii) The matrix should be physically rigid in order to allow good flow properties and able to withstand some mechanical agitation without breaking up.
(iv) The matrix should withstand a reasonable range of pH and temperature.
(v)   The matrix should be easily activatable for coupling of ligands at high density yet be otherwise inert to non-specific binding of proteins.
(vi) Matrix properties should not be substantially altered on functionalization.
(vii) The matrix should be uniform in structure, particularly when functionalized so that ligand molecules can be homogeneously distributed.

For any particular application the importance of the above criteria may differ; for example, incompressibility would be more important at higher pressures or diffusibility more important for purification of very large proteins.

The properties of the commonly available matrices have been described in Chapter 4. Comparative studies of matrices (13) generally with model proteins, have indicated that agarose is one of the better matrix materials. The novice is therefore advised to use agarose or cross-linked agarose unless otherwise indicated by the nature of his or her application. Systematic comparisons of matrices may be carried out at a later stage,

Table 2. Examples of commercially available matrices with spacer arms.

Supplier	Spacer arm description	Support matrix
Bio-Rad	3-aminopropyl and succinylated aminopropyl	agarose
	3,3'-diaminopropyl and succinylated diaminopropyl	agarose
IBF	1,6-diaminohexane	polyacrylamide−agarose
	6-aminohexanoic acid	polyacrylamide−agarose
ICN	1,2-diaminoethane	agarose
	1-diaminohexane	agarose
	3,3'-diaminodipropylamine	agarose
Merck	aminopropyl	controlled-pore glass
	aminohexyl	controlled-pore glass
Miles	aminoalkyl (2,4,6,8,10)	agarose
Pharmacia	1,6-diaminohexane	agarose
	6-aminohexanoic acid	agarose
Pierce	3,3'-diaminopropylamine	agarose
Serva	1,2-diaminoethane	agarose
	3,3'-diaminopropylamine	agarose
	p-aminobenzoyl-3,3'-diamino propylamine	agarose

Some of the activated matrices in *Table 3* also have spacer arms by virtue of the activation chemistry used. Ligands may be coupled using the cross-linking agents described in the text (Section 2) or using the succinic anhydride method (1).

Table 3. Examples of commercially available pre-activated matrices.

Manufacturer/Supplier	Activation method	Support matrix
Pharmacia	CNBr	Sepharose (agarose/cross-linked agarose)
Pharmacia	Sulphonyl chloride	Sepharase (agarose/cross-linked agarose)
IBF	Glutaraldehyde	Ultrogel (polyacrylamide−agarose)
IBF	Glutaraldehyde	Magnogel (magnetized Ultrogel)
Rhom Pharma	Epoxy	Eupergit C (acrylate/acrylamide)
Anachem/Tessek	Epoxy	Separon HEMA (hydroxyethyl methacrylate)
Anachem/Tessek	Vinyl sulphone	Separon HEMA (hydroxyethyl methacrylate)
BioProbe International	FMP	Avid gel (cross-linked agarose)
Pierce	CDI	Reacti-Gel (cross-linked agarose)
Pierce	CDI	Reacti-Gel HW-65F (TSK gel)
Pierce	CDI	Reacti-Gel GF-2000 (Trisacryl GF2000) (Synthetic support from IBF)
Pierce	Tresyl	Selectispher 10 (Silica)
Bio-Rad	N-hydroxysuccinimide	Affi-Gel 10 (cross-linked agarose)
Bio-Rad	N-hydroxysuccinimide	Affi-Gel 15 (cross-linked agarose)
Bio-Rad	N-hydroxysuccinimide	Affi-Prep 10 (synthetic polymer)
Dominick Hunter	Epoxy	Memsep (cellulose, cartridge)
Sterling Organics	Glutaraldehyde	Macrosorb K (Kieselguhr)
Sterogene	Aldehyde	Actigel A (cross-linked agarose)
DuPont	Undisclosed	Perflex (perfluorocarbon)

preferably using similar ligand densities, in order to improve a process or circumvent a problem such as leakage.

A large number of novel matrices have become available in the last five years, which may replace agarose (14). For examples refer to *Tables 2* and *3*. The interested reader should obtain further information from the manufacturers. An example of the way in which a new matrix should be evaluated is given in Section 4.

2.3 Choice of spacer

A spacer molecule may be employed to distance the ligand from the matrix backbone in applications where the small size of the ligand excludes it from free access to protein molecules in the solvent (4,5,7). A spacer molecule may also be effective if the ligand is immobilized through a site near enough to the protein-binding surface to interfere with protein binding.

The length of the spacer arm is crucial and must be determined empirically. The number of methylene groups most often successful is 6−8. The spacer should be hydrophilic and not itself bind proteins, either because of its hydrophobicity or its charged groups.

Some activation chemistries will automatically insert a spacer molecule between the matrix and the ligand (for example, bis-oxirane as illustrated in Section 4). Matrices with spacer arms already attached are commercially available (*Table 2*).

2.4 Activation and coupling chemistry

The detailed methodology of activation of support matrices is beyond the scope of this chapter and the reader is referred to a previous text in this series (1). A brief description of the major methods is given below to allow the reader to grasp the salient points of each method before selecting pre-activated matrices which are available commercially. (*Table 3*).

2.4.1 *Cyanogen bromide*

Cyanogen bromide (CNBr) reacts with hydroxyls in agarose and other polysaccharide matrices to produce a reactive support which can subsequently be derivatized with spacer molecules or ligands containing primary amines (15). The *N*-substituted isourea formed on reaction with the primary amine is positively charged at physiological pH (pK_a ~9.5) thus imparting anion exchange properties to the adsorbent. The isourea derivative is susceptible to nucleophilic attack (e.g. amine-containing buffers, proteins) and slow hydrolysis can occur at extremes of pH resulting in leakage of ligand into the medium. Cyanogen bromide is extremely toxic (releases HCN on acidification) so commercially activated agarose is recommended. Even with these limitations, CNBr activation is one of the most widely used processes in the preparation of affinity adsorbents.

Unprotonated primary amines (except Tris) couple efficiently to CNBr-activated agarose, therefore coupling pH should be greater than the pK_a of the ligand but less than 10 (e.g. pH 7−8 for aromatic amines, pH 10 for aliphatic amines). Borate or carbonate buffers are recommended. The coupling efficiency decreases at pH values above 9.5−10.0 which reflects the sharp decline in stability of the activated complex. CNBr-activated agarose is not stable at elevated temperatures and the coupling should be carried out at +4°C. For details of the coupling procedure see Section 6.

2.4.2 *Bis-oxirane*

Bis-oxiranes (bis-epoxides) react readily with both hydroxyl or amino-containing gels at alkaline pH to yield derivatives which possess a long chain hydrophilic reactive oxirane. These in turn may be reacted with nucleophiles to prepare affinity adsorbents with lower numbers of hydrophilic and ionic groups than those obtained with CNBr activation. The procedure automatically introduces a long chain hydrophilic spacer arm which may be desirable in certain applications. Coupling reactivity is in the order SH > NH > OH and the efficiency is enhanced by raising the temperature to 40°C. Oxirane-coupled ligands are extremely stable. Details of coupling procedures will be found in Section 4.

2.4.3 *Carbonyldiimidazole*

Polysaccharide matrices may be activated using carbonylating agents such as *N,N'* carbonyldiimidazole (CDI) under anhydrous conditions to give reactive imidazole carbonate derivatives (16). These in turn will react with ligands containing primary amino groups at alkaline pH to give stable carbonate derivatives.

CDI is non-toxic, gives high levels of activation and does not introduce ion-exchange groups into the matrix. CDI-activated agarose (available from Pierce) is stable in anhydrous conditions (but not in aqueous solution) and supplied in acetone. Before coupling the acetone should be removed by quickly washing the gel on a sintered glass funnel with ice-cold water (not necessary if the ligand is insensitive to acetone). Coupling is recommended in 0.1 M borate buffer pH 8.5−10 at room temperature.

2.4.4 *Sulphonyl chloride*

Organic sulphonyl halides react with matrix hydroxyl groups to form sulphonyl esters which are themselves excellent leaving groups (17). Nucleophiles (e.g. amino and thiol groups of proteins) will readily displace the sulphonate enabling efficient and rapid coupling. Activation using *p*-toluene sulphonyl (tosyl) or trifluoroethyl sulphonyl (tresyl) chloride is carried out in organic solvents and the resulting matrix is stable (in 1 mM HCl or in the dry state) for extended periods. The reactivity of the sulphone ester is strongly influenced by the substituent attached to the sulphonate. Thus tresyl displacement can occur at neutral pH and 4°C while tosyl-activated matrices need to be coupled in alkaline pH (9−10.5, bicarbonate buffers) at 40°C. An advantage of the tosyl-activated matrix is that tosyl displacement can be followed spectrophotometrically.

2.4.5 *Periodate*

Vicinal diol groups of polysaccharide matrices may be oxidized by the use of sodium *m*-periodate ($NaIO_4$) to generate aldehyde functions. $NaIO_4$ is very soluble in water and can therefore be easily removed from the activated gel by washing. The activated gel is stable at 4°C for several days. The aldehyde groups react with primary amines at pH 4−6 to form a Schiff's base which can be stabilized by reduction with sodium borohydride (or cyanoborohydride).

Details of activation and coupling follow.

(i)   Mix the gel with an equal volume of 0.2 M $NaIO_4$ and place in a tightly closed polythene bottle.

(ii)   Allow to mix gently for about 2 h at room temperature.

(iii)   Wash the gel thoroughly on a sintered glass funnel with distilled water and then with 0.5 M phosphate buffer pH 6.0.

(iv)   Add the gel to an equal volume of 0.5 M phosphate buffer pH 6.0 containing about 25 mM of the desired amine ligand and 0.5 mM sodium cyanoborohydride.

(v)   Allow to mix gently for 3 days at room temperature.

(vi)   Wash the gel exhaustively and reduce any remaining aldehydes by incubation with 1 M sodium borohydride for 15 h at 4°C.

(vii)   Wash the gel extensively in distilled water and then in storage buffer.

Periodate oxidation represents a convenient alternative to other methods, particularly where ligands are sensitive to alkaline pH. The reagents are non-toxic and the alkylamine product is stable.

2.4.6 *Fluoromethyl pyridinium sulphonate*

Facile activation of polysaccharide hydroxyls can be carried out using 2-fluoro-1-methylpyridinium toluene-4-sulphonate (FMP) in polar organic solvents (18). The reaction is carried out in the presence of a tertiary amine such as triethylamine and is completed in 10 min at room temperature. The resulting activated groups (2-alkoxy-1-methylpyridinium salts) can react readily (pH 8−9, aqueous or polar organic) with amino or sulphydryl-containing ligands. Both activation and coupling can be followed spectrophotometrically since 1-methyl-2-pyridine, the leaving group, absorbs strongly at 297 nm. FMP is non-toxic, inexpensive and claimed to form stable, non-ionic linkages (thioether or secondary amine) at high degrees of substitution. FMP-activation is the most recent addition to the methodology of affinity chromatography and appears to be worth evaluating. FMP-activated agarose is available from BioProbe International.

2.4.7 *Glutaraldehyde (19)*

Activation of amino or amide functions (e.g. polyacrylamide or agarose with amine spacers) can be carried out using 25% aqueous (fresh) solutions of glutaraldehyde (0.5 M phosphate buffer, pH 7.5, 20−40°C, overnight). The gel should be washed to remove the aldehyde and then coupling of primary amines carried out in the same buffer but at 4°C. This method is simple, inexpensive and suitable for ligands sensitive to alkaline pH. The linkage is stable and simultaneously introduces a spacer arm. Glutaraldehyde can polymerize on storage and is moderately toxic.

2.4.8 *Cross-linking techniques*

Nucleophilic ligands (e.g. containing amino groups) can be coupled to carboxylated matrices using carbodiimides which promote condensation to generate peptide bonds (20).

Dicyclohexyl carbodiimide is insoluble in water and must be used in organic solvents such as dioxane, dimethyl sulphoxide, 80% (v/v) aqueous pyridine or acetonitrile. A problem encountered in these condensations is the removal of insoluble ureas and other by-products which must be washed from the beads by washing in organic solvents.

Water-soluble carbodiimides are more convenient since their corresponding ureas are much more soluble. An example is described below.

(i)   Prepare carboxylated matrix (e.g. CH-Sepharose) by washing extensively in water and adjusting to pH 4.0−4.5. Do not use amino-, hydroxyl- or phosphate-containing buffers throughout the reaction.

(ii)   Freshly prepare a solution of 1-ethyl-3-(3-dimethyl aminopropyl)-carbodiimide (40 mg in 10 ml of buffer at pH 4.5). Add to 4 ml of matrix.

(iii)   Add an amino-containing ligand (10× concentration of spacer which is usually 10−20 μmol ml^{-1}) and mix by end-over-end rotation at room temperature for at least 1 h but up to 2 h.

(iv)   Terminate the reaction by filtering the matrix and washing out the unreacted carbodiimide and by-products.

2.4.9 *Triazine dyes*

Triazine dyes are 'reactive' dyes containing mono or dichlorotriazinyl groups which may be coupled directly to hydroxyl-containing matrices using the following procedure.

(i)   Wash the matrix (e.g. agarose) with water, weigh out the moist gel (20 g) and suspend it in 50 ml of water.

(ii)   Weigh 200 mg of dye (available from Sigma or ICI) and add 20 ml of water.

(iii)   Mix the matrix and dye solution thoroughly.

(iv)   Add 10 ml of NaCl (20% w/v) solution and incubate at room temperature for 30 min.

(v)   Add 20 ml of Na_2CO_3 (5% w/v) and incubate in a 45°C water bath. (Dichloro-triazines couple in 1 h whereas monochlorotriazines require 40 h).

(vi)   Wash the coupled matrices with warm water, 6 M urea and 1% (w/v) Na_2CO_3 until the washings are colourless. Blocking is not required. Store in 0.1% (w/v) Na_2CO_3 containing 0.02% (w/v) sodium azide.

Refer to Section 3.5 for information on the choice of dyes and their use.

2.4.10 *Blocking reactive groups*

Many commercial pre-activated supports have reactive group concentrations in the range $5-25$ μmol ml^{-1} while coupled ligand concentrations are in the range $1-10$ μmol ml^{-1}. Thus a number of residual activated groups will remain on gels after coupling. These groups may be blocked by the addition of low molecular weight compounds. The most common blocking agent is 1 M ethanolamine adjusted to pH 8.0 and incubated with the matrix at room temperature for 1 h. If other small molecules are used (e.g. glycine) their effect on the level of non-specific binding should be examined. At the end of the blocking period the adsorbents should be washed using the most stringent conditions recommended by the manufacturer before storage in the presence of preservatives (e.g. 0.02% sodium azide or methiolate). Other conditions of storage will depend on the ligand and matrix.

2.5 **Estimation of ligand concentration**

It is essential to determine the success of a ligand immobilization procedure at the time of coupling. This can be done using one of the following methods.

2.5.1 *Difference analysis*

Measure the amount of ligand added to the coupling mixture and that recovered after the washing procedures. The method used for measurement will depend on the nature of the ligand (e.g. spectrophotometry, immunoassay, activity assay).

2.5.2 *Direct measurements*

(i)   Suspend derivatized matrix in 50% (v/v) glycerol, mix thoroughly and measure absorbance against a similarly treated suspension of underivatized matrix. This method is suitable for ligands absorbing at wavelengths different from that at which the matrix absorbs.

(ii)   Alternatively, derivatized gels may be assayed for protein using the Lowry method (Chapter 1). Again, it is necessary to use the underivatized gel as a control.

(iii)   If the immobilized ligand contains a unique group (e.g. phosphate) direct elemental analysis is most appropriate.

(iv)   If the ligand contains a chemically reative group (e.g. thiol) appropriate measurements can be made after reaction.

2.5.3 *Radioactive ligand*

Radioactive ligand may be incorporated into the coupling reaction to give the most sensitive determination of total immobilized ligand.

2.5.4 *Hydrolysis*

This method is suitable only if the hydrolysed ligand can be measured in the presence of hydrolysis products of the matrix. Agarose can be solubilized by heating to 90°C in 50% (v/v) acetic acid or 0.5 M HCl for about 1 h. Alternatively, more vigorous hydrolysis (e.g. 6 M HCl at 110°C *in vacuo* for 24 h) will liberate amino acids from agarose derivatized with protein ligands. Cross-linked agaroses are more resistant to hydrolysis. Immobilized ligands may be hydrolysed using suitable enzymes (e.g. alkaline phosphatase for nucleotides and pronase for proteins). However, release of ligands reflects their accessibility to the digesting enzymes and the method is likely to underestimate ligand concentration.

3. USE OF AFFINITY ADSORBENTS—S.Angal and P.D.G.Dean

3.1 **Initial questions**

Having prepared or obtained an affinity absorbent the first question is whether it will bind and purify the protein of interest. This question can be answered after the following experiment.

(i)   Using 1 ml of the adsorbent prepare a suitable column (1−2 ml capacity plastic columns with 10 ml reservoirs are widely available).

(ii)   Equilibrate the column in a buffer chosen to encourage optimum binding of ligand and protein. Some guidance is provided in a later section, otherwise a probable set of conditions can be chosen for the known properties of the ligand and protein pair.

(iii)   Prepare the crude protein extract, preferably in or dialysed against the equilibration buffer. This should be as concentrated as possible.

(iv)   Apply a small aliquot of the extract to the column. It should contain about 10 times the amount detectable in a specific assay for the protein to be purified and should not exceed the probable capacity of the column. Assume that 1% of the immobilized ligand is likely to be functional where ligands are small and immobilized at high concentrations. For protein ligands the capacity may be 10% of the immobilized ligand.

(v)   Wash the column with 10 ml of equilibration buffer and collect 1 ml fractions.

(vi)   Elute bound proteins with 5 ml of equilibration buffer containing an eluent. If a possible eluent is not known try 1 M NaCl (5 ml) followed by 0.1 M glycine−HCl, pH 2.5.

(vii)   Analyse all collected fractions for total protein content and for the protein of interest. Calculate % purity in each fraction. Four possible results may be obtained.

     (1)   The protein of interest is found in fractions 1−3 of the wash together with most of the contaminating protein. It therefore has no affinity for the adsorbent and the choice of the adsorbent is in question.

     (2)   The protein binds and is eluted by one of the eluents with improved purity. This is the ideal situation and a larger scale purification could follow.

     (3)   The protein of interest is found in fractions 4−10 (i.e. it is retarded by the adsorbent) with improved purity.

(4)   The protein of interest is not found in any of the fractions implying
       that it is still bound to the column or destroyed during the course
       of the experiment.

In case of situation (3) or (4) it is necessary to manipulate the operating conditions
as described in the next section.

3.2 Selection of conditions for operation

These studies should also be carried out on the small scale described above.

3.2.1 *Choice of adsorption conditions*

The buffer chosen to effect adsorption should reflect the conditions required to achieve
a strong complex of the ligand with the protein to be purified. As indicated in Section
1 the binding interaction may be mediated by a variety of molecular forces. If the
interaction is thought to be predominantly hydrophobic an increase in the ionic strength
and/or pH will improve adsorption. An example of such an improvement is the binding
of protein A to murine immunoglobulins. Other interactions may be reinforced by the
addition of divalent metal ions or specific factors able to preserve a particular protein
conformation. Some knowledge of the ligand—protein interaction may suggest other
ways of improvement. (For example, lower temperatures weaken hydrophobic
interaction). Some trial and error experiments are required if nothing is known about
the interaction.

A second method for improving the affinity of the adsorbent for a protein is to increase
the concentration of the immobilized ligand by increasing the amount and concentration
of ligand used at the coupling stage. It may be necessary to start with a matrix containing
a higher level of activated groups.

Stronger binding can sometimes be promoted by allowing longer incubation times
(21). This effect may be due to secondary binding interactions. Consider operation in
batch mode.

Further improvements can be made on scaling up; for example, longer column length,
slower flow rate and lower sample volumes will all allow for increased retardation.

3.2.2 *Choice of washing conditions*

If the ligand—protein complex formed is of high affinity it may be stable to washing
conditions which could desorb non-specifically bound proteins. Washing buffers should
be intermediate between the best adsorption and elution conditions. Thus if a protein
binds in low molarity phosphate buffer and elutes in 0.75 M NaCl, it would be reasonable
to try washing the column in 0.4 M NaCl to discover if any improvement can be made
to the purity.

3.2.3 *Choice of elution conditions*

A protein molecule adsorbed to an affinity column is in equilibrium with the surrounding
immobilized ligand (i.e. it is constantly undergoing desorption and adsorption events).
If both the ligand concentration and the affinity are high it will be 'captured' by a ligand
molecule at a short distance from its original position and show little, if any, movement

through the adsorbent bed. Elution is effected by changing the environment of the ligand−protein complex such that the affinity of the ligand for the protein is lowered.

Most ligand−protein interactions are composed of a combination of molecular forces (e.g. ionic, hydrophobic) and any solvent condition which sufficiently alters these will destabilize the ligand−protein complex. The destabilization should not, however, irreversibly denature the protein (if active protein is required) or the ligand (if column re-use is required). Often the nature of the binding interaction is not defined and elution conditions are found empirically. Continuous gradient elutions for up to 20 column volumes should be employed and compared for recovery and purity.

(i) *Change of ionic strength.* An increase in ionic strength (e.g. continuous gradient) is used to desorb proteins from ligands where ionic interactions predominate. Conversely, a lowering of ionic strength will be required to effect protein elution from adsorbents where hydrophobic interactions predominate. This is the most popular method of elution since it is inexpensive.

(ii) *Change of pH.* This, (generally downward) alters the degree of ionization of a charged group at the binding surface so that it can no longer form a salt bridge with the opposing ions thus reducing the strength of interaction. In principle, an increase in pH should be similarly effective but is not as common in practice. This method is also inexpensive but gradients of pH are less reproducible and stepwise elution is more convenient.

(iii) *Selective elution or affinity elution* (22). This uses molecules which are able to interact either at the ligand binding site or at a different site, such that the binding surface is no longer available for binding (due to conformational change or steric occlusion). This is the mechanism by which affinity elution can be used for 'selective' desorption from ion-exchange matrices. The biospecificity of the eluent does not necessarily imply biospecificity in the binding interaction.

A characteristic of affinity elution is that very low concentrations of eluent are required, often less than 10 mM. Examples include elution of dehydrogenases from dye columns by various nucleotides (1), or elution of glycoproteins from lectins by free sugars (Section 5). Choice of selective eluents will be dictated by the availability and cost for each individual ligand−protein pair.

It should be noted that eluted protein may be difficult to separate from the eluent.

(iv) *Chaotropic agents.* These are used for elution when other methods fail, because of the very high affinity of an interaction. Chaotrophs effect desorption by disrupting the structure of water thus reducing ligand−protein interaction. Potassium thiocyanate (3 M), potassium iodide (2 M) or $MgCl_2$ (4 M) could be tried. (See also Section 6).

(v) *Denaturants.* Urea (8 M) and guanidine−HCl (6 M) may also be effective. For both chaotrophs and denaturants their effect on protein activity should be examined before proceeding to scale-up.

(vi) *Polarity reducing agents.* Dioxane, 10% (v/v) or ethylene glycol, 50% (v/v) should be tried. Similarly, low concentrations of detergents will also effect elution.

(vii) *Other methods of facilitating elution.*

(1) Temperature gradient (23).

(2)   Disruption of a ternary complex (24).

(3)   Electrophoretic desorption (25).

(viii) *Difficult elution or irreversible denaturation.* If either of these is experienced then reduction of the ligand concentration by dilution with underivatized matrix may permit the use of milder elution conditions. Alternatively, a lower concentration of immobilized ligand should be used.

3.3 Estimation of capacity

Having determined the conditions for effective adsorption and elution of the desired protein from an affinity adsorbent, it is necessary to determine adsorbent capacity before conditions for larger scale purification can be finalized. The best method, using a technique known as frontal analysis (26), is as follows.

(i)   Prepare a 1-ml column (6 mm i.d.) of the selected adsorbent (unused) and equilibrate with the chosen buffer at $6-8$ ml h^{-1}.

(ii)   Apply a constant stream of the crude protein mixture to be purified. Concentration of the desired protein P in the mixture is P_0. Collect 1 ml fractions.

(iii)   Monitor effluent for total protein (A_{280}) and assay fractions for the emergence of P.

(iv)   When the concentration of P in the effluent reaches P_0 adsorbent capacity is saturated. Wash the column until A_{280} reaches the baseline.

(v)   Elute with chosen eluent. Measure the yield of P.

(vi)   Calculations (*Figure 1*): the volume at which the concentration of P in the effluent reaches 50% of P_0 is defined as V_e. Similarly, V_0 is defined as the volume at which the unadsorbed protein concentration is 50% of the original. (This assumes that P is an insignificant proportion of the total protein).

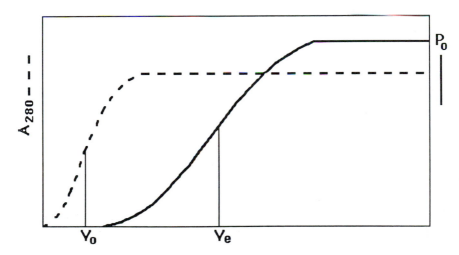

Figure 1. Estimation of the capacity of an affinity adsorbent. Breakthrough curve for total protein (broken line). Breakthrough curve for protein being purified (solid line). Figure prepared by D.Atwal.

The amount of *P* adsorbed to the column is $P_0 (V_e - V_0)$. This is the expected capacity. Ideally this should be the same as the 'working capacity' which is equal to the yield of *P* in the eluted fractions.

Note the following.

(1)   If the working capacity is much lower than the expected capacity, elution conditions need to be improved.

(2)   If assays for *P* cannot be accomplished in the time-frame of the experiment an 'educated guess' will have to be made as to the volume required. If the guess is wrong the experiment should be repeated, but on fresh unused adsorbent.

(3)   If the flow rate is too high some *P* will appear in the effluent throughout the experiment.

(4)   Capacity experiments should be carried out using conditions similar to those to be used in the actual purifications. Changes in the constituents of the feedstream will invalidate the capacity experiment, particularly if the adsorbent binds several proteins whose proportions change.

(5)   The capacity experiment should be repeated on the once-used adsorbent to determine the re-useability of the column. This is particularly important for expensive affinity columns which are to be used repeatedly at more than 80% of the working capacity.

3.4 Ligand efficiency

Ligand efficiency may be defined as:

$$\frac{\text{Working capacity}}{\text{Theoretical capacity}} \times 100$$

The theoretical capacity is equal to the immobilized ligand concentration × the number of binding sites per molecule of ligand. In practice, ligand efficiencies of 1% or less are common for small ligands (4) indicating immobilization of ligand molecules in positions or orientations which are inaccessible to the interacting protein. On the other hand, macromolecular ligands, such as monoclonal antibodies, often show efficiencies in excess of 10%. This observation probably reflects the much lower density of immobilization achievable with larger molecules. Thus for successful affinity chromatography the aim should be to maximize ligand efficiency, particularly for expensive ligands, by experimenting with a range of ligand concentrations at the coupling stage.

3.5 Application to protein purification

3.5.1 *Operational considerations*

Previous sections of this chapter have dealt with the factors to consider in the design and preparation of affinity adsorbents as well as detailing the types of preliminary experiments required to be done before adaption of such adsorbents for protein purification. This section briefly discusses the operational factors to be considered. (See ref. 27 for a recent review.)

(i) *Mode of operation.* Batch mode is suitable for higher affinity systems, particularly when the volume of the starting material is high and concentration is not facile because of sensitivity of the protein of interest or because of high contaminant concentrations. Care should be taken to avoid excessively long incubation times which can contribute to secondary interactions.

Column mode is preferable for lower affinity systems since it is less labour-intensive and more flexible.

(ii) *Column dimensions.* The volume of affinity matrix required will depend on the empirically found capacity and the type of separation required. For expensive ligands it is desirable to use more than 80% of the binding capacity. If the choice of ligand and elution conditions has resulted in good selectivity, resolution of proteins will depend on the adsorbent itself and not to any great degree on column length. Thus short, fat columns are quite suitable (e.g. diameter:length = 1:1) for higher affinity systems.

(iii) *Flow rate.* This will depend on the porosity of the matrix and the size of protein to be purified and should generally be as slow as practicable. If the flow rate is too high the protein of interest will appear in the breakthrough before the column is saturated and will tend to cause tailing of peaks during the elution phase. As a guide, for agarose-based adsorbents, linear flow rates of 10 ml cm^{-2} h^{-1} can be recommended. Faster flow rates can be used during the washing and re-equilibration steps.

(iv) *Ligand leakage.* A serious limitation of affinity chromatography is leakage of ligand into the feedstream or eluate (28). With high affinity systems (e.g. hormone − receptor interactions) the isolation of minute amounts of protein is compromised by the release of ligand from the matrix.

In the case of therapeutic proteins the presence of ligand in the final product is likely to result in serious consequences (29). Leakage may be due to instability of the immobilized ligand, instability of the linkage to the matrix or dissolution of the matrix itself. All these factors will need to be investigated and their effects minimized in the development of therapeutic proteins.

Leakage of ligands in any substantial amount will lower the capacity of an adsorbent, which is unacceptable if the adsorbent is expensive and re-useability is essential for economy.

Ligands coupled to CNBr-activated matrices have been reported to show a greater degree of leakage than those coupled using other methods. However, since the instability of linkage is not the sole cause of leakage further work must be done on individual matrix − ligand systems where any degree of leakage is unacceptable.

(v) *Cleaning and storage of affinity adsorbents.* The robustness of an affinity adsorbent will depend on the nature of the ligand. For many systems, particularly those involving protein ligands, cleaning with extremes of pH or heat-sterilization is not possible. Affinity adsorbents are therefore often cleaned with high salt concentrations (2 M KCl) and stored in the presence of bacteriostatic agents (e.g. 0.02% w/v sodium azide).

3.5.2 *Selected examples of use of affinity adsorbents*

Some of the most commonly used ligands in affinity chromatography are lectins and antibodies whose use is described in detail in Sections 5 and 6. A brief summary of the use of other commonly employed ligands is given below. It should be used as a quick reference guide. More detailed information may be sought from refs 1 − 11 and manufacturer's publications.

(i) *Protein A* (30). Binds specifically to the Fc region of immunoglobulins from various species. Binds only weakly to murine IgG1, horse IgGc, chicken IgG, most IgA and IgMs. Does not bind human IgG3, or rat IgG2a and 2b. Binding is enhanced in the presence of high salt concentration, for example 3 M NaCl, and at high pH (e.g. 8 − 9). Usual binding buffer is phosphate or Tris based. Elution should be carried out with a decreasing pH gradient using 0.1 M citric or 1 M acetic acids. The ligand itself is stable to 6 M guanidine − HCl, which should be used to clean the column from time to time. Store in 70% ethanol.

(ii) *Protein G* (31). Complements protein A in that it can bind human IgG3, rat IgG2a and 2b as well as other IgGs mentioned above (with the exception of chicken IgG). Binding should be carried out in physiological buffers. Elution requires 0.1 M glycine − HCl pH 2.5. Eluted proteins should be neutralized immediately. Clean the column with 6 M guanidine − HCl at low pH and store in 70% ethanol.

(iii) *Cibacron blue 3 G-A* (32). Binds human serum albumin, fibroblast interferon, lipoproteins, nucleotide-requiring enzymes (at least some are bound through their co-enzyme binding sites). Binding is usually in low molarity Tris − HCl pH 7 − 8.5 (but depends on the stability of the protein of interest). Binding is enhanced at lower pH, probably at least partially due to the ion-exchange effect of the sulphonate groups on the dye. Elution is often successful using a salt gradient to 1 M KCl or alternatively affinity elution is possible. More stringent conditions are required for complete elution of human serum albumin (e.g. 0.5 M potassium thiocyanate). Columns can be cleaned with 8 M urea. Lipoproteins (e.g. from serum feeds) may clog the column surface and are best dealt with by unpacking the column into distilled water and decanting the lipo-protein particulate matter followed by several urea washes at extremes of pH. Leakage of ligand molecules is easily apparent due to their colour, particularly after storage, and adsorbents should be thoroughly washed to remove leached or non-covalently bound dye. A blank run with equilibration and elution buffers is recommended.

(iv) *Other triazine dyes.* Since the discovery of the interaction of Cibacron blue with various enzymes there has been a considerable expansion in the applicaton of other reactive dyes to protein purification. Dye-ligand chromatography has become popular because the ligands are inexpensive, easy to couple to matrices and extremely stable.

Reactive dyes have been termed 'pseudo-ligands' or 'biomimetic ligands' as they do not in general have a direct relationship to biological ligands. Numerous examples of protein purification using dye-ligands have been published (33 and references within). However, it is not possible at present to predict if a particular protein will bind to a given dye column. It is therefore necessary to use systematic screening procedures (34,35) to identify suitable adsorbents from a collection (e.g. kits available from Cambio, Sigma or Amicon). The general procedure for testing any dye adsorbents is that given for Cibacron blue above. Dye-ligand adsorbents can be very effectively used in tandem (35).

(v) *Phenyl boronate (PBA).* PBA forms covalent complexes with *cis*-diols and may therefore be used to purify some glycoproteins (1). PBA-agarose is available at two

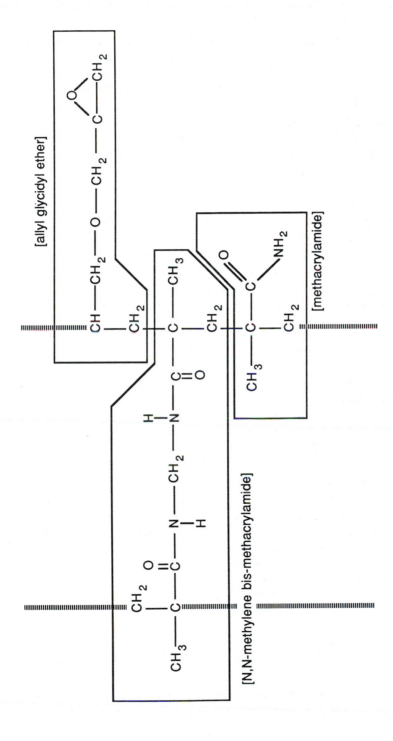

Figure 2. Chemical structure of Eupergit oxirane acrylic beads. The three components of the polymer are indicated in outline.

ligand densities (10 and 30 μmol ml^{-1}) from Amicon. Binding is effected in 50 mM Tris−HCl pH 8 containing 20 mM MgCl$_2$. Elution requires sorbitol (0−50 mM gradient) in 50 mM Tris−HCl pH 8. A further wash at pH 5 using acetate buffer will clean and re-equilibrate the matrix.

(vi) *Lysine.* Binds plasminogen, plasmin and plasminogen activator (36). Binding is effected in physiological buffers and is followed by washing with 0.5 M NaCl in the same buffer. Specific elution may be carried out using 0.2 M ϵ-aminocaproic acid in distilled water.

(vii) *Other applications.* The reader is referred to other texts in this series for additional practical information on applications (1,37). Regular literature searches are prepared by Sturgen and Kennedy (38). Useful monographs and technical documents are also available from suppliers (e.g. Pharmacia-LKB, IBF, Bio-Rad, Pierce Amicon).

4. PREPARATION OF ADSORBENTS USING OXIRANE ACRYLIC BEADS AND THEIR USE—M.Cusack and R.J.Beynon

Traditionally, CNBr-activated Sepharose has been used in affinity chromatography, offering good capacities for proteins and low molecular weight ligands, good flow properties and providing a support matrix that is relatively inert. More recently, other matrices have become available, offering new advantages for protein purification. Here we describe the use of one of these relatively new matrices, Eupergit C. In addition to simple coupling protocols, this matrix exhibits good chemical and physical stability and is suitable for a wide variety of applications. Eupergit C, an oxirane acrylic matrix, is available from Rohm Pharma Gmbh.

4.1 Properties of Eupergit C

4.1.1 *Chemical structure*

Eupergit C is formed by copolymerization (*Figure 2*) of methacrylamide, methyl bis-acrylamide, glycidyl methacrylate and allyl glycidyl ether (1-allyloxy-2,3-epoxy-propane). The product is polymerized, initially as microbeads that are subsequently assembled into larger bead structures. The matrix is virtually electroneutral and pre-dominantly hydrophilic; as such it has neither the properties of an ion-exchanger nor a hydrophobic chromatography matrix. Some hydrophobic character may be contributed by the methyl groups that are derived from the methacrylamide moieties (39).

4.1.2 *Physical properties*

The matrix, which is electroneutral and predominantly hydrophilic, is very resistant to shape changes and, when hydrated, undergoes no significant swelling or contraction over the range of pH values (pH 2−12) and ionic strength values (I = 0.01−1.0) that would normally be used in biospecific chromatography. Eupergit beads are resistant to pressures of 300 bar and yield good flow rates even at low pressure. For example, a column of 28 × 2 cm, under a pressure of 2 bar, gives a flow rate of 100 ml min^{-1}. Compression of column packings does not occur under normal conditions. The beads are very tolerant of stirring and other forms of agitation but magnetic stirrers or stirrers that impinge on the sides of the reaction vessel can damage the beads. Four variants

Table 4. Properties of Eupergit oxirane acrylic beads.

Form	Particle size (μm)	Exclusion limit (M_r)	Surface area ($m^2 g^{-1}$ dry wt)	Binding capacity (mg albumin per g wet wt)
Eupergit C250L	200−250	1 000 000	n.a.	15
Eupergit C	140−180	200 000	180	48
Eupergit C 30N	30−50	200 000	40	52
Eupergit C 1z[a]	0.6−1.4	n.a.	6	5

[a]Eupergit C 1Z is in the form of compact, non-porous spheres. It is very suitable for packing into HPLC columns and the lack of pores explains the lower capacity for protein binding.
n.a.; not applicable.

of Eupergit C are available; these differ in particle size, exclusion limit, surface area and hence capacity. (*Table 4*). The smallest of the beads consists of approximately 1-μm particles which are suitable for packing into HPLC columns. However, these beads are not porous and thus have a lower surface area than the other forms.

The underivatized beads are stable at −20°C and are best stored in a container sealed in a plastic bag containing dry silica gel. The bag is allowed to warm to room temperature before the beads are weighed, to prevent condensation on the inside of the container. The water regain value of the dry beads is about 2.5 ml g^{-1} Eupergit C.

4.1.3 *Chemistry of ligand binding*

The oxirane groups of the glycidyl moieties are the sites of ligand immobilization (*Figure 3*) and are present at a level of approximately 1 mmol g^{-1} dry beads. These interact with a number of different functional groups as shown in *Figure 4*. The conditions that are optimal for reaction depend upon the nature of the functional groups used for coupling. Maximal binding through tryptophan residues is observed at low pH values whereas primary amines bind preferentially at higher pH values (40).

Proteins are most often bound at pH values around neutrality, although the choice of pH may be largely determined by the need to maintain the protein in a stable conformation during the coupling procedure, which is fairly long (16−48 h is typical). The selection of buffer is clearly important as the oxirane groups react with many 'biological' buffer systems, including amines. In particular, care should be taken to ensure that the underivatized matrix is not exposed to ammonium salts as these will be highly reactive and inactivate the beads rapidly. Other buffers that should be avoided during the coupling step include the 'Good' buffers (Chapter 1, Table 1B) and Tris, although the selection of a buffer for subsequent chromatographic steps can be more relaxed.

The manufacturers recommend that proteins be coupled to the matrix in a potassium phosphate buffer at high concentrations (1.0 M) and at near-neutral pH values. However, such a high ionic strength (I = 2.4) may not be required in all instances. Indeed, some data suggest that the optimal ionic strength is dependent upon the protein being coupled and collated data suggest that effective coupling can be achieved over a wide range of buffer concentrations, although experiences to date seem to have been limited to phosphate buffers. Note that the coupling of immunoglobulins is more efficient at solutions of lower ionic strength, whereas albumin and a number of other proteins have routinely been coupled at higher values of I (see below).

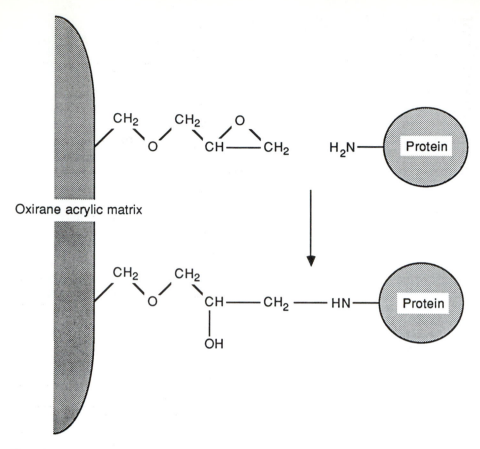

Figure 3. Coupling of oxirane acrylic beads to proteins via amino groups on proteins. The reaction between the oxirane group and primary amino groups (α and ϵ amino groups on proteins) is shown, although the oxirane group is able to react with other moieties on proteins and other ligands.

4.2 **Use of Eupergit C**

4.2.1 *Examples of existing applications*

To date, there have been relatively few applications of oxirane acrylic beads. (*Table 5*). Even fewer applications have established rigorously the optimal conditions for coupling of protein that retains a high degree of biological activity. In one study, the coupling of three proteins, albumin ($M_r = 69\,000$), gamma globulin ($M_r = 150\,000$) and beta galactosidase ($M_r = 520\,000$) was compared (41). All three proteins were coupled effectively at pH values betwen 6 and 9, in a phosphate buffer. Albumin was bound optimally at pH 7.6 at an ionic strength greater than 2 whereas optimal binding of gamma globulin was attained at much lower ionic strength ($I = 0.5$). Coupling of the three proteins was inversely related to their molecular weight, implying that a greater proportion of the surface of the macroporous bead structure was inaccessible to larger proteins. Effective coupling of small proteins has been achieved; aprotinin (bovine pancreatic trypsin inhibitor, $M_r = 6500$) has been bound to Eupergit C. The product

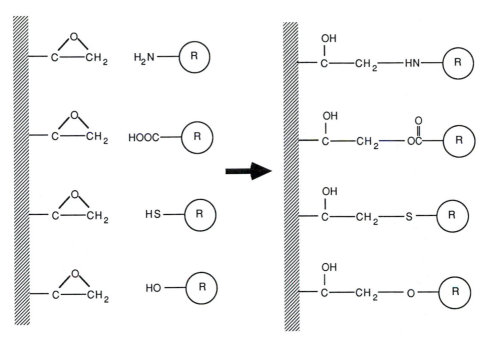

Figure 4. Interactions of oxirane beads with different functional groups.

Table 5. Representative applications of oxirane acrylic beads.

Protein	Buffer	pH	Time (h)	T (°C)	References
Mouse monoclonal antibodies	0.1 M phosphate	8.0	24	RT	43,44
Mouse monoclonal antibodies	0.1 M phosphate	n.s.	24	RT	45
Concanavalin A	1.0 M phosphate	5.0	30	RT	46
Esterase	1.0 M phosphate	7.5	24	RT	47
Dehydrogenases	0.2 M phosphate	8.0	48	n.s.	48
Penicillin acylase	0.1 M phosphate	7.2	40	RT	49
Albumin β-galactosidase, γ-globulin	0.1 M phosphate	6.0−8.0	12−80	RT	41
alcohol dehydrogenase	0.5 M phosphate	7.5	24 h	RT	50
Albumin, trypsin papain, chymotrypsin trypsin inhibitor	Variable	3−10	48	30	40

n.s.; not specified.
RT, room temperature.

is an effective chromatographic medium for plasmin and has been employed in assays for tissue plasminogen activator (42).

4.2.2 Procedures for coupling of ligands
Although coupling of materials to oxirane beads is facile, it may be worthwhile

Method Table 1. General procedure for small-scale exploratory coupling of ligands to oxirane acrylic beads.

1.	Dissolve the ligand in an appropriate buffer, at as high a concentration as feasible. Inorganic buffers such as phosphate or pyrophosphate are suitable. The pH of the buffer should be considered carefully from the point of view of stability of the ligand and the oxirane groups. At pH values > 10, conversion of the oxirane moiety to the diol derivative is significant.
2.	Add 1.0 ml of the ligand solution to 100 mg of dry Eupergit C beads, weighed into a 1.5-ml Eppendorf tube.
3.	Mix the suspension of beads by inversion or rotation for $16-48$ h at room temperature or a temperature commensurate with the stability of the ligand.
4.	At various times during the incubation period, centrifuge the suspension of beads briefly in a bench centrifuge (10 000 g for 10 sec) to sediment the beads, and remove small samples of the resulting supernatant to determine residual ligand. The choice of assay is determined by the nature of the ligand and can be as simple as absorbance or as complex as an enzyme assay or immunoassay. Note that Eupergit C contains a small proportion of acetone that can interfere with simple E_{280} determinations of protein.

Method Table 2. General procedure for preparative-scale coupling of ligands to oxirane acrylic beads.

1.	Dissolve the ligand in an appropriate buffer, as indicated by previous exploratory studies. In the first instance, 0.1 M phosphate buffer, pH 7.5 may prove acceptable for most proteins or low molecular weight ligands.
2.	Add 4 ml of the ligand solution to every 1 g of dry oxirane acrylic beads. The beads will take up all of the buffer; this increases the efficiency of coupling.
3.	Allow the moist beads to stand for $16-48$ h at room temperature if possible, but at 4°C if the ligand is unstable.
4.	Suspend the beads in a suitable buffer (e.g. 0.01 M phosphate buffer, pH 7.5) and rinse on a sinter funnel (pore size 1) with at least 10 vol of the same buffer.
5.	Block unreacted oxirane groups on the beads by reacting overnight with 4 ml g^{-1} beads of aqueous mercaptoethanol (5% v/v), 1 M glycine or 1 M ethanolamine, pH $7-8$ for 16 h at room temperature.
6.	Wash the derivatized, blocked beads extensively on a sintered glass funnel with double distilled water, and finally with 10 vol of the chromatography buffer.
7.	Store the derivatized beads at 4°C, with a suitable bacteriostat if required (e.g. a drop of toluene, 0.02% w/v sodium azide or 0.1% v/v formaldehyde).

attempting a small-scale coupling to optimize binding conditions. *Method Table 1* gives a simple small-scale coupling method that has proven valuable in our hands. The determination of free and bound ligand is very dependent upon the ligand that is being coupled. Note that the method should be able to differentiate between ligand that is coupled covalently to the matrix and ligand that is bound non-covalently; the latter

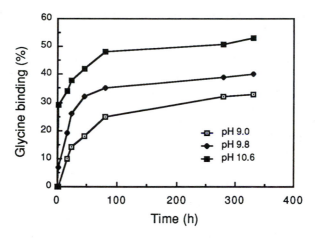

Figure 5. Coupling of glycine to Eupergit C beads at different pH values. Small samples (100 mg) of Eupergit C 250 N beads were incubated at room temperature with 1.0 ml of 50 mM sodium phosphate buffer containing 0.1 M [^{14}C]glycine, sp. act. 40 μCi mmol^{-1}, adjusted to the indicated pH values (most of the buffering capacity is provided by glycine). At the indicated times, samples (10 μl) were removed and the unbound glycine was determined by scintillation counting. Data are expressed as bound glycine (% of total); 100% binding is equivalent to a capacity of 1 mmol g^{-1} dry weight of beads. Key: pH 9.0, ($\square$); pH 9.8, ($\bullet$); pH 10.6 ($\blacksquare$).

behaviour often manifests itself as a rapid binding (in the first few minutes). The volume of liquid added to the beads is sufficiently large that the suspension of beads can be centrifuged; the supernatant can then be sampled for determination of unbound ligand.

A larger scale coupling can be achieved by simply scaling up the method in *Method Table 1*. However, *Method Table 2* gives a method that can be used for larger quantities of material but which differs in that a considerably lower volume of ligand solution is added to the dry beads. This increases the local concentration of the ligand and may make the coupling more efficient. No surplus liquid will be apparent in this method.

Figure 5 shows the effect of pH on the binding of a simple ligand, glycine, to the beads. Note that optimal binding is attained at high pH values, suggesting that the uncharged amino group is an effective reactant. However, the binding of another ligand, 4-aminobenzamidine, was largely unaffected by the coupling conditions between pH 4 and 10 (data not shown), suggesting that this simple explanation may not be satisfactory. Similarly, the binding of simple ligands to Eupergit C is largely insensitive to ionic strength. This implies that the ionic strength effects on protein coupling are largely due to an influence on the protein (either by enhancing weak hydrophobic interactions between the matrix and the protein or by altering the aggregation state of the protein and hence, accessibility to the pores of the matrix).

To examine further the behaviour of the oxirane acrylic beads we prepared two affinity matrices suitable for the purification of trypsin-like serine proteases. The ligands were 4-aminobenzamidine (4-BA), which was coupled to the beads at a level of 0.5 mmol g^{-1} dry weight, and soybean trypsin inhibitor (STI), a protein of molecular weight 23 000. The STI−Eupergit was coupled such that the capacity was at least 5 mg trypsin g^{-1} wet beads. *Figure 6* indicates that trypsin binds rather more slowly to the STI−Eupergit than the BA−Eupergit, as might be expected for highly specific

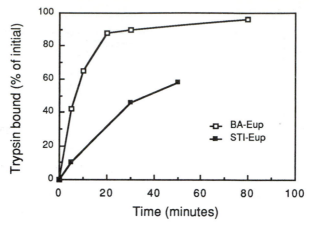

Figure 6. Time dependence of binding of trypsin to BA-Eupergit and STI-Eupergit. Benzamidine (BA)-Eupergit and soybean trypsin inhibitor (STI)-Eupergit were prepared as described in the text. A sample of each affinity matrix (total packed volume 50 μl) was incubated on a bottle roller in a final volume of 1.0 ml of 0.1 M Hepes buffer, pH 7.5, containing $100-150$ μg of trypsin. At the indicated times, the incubation mixture was centrifuged briefly and the residual trypsin in 10 μl of the supernatant was assayed towards Z-Phe-Arg-N-methyl-coumarylamide. Key: binding to BA-Eupergit, ($\square$); binding to STI-Eupergit, ($\blacksquare$).

interactions with a large ligand. Chromatography of semi-purified trypsin preparations indicates that they are effective affinity matrices (*Figure 7*). However, it is interesting to note that much less non-trypsin material is eluted from BA − Eupergit when the column is washed with loading buffer. Further, the specific activity of the trypsin eluted at low pH from the BA − Eupergit is much lower than that from STI − Eupergit. We suggest that this might be due to the fact that the BA − Eupergit may also be acting as a non-specific ion-exchanger; an important pitfall to be wary of when dealing with a matrix that can achieve such a high degree of substitution (0.5 mmol g^{-1} 4-BA in this instance). Such behaviour might also explain the differences in speed of binding of trypsin to the matrices (*Figure 6*).

4.3 Conclusions

The pre-eminence of existing methods for immobilization of affinity ligands onto matrices is such that newer methods have yet to acquire wide popularity. However, the mechanical and physical stability of oxirane acrylic polymers suggests that they deserve a close look in the selection of a suitable matrix. The chemistry of coupling is facile and moreover, safe, and the ease with which the resulting materials may be used in FPLC and HPLC applications suggests that they have potential in the high resolution protein purification methodologies that are coming to the fore.

5. LECTIN AFFINITY CHROMATOGRAPHY—C.Sutton

Lectins are a family of carbohydrate binding proteins that are produced by slime moulds, plants and animals. In each they have adapted for different roles. Despite this diverse range of sources lectins share common structural properties. These properties have been reviewed extensively (51,52) and will only be discussed briefly here.

a) BA-Eupergit

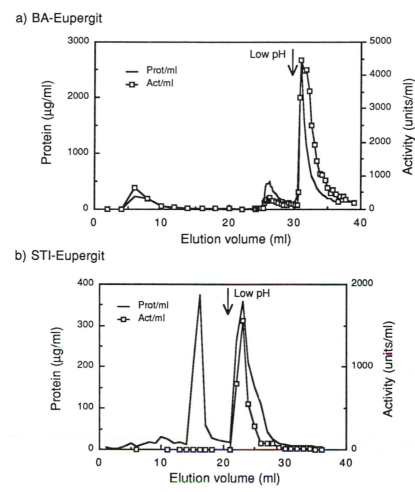

b) STI-Eupergit

Figure 7. Chromatography of partially-pure trypsin on BA-Eupergit and STI-Eupergit. BA-Eupergit and STI-Eupergit were prepared as described in the text. For both columns (volume ~ 1.0 ml) the trypsin was applied at a flow rate of 0.1 ml min^{-1} in 0.1 M Hepes buffer, pH 7.5 and the column was washed with the same buffer until baseline absorbance was attained. Bound material was eluted with 0.1 M sodium citrate buffer, pH 2.0, indicated by an arrow on the elution profile. For the BA-Eupergit column, a total of 8.4 mg of trypsin was applied in two successive applications (Ve = 5 ml and Ve = 25 ml). For the STI-Eupergit, 1.9 mg of trypsin was loaded in a single application (Ve = 12 ml). The protein concentration of each fraction (continuous line) was determined by absorbance at 280 nm and the trypsin activity was assayed with the fluorogenic substrate Z-Phe-Arg-N-methyl-coumarylamide (□). The unit of activity is arbitrary.

Lectins were originally recognized for their ability to agglutinate different human erythrocyte types (A, B and O) by binding to specific receptors on the surface and hence are often called agglutinins. Others initiate proliferation and/or morphological transformations of lymphocytes in a dose-dependent manner and such lectins are referred to as mitogens. The specificity of lectins was determined initially by the ability of particular sugars to inhibit these biological responses.

Lectins are made up of one or more subunits in a multimer—most frequently a

Table 6. Biochemical characteristics and specificities of some commercially available lectins.

Lectin (abbreviation)	Mol. wt	No. of subunits	Simple sugar specificity	Metal ions required
Concanavalin A (Con A)	55 000[a]	2 4	α-D-Man > α-D-Glc > α-D-GlcNAc	Ca^{2+} Mn^{2+}
Lentil lectin (LCA)	49 000	2	α-D-Man > α-D-Glc > α-D-GlcNAc	
Soybean lectin (SBA)	110 – 120 000	4	α-D-GalNAc > β-D-GalNAc > α-D-Gal	Ca^{2+} Mn^{2+}
Castor bean lectins (RCA$_{60}$) (RCA$_{120}$)	60 000 120 000	2 4	D-GalNAc β-D-Gal > D-Gal	
Wheatgerm agglutinin (WGA)	36 000	2	$(\beta$-D-GlcNAc$)_3$ > $(\beta$-D-GlcNAc$)_2$	Ca^{2+} Mn^{2+} Zn^{2+}
Peanut agglutinin (PNA)	120 000	4	β-D-Gal-(1 – 3)-D-GalNAc > D-GalNH$_2$ = α-D-Gal	
Pea lectin	49 000	4 ($\alpha\beta$)	α-D-Man	
Phytohaemagglutinin (PHA)	128 000	4	D-GalNAc	Ca^{2+}, Mn^{2+}
Bandeirea simplicifolia BS-I BS-II	114 000 113 000	4 4	α-D-Gal α-D-GlcNAc	Ca^{2+}, Mg^{2+} Ca^{2+}, Mg^{2+}
Pokeweed mitogen (PWM)	32 000		(D-GlcNAc)$_3$	
Ulex europeus agglutinin (UEA1)	170 000		α-L-Fucose	
Dolichos biflorus agglutinin (DBA)	111 000	4	α-D-GalNAc	
Jacalin	40 000	4	α-D-Gal	
Helix pomatia lectin	79 000	6	α-D-GalNAc < α-D-GlcNAc < < α-D-Gal	
Limulin	400 000	18	α-*N*-acetylneuraminic acid	

[a]Con A exists as a dimer (55 000) below pH 5.6, as a tetramer (110 000) between pH 5.6 and 7.0 and as aggregates above pH 70.

tetramer. If all the subunits are identical the lectin has multiple binding sites recognizing a single, specific saccharide. However, if there are two functionally distinct subunits the lectin may have the ability to bind two different saccharides, mono or multivalently. That two subunits can be combined in multimer lectins in different proportions may be important in conferring a specificity for different functions. Some important structural characteristics of the more common lectins are summarized in *Table 6*.

5.1 Materials for lectin chromatography

5.1.1 *Selection of a lectin*

The affinity of lectins for specific carbohydrate moieties has made them particularly useful for the purification of distinct groups of glycoproteins. For example, many receptors bind to wheatgerm agglutinin (WGA) whereas cytokines and other growth factors bind preferentially to concanavalin A (ConA) or lentil lectin (LCA). In order to choose a lectin to purify a glycoprotein two factors have to be considered.

(i)   The availability of the lectin; whether it has to be prepared from its natural source or is available commercially and if so whether or not it is available immobilized.

(ii)   The nature of the oligosaccharide linked to the protein of interest.

(i) *Availability of lectins.* For lectins to function as purification media it is important that they are coupled to a solid phase allowing specific binding of glycoproteins and the subsequent dissociation using selected elution conditions. The quantities of lectin required can vary significantly from milligram to gram amounts depending on the scale of the purification. This limits the choice of source of lectin considerably; predominantly to the seeds of legume plants, in which these proteins are particularly abundant, comprising up to 3% of the seed total protein. A large number of lectins are available commercially (e.g. from Sigma, BDH, Pharmacia-LKB, Pierce, BioInvent). The more common lectins can be purified from the original sources using either readily available adsorbents or specific affinity chromatography methods (see *Table 7*). A number of immobilized sugar columns that can be used for purification of specific lectins are available from Pierce and Sigma. Having obtained the pure lectin of interest it can be immobilized onto activated supports.

(ii) *Importance of oligosaccharide structure in lectin chromatography.* Glycoproteins are produced by eukaryotic cells for a diverse range of functions. Lectin chromatography can be used for the isolation of glycoproteins, not only from their natural sources but also when produced by recombinant DNA techniques. Some proteins secreted by yeast or mammalian cell lines may have an oligosaccharide moiety, though the natural form may not. Even those that are usually glycosylated may have aberrant sugar chains and significant microheterogeneity in their saccharide sequence. Whether a protein is native or cloned, the oligosaccharide patterns, though complex, can be divided into groups (60). The characteristics of each group greatly influences the binding of lectins and is an important parameter in choosing a lectin for chromatography. The sugar moiety is bound covalently to the protein by one of two main linkages.

(1)   *O*-glycosylation linkage, of which the predominant form is *N*-acetylgalactosamine covalently bound to L-serine or L-threonine. Ser/Thr linked sugar chains are typically found in mucins (see *Figure 8*). Of the common lectins, peanut lectin with a sugar specificity for galactose preferentially binds to mucin-type chains.

(2)   *N*-glycosylation linkage between *N*-acetylglucosamine and asparagine. This group can be further subdivided on the basis of oligosaccharide composition and complexity into high-mannose-type, complex-type and hybrid-type (*Figure 8*). Many of the most frequently used lectins bind to *N*-linked sugar chains. For example castor bean agglutinin, a galactose-binding lectin, (compare with peanut lectin) primarily binds to complex-type and hybrid-type oligosaccharides.

Lectin Recognition	Carbohydrate Structure	Class
WGA PNA SBA	NeuAc2 — 3Gal1 —β— 3GalNAc1 —α— Ser/Thr with NeuAc 2—α—6 branch on GalNAc	Mucin-type
	(Man1 —α2)Man1 —α 6 Man1 —α (Man1 —α2)Man1 —α3 6 Man1 —β— 4GlcNAc1 —β— 4GlcNAc — Asn (Man1 —α2)Man1 —α3	High mannose-type
Con A LCA Pea Lectin	(NeuAc)2 — 6Gal1 —β— 4GlcNAc1 —β4 (NeuAc)2 — 6Gal1 —β— 4GlcNAc1 —β2 Man1 —α6 Man1 —β— 4GlcNAc1 —β— 4GlcNAc — Asn (NeuAc)2 — 6Gal1 —β— 4GlcNAc1 —β— 2Man1 —α3 with Fuc 1 —α— 6 branch	Triantennary complex-type

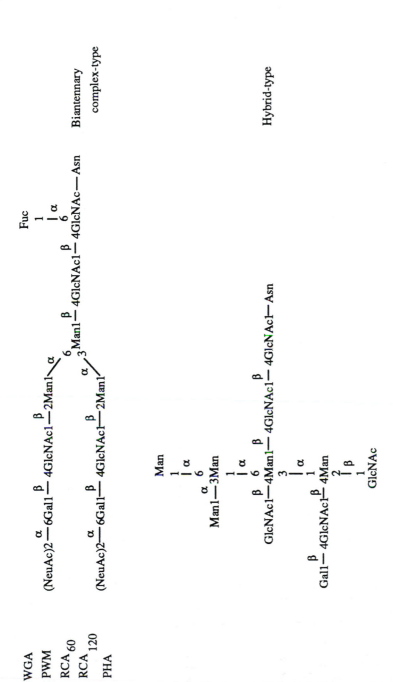

Figure 8. Examples of the various types of sugar chains of glycoproteins. Mucin-type, ref. 61; high mannose- type, ref. 62; triantennary complex-type, ref. 63; biantennary complex-type, ref. 65; hybrid-type, ref. 64.

273

Table 7. Purification of lectins.

Lectin (Abbreviation)	Source	Chromatography adsorbent	Eluent	References
Concanavalin A (ConA)	Jack bean (*Canavalia ensiformis*)	Sephadex G-100 Derivatized maltose Bio Gel P-150	20 mM glycine−HCl pH 2.0 0.1 M α methyl mannoside in PBS	53 54
Lentil lectin (LCA)	Common lentil (*Lens culinaris*)	Sephadex G-100 Derivatized maltose Bio Gel P-150	0.1 M glucose 0.1 M α methyl mannoside in PBS	55 54
Soybean lectin (SBA)	Soybean (*Glycine max*)	D-galactosamine Sepharose	2.5% lactose	56
Castor bean agglutinin (RCA)	Castor bean (*Ricinus communis*)	Agarose A 0.5 Derivatized lactose Bio Gel P-150	0.2 M galactose in 5 mM phosphate buffer pH 7.2 plus 2 M NaCl 0.2 M lactose in PBS	57 54
Wheatgerm agglutinin (WGA)	Wheatgerm (*Triticum vulgaris*)	Ovomucoid Sepharose Derivatized di-N-acetylchitobiose Bio Gel P-160	0.1 M acetic acid 0.1 M N-acetylglucosamine in PBS	58 54
Peanut agglutinin (PNA)	Peanut (*Arachis hypogea*)	Derivatized lactose Bio Gel P-150	0.1 M lactose in PBS	54
Pea lectin	Garden pea (*Pisum sativum*)	Sephadex G-150	10 mM glycine−HCl pH 2.0	59
Bandeirea simplicifolia lectins (BSI) and (BSII)	*B.simplicifolia*	Derivatized melibiose Bio Gel P-150	0.1 M melibiose in PBS	54

Frequently, however, this latter parameter is unknown. Therefore, in order to find the appropriate lectin it is necessary to screen a range of lectins for binding of the glycoprotein. Initially the use of immobilized lentil lectin, wheatgerm agglutinin and soybean agglutinin can give an indication of the oligosaccharide-type present.

5.1.2 *Coupling to the matrix*

(i) Crosslinking of lectins to CNBr-activated Sepharose-4B. Protocols for coupling ligands to CNBr-activated Sepharose are described in Section 6. It is important to note, however, the specific requirements for lectin immobilization outlined at the end of this section.

(ii) Crosslinking of lectin to Affigel 10 (Bio-Rad). Affigel matrices are available pre-swollen and only require washing on a sintered glass filter with the appropriate coupling buffer prior to addition of the lectin.

(1) Prepare the lectin $(5-7 \text{ mg ml}^{-1})$ in 0.1 M NaHCO$_3$ pH 8.4, containing 100 μM appropriate divalent metal ions (see Note A below) and 5% (w/v) appropriate saccharide (see Note B below).

(2) Add 1 ml of lectin solution per gram of moist Affigel cake previously washed in 0.1 M NaHCO$_3$ pH 8.4 containing appropriate components (see Notes A and B).

(3) Incubate the mixture overnight at 4°C on a roller rig or end-over-end mixer.

(4) Inactivate unreacted sites by adding 1 M glycine, glycine ethyl ester or 0.1 M ethanolamine for 2 h at room temperature.

(5) Wash the lectin−Affigel matrix with coupling buffer and store in PBS pH 7.2 containing 0.05% sodium azide.

Note A.

A number of lectins (see *Table 8*) can only bind saccharides in the presence of divalent

Table 8. Saccharide and divalent metal ion requirements for immobilization of lectins.

Lectin	Saccharide	Divalent metal ion[a]
ConA	α-D-mannose α-D-glucose or α-methyl mannoside	MnCl$_2$ and CaCl$_2$
LCA	α-D-mannose α-D-glucose or α-methyl mannoside	MnCl$_2$ and CaCl$_2$
SBA	D-*N*-acetyl galactosamine	−
WGA	*N*-acetyl glucosamine	MnCl$_2$ and CaCl$_2$
PHA	D-*N*-acetyl galactosamine	MnCl$_2$ and CaCl$_2$
RCA$_{60}$	D-*N*-acetyl galactosamine	−
RCA$_{120}$	β-D-galactose	−
BS-I	β-D-galactose	MgCl$_2$ and CaCl$_2$
BS-II	α-D-glucosamine	MgCl$_2$ and CaCl$_2$
PNA	D-galactose	−
Pea lectin	α-D-mannose or methyl mannoside	−

[a]Other transition metal ions can replace Mn^{2+}, e.g. Cd^{2+}, Cr^{2+}, Ni^{2+} or Zn^{2+}.

metal ions. The addition of divalent metal ions results in a transformation in the three-dimensional structure of the lectin leading to the formation of a saccharide binding site. As the lectin is to be immobilized the ability to undergo this transition may be limited. Therefore, by including the appropriate divalent metal ions (see *Table 8*), prior to coupling, it is possible to ensure that the lectin is in the correct form to bind glycoproteins. It also assists the binding of sugar, to the lectin, added to the coupling solution (see Note B).

Note B.

Lectin columns can be prepared when the coupling buffer contains an appropriate sugar of the correct specificity (see *Table 8*). The sugar serves to protect the lectin binding sites which might otherwise be cross-linked to the activated group on the matrix. Addition of the sugar therefore ensures that the maximum number of sites are available for glycoprotein binding. In the particular case of immobilizing ConA the presence of α-methyl mannoside, mannose or glucose at pH 7.0 or above prevents a time-dependent aggregation which would decrease the efficiency of coupling.

Note C.

Lectin solutions greater than 10 mg ml^{-1} should be avoided. Lectins have two or more binding sites. An excessive number immobilized to the matrix may cause steric hindrance of the affinity-binding glycoproteins. This may be less of a problem with matrices which have activated groups on the end of a spacer arm (e.g. Affigel 10) .

A number of the more common lectins can be purchased immobilized on agarose, from commercial sources (Sigma, BDH, Pharmacia-LKB, Pierce, BioInvent and Bio-Rad).

5.2 Procedures for using immobilized lectins

5.2.1 *Preparation of samples for lectin chromatography.*

For the practical application of lectin chromatography glycoproteins can be considered in two main categories.

(i)   Aqueous soluble glycoproteins: those which are secreted by cells found in biological fluids or culture medium (e.g. hormones) and structural proteins (e.g. collagen).

(ii)   Membrane-bound glycoproteins including transport proteins and cell surface recognition markers (e.g. receptors).

(i) *Preparation of soluble glycoproteins.* There are a number of requirements for lectin chromatography of glycoproteins particularly if the sample is a crude extract or culture medium.

(1)   Binding of glycoproteins to lectin columns is pH-dependent, therefore samples are preferably buffered (phosphate or Tris−HCl) at a physiological pH (6.8−7.5).

(2)   One of the advantages of lectin chromatography is the ability to run columns in the presence of relatively high salt concentrations, as binding between the sugar residue on the glycoprotein and the lectin is not due to ionic interaction. Salt also prevents non-specific binding of proteins with the matrix.

(3)   In *Table 8* lectins are listed which require divalent metal ions for glycoprotein binding, which should therefore be added to samples prior to chromatography. Final concentrations of 1 mM of the chloride salts of these ions are suitable.

(4)   Free sugars, present in the cytoplasm or culture medium should be removed to avoid competition with the glycoproteins for lectin binding sites.

(5)   A bacteriostatic agent such as sodium azide (0.05%) should be included, to prevent microbial contamination of the matrix. However, where detection of the glycoprotein of interest is dependent on a biological assay, alternative reagents such as antibiotics should be used (e.g. gentamycin at 100 μg ml^{-1}).

These requirements indicate that the sample should be modified to introduce or remove the various components. This can be performed through dialysis, molecular exclusion or after prior concentration by ion-exchange chromatography or ammonium sulphate or PEG fractionation.

(ii) *Preparation of membrane-bound glycoproteins.* The requirements described for aqueous soluble glycoproteins are equally important for membrane-bound glycoproteins. However, modification of sample buffer conditions is made easier by washing the cells or membrane pellets in the appropriate buffer prior to or simultaneously with solubilization of the glycoprotein. Solubilization of membrane-bound glycoproteins together with purification on lectin columns, has been comprehensively examined by Lotan and Nicolson (65). Most membrane proteins have a low solubility in aqueous buffer systems. A number of different methods are available for disrupting protein−lipid and protein−protein interactions in the membrane.

(1)   Strong denaturants—urea or guanidine hydrochloride.

(2)   Chaotrophs—thiocyanate or iodide.

(3)   Organic solvents—ethanol or isopropanol.

(4)   Detergents (66)—see *Table 9*.

Agents in groups (1), (2) and (3), although successful in disrupting hydrophobic interactions between lipids and proteins, may result in denaturation of the protein and prevent detection in a specific assay. Detergents are by far the most successful membrane solubilizing agents; selected uses are shown in *Table 9*.

The efficiency of solubilization of a particular glycoprotein from membranes is a unique property of that glycoprotein and therefore extraction of each will require some initial development. A major consideration, however, in deciding an appropriate detergent should be the effect of particular detergents on the efficiency of glycoprotein binding to particular immobilized lectins (see *Table 10*). Some general points can be discerned from these results.

(1)   Chromatography with ConA is inefficient in the presence of detergents.

(2)   Non-ionic detergents (Triton X-100, Nonidet P-40) have a less detrimental effect on lectins than other detergents.

(3)   SDS will reduce binding, probably by unfolding the immobilized lectin.

(4)   Detergents are preferentially used at less than 1% (w/v or v/v) for optimum solubilization and binding efficiency.

Disruption of cells by detergents results in the release of lysosomal proteases that will rapidly degrade or modify many membrane-bound glycoproteins. At the same time

Table 9. Selected examples of detergents used to solubilize glycoproteins prior to lectin chromatography.

Detergent	Concentration	Protein	Cell source	Lectin chromatography	References
Non-ionic					
Triton X-100	1% (v/v)	NGF receptor	human melanoma	WGA	67
	0.5%	glycophorin	human erythrocytes	WGA	68
	1%	IFN-γ-receptor	human foreskin fibroblasts	RCA120	69
		acetylcholinesterase	rat brain	LCA, WGA and/or RCA	65
Nonidet P-40	2.5%	PDGF receptor	porcine uterus	WGA	70
		H-2K, H-2D Thy-1	mouse T-cells	ConA	71
Octyl β-D-glucoside	60 mM	PDGF receptor	mouse 3T3 cells	WGA	72
Anionic					
Sodium deoxycholate	1% (w/v)	HLA antigen	human lymphoblastoid cell line 1M1	LCA	73
	6.6%	viral envelope glycoprotein	influenza, sendai, mouse mammary tumour viruses	LCA	74
Cationic					
Dodecyl trimethyl ammonium bromide	1.4%	rhodopsin	bovine retina	ConA	75

Table 10. Effect of increasing concentrations of detergents on the glycoprotein binding efficiencies of immobilized lectins[a] (65).

Detergent (%)	Relative binding efficiencies[b] (%)				
	R. communis agglutinin	Peanut agglutinin	Soybean agglutinin	Concanavalin A	Wheat germ agglutinin
Nonidet P-40					
0.1	100	103	100	88	95
1.0	98	87	100	78	90
2.5	100	92	100	80	92
Triton X-100					
0.1	105	92	100	85	90
1.0	105	90	100	76	76
2.5	100	90	99	74	79
Deoxycholate					
0.1	100	90	53	60	85
1.0	102	85	34	46	39
2.5	105	37	10	20	10
SDS					
0.05	100	82	22	70	89
0.1	70	64	11	26	13

[a]Lectins immobilized on glutaraldehyde-substituted polyacrylichydrazido-agarose were treated for 30 min in solutions of the various detergents at the indicated detergent concentrations. The binding of [^{3}H]fetuin or [^{3}H]asialofetuin to the immobilized lectins was compared in the absence and in the presence of the detergents, and the binding efficiencies were calculated from the equation:

$$\frac{\text{c.p.m. eluted with saccharide}}{\text{c.p.m. removed without saccharide} + \text{c.p.m. eluted with saccharide}} \times 10$$

[b]Calculated according to the equation:

$$\frac{\text{binding efficiency in detergent}}{\text{binding efficiency in absence of detergent}} \times 100$$

as using the detergent it is therefore useful to include one or more protease inhibitors. Those most commonly used are dithiothreitol (inhibition of thiol-dependent proteases), EDTA or EGTA (metalloproteinases), PMSF (serine proteases), pepstatin (acid proteases) and leupeptin (thiol-dependent proteases).

5.2.2 *Preparation of the adsorbent and chromatography*

Before use the immobilized lectin should be washed with an equilibration buffer (e.g. 20 mM phosphate buffer pH 7.2, containing 1 M sodium chloride). This is best performed by packing the gel into a column and washing with five column volumes of equilibration buffer. The size of column used will be dependent on the capacity of the lectin for the glycoproteins being purified. Optimally chromatography should initially be carried out on a small scale (1−5 ml of immobilized lectin) to assess capacity (and elution conditions) before scaling up (*Method Table 3*).

(i) *Sample application.* The sample should be prepared according to the guidelines above and applied directly to the equilibrated column. The column should then be washed with equilibration buffer (two column volumes at least, but preferably five volumes) to remove non-binding or isocratically eluting proteins. After the sample has been applied and the immobilized lectin washed it is advantageous to reverse the flow through the

Method Table 3. Method for using lectin columns.

1.	Equilibrate the gel with five column volumes of 20 mM phosphate buffer pH 7.2 containing 1 M sodium chloride (equilibration buffer). Include divalent metal ions (0.1 mM) and detergents (0.5%) as appropriate for lectin binding efficiency and sample stability respectively.
2.	Apply sample in equilibration buffer.
3.	Wash the column with five column volumes of equilibration buffer.
4.	Elute with equilibration buffer containing specific sugar (see *Table 11*, 1st option) using 0.1 M increments; five column volumes per increment.
5.	Elute with five column volumes of 0.5 M specific sugar containing 50% ethylene glycol in equilibration buffer.
6.	Regenerate column with 10 column volumes of equilibration buffer.

column. This procedure avoids unnecessary interaction with unused immobilized lectin as the glycoproteins elute from the column.

(ii) *Elution of bound glycoproteins.* A number of approaches are available for elution of bound glycoproteins. Various eluents can be applied either stepwise or by gradient.

(1) Most commonly glycoproteins are eluted with the sugar or an analogue for which the lectin has an affinity. For example glycoproteins binding to ConA are frequently eluted with α-D-methyl mannoside (0−0.5 M) in equilibration buffer and those binding to soybean agglutinin eluted with α-D-*N*-acetyl galactosamine (0−0.5 M) in equilibration buffer (*Table 11*).

(2) Hydrophobic interaction plays a significant role in binding many glycoproteins to lectin columns, therefore, ethylene glycol (0−50%), in addition to the specific sugar, can be incorporated into the elution buffer.

(3) Glycoproteins can be eluted from immobilized lectins by changing the pH to acid (but not below pH 3.0), or alkali (but not above pH 10.0). This method is not always suitable particularly if the pH stability of the glycoprotein of interest is unknown.

(4) Borate buffers can be used to elute glycoproteins as borate ions form complexes with some polysaccharides. Such glycoproteins probably have a high proportion of sugar residues and may purify equally well on phenyl boronate matrices.

Elution from a lectin column may be time-dependent and improved yields can be achieved by overnight incubation of the column, with glycoproteins bound, in the elution buffer.

5.2.3 *Regeneration and storage of adsorbents*

After chromatography, replace the elution buffer in the gel with 0.1 M acetate buffer pH 6.0. Include 100 μM of the appropriate divalent metal ions ($CaCl_2$ and $MnCl_2$ or $MgCl_2$; see *Table 6*) and a bacteriostatic agent, such as 0.05% (w/v) sodium azide. Storage of the lectin adsorbent under these conditons at 4°C should ensure stability for at least 18 months.

Denaturants (guanidine hydrochloride, urea) and proteinases should be avoided. Those

Table 11. Buffers for eluting glycoproteins from lectin columns.

Lectin	Elution conditions		
	1st Option	2nd Option	3rd Option
ConA	0.1–0.5 M α-methyl mannoside	0.1–0.5 M α-methyl glucoside	sugar + 50% ethylene glycol
LCA	0.1–0.5 M α-methyl mannoside	0.1–0.5 M α-methyl glucoside	sugar + 50% ethylene glycol
SBA	0.1–0.5 M N-acetyl galactosamine	sugar + 50% ethylene glycol	0.1–0.5 M galactose
WGA	0.1–0.5 M N-acetyl glucosamine	N,N′,N″ triacetyl chitotriose	sugar + 50% ethylene glycol
PHA	0.1–0.5 M N-acetyl galactosamine	sugar + 50% ethylene glycol	
RCA$_{60}$	0.1–0.5 M N-acetyl galactosamine	sugar + 50% ethylene glycol	
RCA$_{120}$	0.1–0.5 M Lactose	0.1–0.5 M galactose	sugar + 50% ethylene glycol
BS-I	0.1–0.5 M melibiose	0.1–0.5 M galactose	sugar + 50% ethylene glycol
BS-II	0.1–0.5 M melibiose	0.1–0.5 M glucosamine	sugar + 50% ethylene glycol
PNA	0.1–0.5 M lactose	0.1–0.5 M galactose	sugar + 50% ethylene glycol
Pea lectin	0.1–0.5 M α-methyl mannoside	sugar + 50% ethylene glycol	

lectins dependent on divalent metal ions should not be exposed to chelating agents, such as EDTA, and pH of buffering solutions should not exceed the range of 3−9.

6. IMMUNOPURIFICATION—C.R.Hill, L.G.Thompson and A.C.Kenney

Immunopurification is one of the most selective and powerful purification methods available. Antibodies can be found that can distinguish between very similar antigens (76) and overcome separation difficulties that no other method can resolve. Monoclonal antibodies can be obtained by immunization with relatively impure antigen, thus, they can be obtained before alternative purification methods have been developed and so save time and effort in obtaining pure antigen. There may be problems associated with the removal of a particular contaminant from a relatively pure protein (for example, it is often difficult to remove the last traces of albumin from proteins purified from serum or serum-containing cell culture supernatants). In this case immunopurification can be used to subtract unwanted trace contaminants.

A recent study (77) showed that affinity chromatography is rarely used early in a purification procedure, where the high degree of purification often achieved can be exploited to the full. Perhaps the main reason for this is the desire to protect a valuable material from both proteolytic attack and fouling by contaminants in the crude protein solution.

Immunopurification has been perceived as one of the most expensive affinity methods, particularly when using monoclonal antibodies. However, the expansion of monoclonal antibody technology, including methods for large-scale economic production (78,79) has significantly reduced the costs of immunopurification, allowing the development of production scale applications (80−84).

In general antibodies as a class of proteins are particularly resistant to proteolytic attack and are cleaved quite selectively by only a limited number of enzymes. This can give immunopurification a distinct advantage over other affinity methods employing immobilized proteins and enables the method to be used early in a sequence of purification steps, where its power can best be exploited. Indeed immunopurification can often be considered the method of choice for purification of any protein for which a suitable monoclonal antibody is available. Furthermore, monoclonal antibodies are often raised against a new protein of interest early in a research programme to facilitate assay of the protein and identification by immunoblotting, thus antibodies are often available for use in purification.

Given the availability of a monoclonal antibody it is usually a simple task to develop an immunopurification method that gives good results at least for research purposes. Many pre-activated support matrices are readily available (85,86) so workers do not need to be exposed to the, often highly toxic, chemicals needed for the activation. These matrices often can be simply mixed with the antibody in a suitable buffer, then washed and blocked following a straightforward protocol. Some care is necessary to find a suitable eluent to remove the antigen of interest from the column; the stability of both the antigen and the antibody need to be borne in mind when selecting the eluent. Fortunately, most antibodies are remarkably robust to extremes of pH and ionic strength, and it is usually the nature of the antigen that dictates the composition of the eluent.

Table 12. Criteria used for the selection of monoclonal antibodies for development of an immunopurification reagent.

Stage of development	Selection criteria	Typical number of cell lines in panel
1. Screening of hybridoma	Specificity Affinity	50−200
2. Hybridoma culture	Growth kinetics Nutritional requirements Cell line stability	10−15
3. Monoclonal antibody purification	Yield of antibody Stability of antibody	5−8
4. Immunopurification reagent development	Immobilization yield Elution buffer Capacity for antigen Yield of antigen Purity of antigen Re-useability of reagent Leakage of IgG	4−6

6.1 Antibody selection

The development of a successful immunopurification reagent is dependent upon the selection of a monoclonal antibody with specific properties. Production of monoclonal antibodies can often result in a large panel of antibodies. Several screening procedures are required to identify those antibodies with appropriate properties for development into an immunopurification reagent.

The steps in the development of an immunopurification reagent, at which decisions regarding the selection of an antibody can be made, are listed in *Table 12*. Also given are a number of important criteria to be considered during the selection process.

6.1.1 *Stage 1*

A key step in the selection is identification of an antibody with an appropriate affinity and specificity for the antigen (87−89). A high affinity antibody should be chosen to allow efficient recovery of a dilute antigen from a complex mixture. However, with a high affinity antibody it may be difficult to recover the antigen. Thus, an antibody with an intermediate affinity may be more desirable (90).

It is not practicable to grow 50−200 cell lines in sufficient quantity to purify antibody from each. Screening assays that can be applied during the hybridoma development phase are therefore required, preferably using the supernatants from clones grown in 96-well microtitre plates. At its simplest this may involve determining their relative titres by titrating the antibodies against antigen immobilized on a microtitre plate.

The development of specific assays for screening hybridoma cell lines to identify antibodies with a suitable affinity for immunopurification has been described by Rubinstein and coworkers for immunopurification of α and β interferon (87,99). The objective at this stage should be to identify perhaps 10−15 antibodies for further development.

6.1.2 *Stage 2*

The next stage in the selection procedure requires production of 10 – 100 mg of each antibody. This may be achieved by growth either as an ascites tumour or in suspension cell culture. If cell culture is used, information on the stability of the cell lines, their growth kinetics, and nutritional requirements can be generated. Those cell lines that exhibit low growth rates or poor antibody yields can be eliminated at this stage. Monoclonal antibodies (MAbs) from the remaining cell lines can then be purified using a variety of techniques (37).

6.1.3 *Stage 3*

At this stage information relating to yield and purity of antibody together with its isoelectric point, solubility and stability can be obtained. Those MAbs that are difficult to purify or are unstable (91) can be eliminated at this stage.

6.1.4 *Stage 4*

Further selection will be based on properties after immobilization on a suitable matrix. Important criteria to be considered at this stage include yield of immobilized antibody, elution conditions required to recover antigen, capacity for antigen, yield and purity of antigen recovered, re-useability of the reagent and leakage of immobilized antibody.

6.2 Immobilization of antibodies on CNBr-activated Sepharose

For most research purposes, CNBr-activated Sepharose (Pharmacia-LKB) will be adequate. The material is readily available and convenient to use; the detailed use of this material is described later. However, it is important to draw attention to the situations where this material may be less useful. For these cases the reader is directed to suitable alternatives (85,86). There are three major limitations of CNBr-activated Sepharose.

(i)   It is a relatively soft gel that cannot support high flow rates, or be used effectively in large columns.

(ii)   It has an exclusion limit of 20×10^6, thus severely limiting the surface area accessible to high molecular weight antigens (e.g. IgM).

(iii) The covalent bonds formed between the ligand and the matrix are relatively labile (92), thus leakage of the immobilized antibody into the antigen can be a problem.

Clearly these limitations will not affect the majority of research applications.

(i) *Preparation of the antibody.*

(1)   Dialyse the antibody at 4°C against fresh coupling buffer [either 0.1 M sodium hydrogen carbonate (NaHCO$_3$), pH 8.3 containing 0.5 M sodium chloride, or 0.1 M sodium borate buffer, pH 8]. As a guide, the amount of antibody required for immobilization is approximately 1 – 10 mg ml^{-1} gel for monoclonal antibodies and about 10 – 20 mg ml^{-1} gel for polyclonal antibodies. These quantities vary according to the avidity of the antibodies for their antigens. Excess antibody may result in reduced efficiency due to steric hindrance of the immobilized antibody.

(2)   Adjust the concentration of antibody in coupling buffer to 5 – 10 mg ml^{-1}. This

can be estimated by measuring A_{280} using an average extinction coefficient for antibodies of 1.4 for 1 mg ml^{-1} protein using a 1 cm path length cell.

(ii) *Preparation of CNBr-activated Sepharose-4B for coupling antibody.*

(1)   Chill a small volume ($\sim$ 10 gel volumes) of the chosen coupling buffer to 4°C and prepare a solution of 1 mM HCl ($\sim$ 200 ml g^{-1} gel).

(2)   Weigh out the desired amount of CNBr-activated Sepharose-4B powder immediately prior to use (0.3 g produces $\sim$ 1 ml of swollen gel). The powder should not be allowed to stand for any length of time in the open laboratory as it absorbs moisture from the atmosphere, which can destory the CNBr-activated sites.

(3)   Sprinkle the dry powder slowly onto 10 volumes of 1 mM HCl contained in a sintered glass funnel.

(4)   Stir the gel gently with a spatula to ensure that all the particles are dispersed. The gel particles are extremely fragile and should be treated gently. Shear forces caused by magnetic stirrer bars and virgorous manual stirring can cause the particles to fracture, producing fines which adversely affect ultimate flow rates and can block sinters and column support nets.

(5)   Allow the gel to swell at room temperature for approximately 15 min and then flush through with the remaining 1 mM HCl under suction, taking great care not to allow the gel to dry, as this can also break down the matrix. A total of 200 ml of 1 mM HCl per gram of dry gel is sufficient for this washing step.

(6)   Finally wash the gel with 10 gel volumes of chilled (4°C) coupling buffer.

(iii) *The coupling reaction.*

(1)   Mix one volume of washed gel with five volumes of antibody solution in coupling buffer at 4°C. It is important that both reactants are at 4°C prior to mixing, as at higher temperatures the CNBr-activated groups react faster with the amino groups on proteins. This can cause dense coupling of antibody molecules to the external surfaces of the gel and result in reduced efficiency.

(2)   Incubate the coupling mixture at 4°C for 16−20 h; continual mixing is necessary to prevent the gel settling and a rolling motion rather than an end-over-end motion is more gentle on the gel beads.

(iv) *Blocking and washing the Sepharose gel.* When the coupling reaction is complete it is advisable to test the success of the reaction before proceeding.

(1)   Allow the mixture to settle for a few minutes.

(2)   Take a sample ($\sim$ 1 ml) of the supernatant, centrifuge to clarify and measure the A_{280} of the supernatant; an absorbance close to zero suggests a successful coupling reaction.

(3)   Add this supernatant back to the reaction mixture.

(4)   Filter the coupling reaction mixture by gentle suction in a sintered glass funnel, and wash the gel with a further five volumes of fresh coupling buffer.

(5)   Collect all filtrates in one vessel and determine protein content. Do not allow the gel to dry.

Any remaining CNBr-activated groups on the matrix are then blocked by adding an excess of a primary amino-containing compound. A variety of compounds may be used for this purpose (e.g. ethanolamine, glycine) and each should be considered in the light of the property it may impart to the final reagent. For example, ion-exchange properties or hydrophobic character could enhance the non-specific binding of contaminant proteins to the reagent. The most commonly used blocking agent is 1 M ethanolamine adjusted to pH 8.0 with HCl. Care must be taken in the preparation of this solution, since acidification requires a high volume ratio of concentrated hydrochloric acid to be added to the ethanolamine, and the reaction is exothermic. The following procedure should be followed.

(1)   Incubate the antibody matrix with blocking solution at room temperature for $1-2$ h with continual rolling.

(2)   Remove excess ethanolamine from the gel using a sintered glass funnel and gentle suction, taking care not to dry the gel out.

(3)   Wash the gel in three alternating cycles of high and low pH buffers (0.1 M sodium acetate buffer, pH 4 containing 0.5 M sodium chloride; and 0.1 M sodium carbonate buffer, pH 8.3 containing 0.5 M sodium chloride). These washing cycles promote the release of non-covalently bound antibodies. Use a sintered funnel for the washing and allow the gel to stand at the appropriate pH for $5-10$ min during each cycle. Use 10 gel volumes of each wash solution.

(4)   A further series of washing cycles to minimize eventual antibody leakage can be introduced at this point. For this, the equilibration buffer and elution buffer (see below) are used.

(5)   Store the gel at 4°C in equilibration buffer with 0.1% w/v sodium azide added. Once swollen, the gel should never be frozen.

A convenient means of aliquoting known volumes of gel is as follows. Allow the final mixture to settle for approximately 72 h before adjusting the volume of liquid on top of the gel to equal that of the gel. Each time gel is required from the container it can be gently mixed to form an even slurry and two times the desired volume of settled gel removed.

6.3 Procedures for immunopurification

There are two ways of employing immunopurification. Firstly, in a positive fashion, to purify the antigen from a crude mixture, and secondly, in a negative (subtractive) fashion, to remove a specific contaminant from a partially purified antigen solution. There are also two ways of executing the method, that is either by a batch method or a column procedure.

6.3.1 *Positive immunopurification*

In this method the immobilized antibody is specific for the protein to be purified. It is therefore important to determine under what conditions the antigen binds to the immunoadsorbent and what conditions are necessary to effect dissociation of the complex. All conditions applied in this step should be compatible with both the immunoadsorbent and the antigen.

(i) *Equilibration.* The equilibration buffer is one in which the antigen binds to the immobilized antibody. It is usually a neutral, low ionic strength buffer such as 40 mM sodium phosphate buffer pH 7.2, or phosphate buffered saline. The immunoadsorbent is stored in this buffer containing 0.1% sodium azide, and equilibrated in fresh buffer, without azide, when required for use. The immunoadsorbent should be equilibrated to the same temperature as the equilibration buffer and the environment in which the column is to be used; otherwise air bubbles can form and be trapped if the matrix is already packed in a column. Pack the immunoadsorbent in an appropriate column. Equilibrate in 10 or more volumes of equilibration buffer (to remove traces of azide used in storage and to flush out any ligand, which may have leached on storage). The column is now ready for sample application.

(ii) *Sample loading.* Ensure the crude antigen sample is in equilibration buffer, for example by dialysis, and allow it to equilibrate to the same temperature as the packed column. If the antigen solution contains any particulate matter (i.e. protein precipitate or cell debris) clarify by centrifugation or filtration immediately prior to loading or pass via a 0.22 μm filter onto the column. This procedure protects the immunoadsorbent and prevents column blockage. The flow rate should be as slow as convenient and no faster than that recommended for the Sepharose-4B matrix (~ 30 ml cm^{-2} h^{-1}). A slow flow rate allows all antigen molecules passing through the column to saturate all available binding sites, whereas a faster flow rate risks losing some antigen in the flow-through before capacity is reached.

The capacity of an immunoadsorbent is defined as the maximum amount of antigen that can be bound and recovered per unit volume and should be experimentally determined for each immunoadsorbent.

Theoretically each monoclonal IgG antibody immobilized can bind two antigen molecules. However, in practice only a small percentage of antigen binding sites remain available for binding antigen after covalent immobilization to a matrix (93). Typically this figure will be approximately 10% of available antigen binding sites. However, this can vary over a wide range ($< 1 - 50$). A number of factors will affect the apparent efficiency of the matrix. These include coupling chemistry, matrix pore size, antigen size and antibody structure. For example an antibody that contains a reactive amino group in the antigen combining site may well result in a matrix with a very low overall efficiency.

(iii) *Washing.* Once the antigen has bound to the immunoadsorbent it is necessary to wash off the non-specifically bound contaminant molecules prior to elution. This may be done by passing equilibration buffer through the column until the absorbance at 280 nm of the effluent is about 0 (~ 10 gel volumes). If the Ab:Ag complex is sufficiently strong, inclusion of approximately 0.5 M sodium or potassium chloride in the equilibration buffer may result in a purer product.

(iv) *Elution.* Elution of bound antigen from the immunoadsorbent is effected by breaking the bonds which form the complex. These bonds will be composed of a mixture of weak physical forces such as coulombic salt bridges, hydrogen bonds and Van der Waal's forces.

Elution is usually achieved by a change in ionic strength, pH, dielectric constant, surface tension or temperature. However, to preserve the antigen and immunoadsorbent activity the mildest conditions should be tried before the more stringent conditions. Thus the order in which to try eluents can be summarized as increased ionic strength followed by extremes of pH, chaotropic agents, protein deforming agents and organic

Table 13. Some elution conditions for immunopurification in the order in which they should be tried.

Elution conditions	References
Extremes of pH	
0.1 M Glycine−HCl, pH 2.5	94
1.0 M Propionic acid	95
0.15 M NH$_4$OH, pH 10.5	96
0.1 M NaCaps, pH 10.7	81
Chaotropic salts	
4.0 M MgCl$_2$, pH 7.0	97
2.5 M NaI, pH 7.5	98
3.0 M NaSCN, pH 7.4	99
Organic chaotropes	
8 M Urea, pH 7.0	100
6 M Guanidine−HCl	101
Organic solvents	
50% (v/v) Ethanediol, pH 11.5	102, 103
Dioxane/acetic acid	102, 103

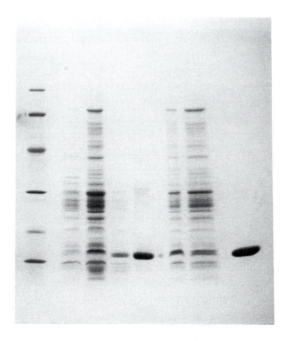

Figure 9. Positive immunopurification of interleukin 2 (IL-2). Coomassie Blue-stained 7.5−15% SDS−PAGE run under reducing conditions. Tracks from **left to right**: mol weight standards, sample load flow-through (and 5 volumes), column wash (1 and 5 volumes), eluted IL-2 (1 and 5 volumes), crude IL-2 *E.coli* lysate (1 and 5 volumes), IL-2 standard. Photograph provided by D.Brady.

solvents (*Table 13*). Since an elution step requires only two gel volumes of elution buffer, elution can be achieved in a relatively short time; thus if the elution condition is particularly harsh a rapid change over to the re-equilibration step and/or a rapid buffer

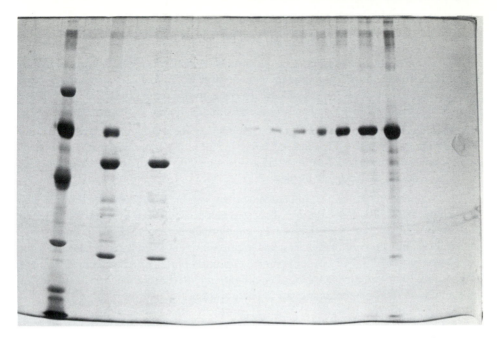

Figure 10. Negative immunopurification of BSA contamination from a monoclonal antibody. Siler stained 7.5−15% SDS−PAGE run under reducing conditions. Tracks from **left to right**: mol. wt standards, sample before purification, sample after purification, BSA standards: 0, 5 ng, 10 ng, 25 ng, 50 ng, 100 ng, 500 ng, 1 μg. Photograph provided by A.Nash.

change for the purified antigen can reduce damage caused by elution. This can be achieved either by collecting the eluent into tubes containing a neutralizing buffer or by rapid gel filtration.

6.3.2 *Example of positive immunopurification of interleukin 2 (IL-2)*

A monoclonal antibody to IL-2 was immobilized on CNBr-activated Sepharose at a concentration of 8 mg IgG ml^{-1} packed gel. The immunoadsorbent was packed into a suitable column and equilibrated with 0.1 M sodium phosphate buffer, pH 7.5 containing 0.5 M sodium chloride. Crude IL-2 derived from *E.coli* lysate was loaded onto the column at 100 ml cm^{-2} h^{-1}. The column was washed with 10 gel volumes of 0.1 M sodium phosphate buffer, pH 7.5 containing 1.0 M potassium chloride. IL-2 was eluted from the immunoadsorbent with five gel volumes of 0.1 M sodium acetate buffer, pH 4.0 containing 1.0 M potassium chloride. The eluent was neutralized by collecting directly into 2.5 gel volumes of 2 M Tris−HCl, pH 7.5 and the immuno-adsorbent was immediately re-equilibrated in equilibration buffer containing 0.1% azide.

Electrophoretic analysis of various fractions can be seen in *Figure 9*. A purification of about 1000-fold was achieved in this single step.

6.3.3 *Negative (subtractive) immunopurification*

In this method the immobilized antibody is specific for a particular contaminant. The principles described in the previous section for positive immunopurification still apply but the emphasis is on different parts of the method.

(i) *Equilibration.* The main criteria for selection of an equilibration buffer are suitability to allow or promote the antigen binding to the immunoadsorbent and compatibility with the desired protein. This buffer will usually be a neutral, low ionic strength buffer.

The column should be packed and equilibrated in this buffer as described previously.

(ii) *Purification.* Prior to application to the immunoadsorbent column, dialyse the extract into equilibration buffer. To be sure of removing all of the contaminant it is important not to exceed the capacity of the immunoadsorbent and to use an appropriate flow rate to allow contaminant antigens to bind.

(iii) *Column regeneration.* The immunoadsorbent can be recycled many times if it is regenerated carefully after each use. This step can be compared with the elution step for positive immunopurification, since the same principles apply. However, unless the contaminant antigen is required, it is less important to consider its compatibility with the elution buffer. Apply $2-5$ gel volumes of elution buffer to the immunoadsorbent followed by 10 or more gel volumes of equilibration buffer. Store the immunoadsorbent in equilibration buffer containing 0.1% sodium azide.

6.3.4 *Example of negative immunopurification*

A monoclonal antibody to bovine serum albumin (BSA) was immobilized onto CNBr-Sepharose at 5 mg IgG ml^{-1} packed gel. The immunoadsorbent was packed into a suitable column with PBS. Two gel volumes of elution buffer (0.1 M glycine$-$HCl pH 2.5) were passed through the column as a pre-wash. The column was then equilibrated with 10 or more gel volumes of PBS.

A sample of a monoclonal antibody containing BSA was applied to the immunoadsorbent column at 10 ml cm^{-2} h^{-1}. The unbound proteins were collected and analysed by SDS-electrophoresis, together with the sample applied (*Figure 10*). An estimate from the electrophoresis gel of BSA clearance shows a reduction in BSA from 10% (w/w) to less than 0.05%, demonstrating at least a 200-fold reduction in BSA in a single pass.

7. ACKNOWLEDGEMENTS

M.Cusack and R.J.Beynon are grateful to Rohm Pharma for making available literature and oxirane acrylic beads. M.Cusack is grateful to the SERC for a studentship. Part of the work described in this chapter was supported by grants from the SERC.

8. REFERENCES

1. Dean,P.D.G., Johnson,W.S. and Middle,F.A. (eds) (1985) *Affinity Chromatography: A Practical Approach.* IRL Press, Oxford.
2. Jakoby,W. (ed.) (1984) *Methods in Enzymology.* Academic Press, New York, Vol. 104C.
3. Scouten,W.H. (1981) *Affinity Chromatography.* Wiley-Interscience, New York.
4. Lowe,C.R. (1979) *An Introduction to Affinity Chromatography.* Elsevier Biomedical, Amsterdam.
5. Lowe,C.R. and Dean,P.D.G. (1974) *Affinity Chromatography.* John Wiley & Sons, London.
6. Jackoby,W. (ed.) (1974) *Methods in Enzymology.* Academic Press, New York, Vol. 34.
7. Turkova,J. (1978) *Affinity Chromatography in J. Chromatography Library.* Elsevier, Amsterdam, Vol. **12**.

8. Jennissen,H.P. and Muller,W. (eds) (1988) *Die Makromol. Chem. Macromol. Symp.*, **17**.
9. Chaiken,I.M., Wilchek,M. and Parikh,I. (eds), (1983) *Affinity Chromatography and Biological Recognition*. Academic Press, New York.
10. Gribnau,T.C., Visser,J. and Nivard,R.T.F. (eds) (1982) *Affinity Chromatography and Related Techniques*. Elsevier, Amsterdam.
11. Hoffman-Ostenoff,O., Breitenbach,M., Koller,F., Kraft,D. and Scheirier,O. (eds) (1978) *Affinity Chromatography*. Pergamon Press, New York.
12. Fowell,S.J. and Chase,H.A. (1986) *J. Biotechnol.*, **4**, 355.
13. Angal,S. and Dean,P.D.G. (1977) *Biochem. J.*, **167**, 301.
14. Groman,E.V. and Wilchek,M. (1987) *TIBTECH*, **5**, 220.
15. Kohn,J. and Wilchek,M. (1982) *Biochem. Biophys. Res. Commun.*, **107**, 878.
16. Bethell,G.S., Ayers,J.S., Hancock,W.S. and Hearn,M.T.W. (1979) *J. Biol. Chem.*, **254**, 2572.
17. Nilsson,K. and Mosbach,K. (1980) *Eur. J. Biochem.*, **112**, 397.
18. Ngo,T.T. (1986) *Bio/Technology*, **4**, 134.
19. Weston,R.D. and Avrameas,S. (1971) *Biochem. Biophys. Res. Commun.*, **45**, 1574.
20. Avrameas,S., Ternynck,T. and Guesdon,J. (1978) *Scand. J. Immunol.*, **8**, 7.
21. Lowe,C.R., Harvey,M.J. and Dean,P.D.G. (1974) *Eur. J. Biochem.*, **41**, 341.
22. Scopes,R.K. (1982) In *Affinity Chromatography and Related Techniques*. Analytical Chemistry Symposium Series Vol. 9, p. 333.
23. Harvey,M.J., Lowe,C.R. and Dean,P.D.G. (1974) *Eur. J. Biochem.*, **41**, 353.
24. Hughes,P., Lowe,C.R. and Sherwood,R.F. (1982) *Biochim. Biophys. Acta*, **700**, 90.
25. Morgan,M.R.A., Slater,N.A. and Dean,P.D.G. (1978) *Anal. Biochem.*, **92**, 144.
26. Nicol,L.W., Ogston,A.G., Winzor,D.J. and Sawyer,W.H. (1974) *Biochem. J.*, **143**, 435.
27. Osterman,L.A. (1986) *Methods of Protein and Nucleic Acid Research, Part 3—Chromatography*. Springer Verlag, New York, p. 308.
28. Tesser,G.I., Fisch,H.-U. and Schwyzer,R. (1974) *Helv. Chim. Acta*, **57**, 1718.
29. Bristow,A. (1989) In *Protein Purification Applications: A Practical Approach*. Harris,E.L.V. and Angal,S. (eds), IRL Press, Oxford.
30. Kruger,N.J., and Hammond,J.B.W. (1988) In *Methods in Molecular Biology*. Walker,J.M. (ed.), Vol. III, p. 363.
31. Frederiksson,G., Nilsson,S., Olsson,H., Bjorck,L., Akerstrom,B. and Belfrage,P. (1987) *J. Immunol. Methods*, **97**, 65.
32. Dean,P.D.G. and Watson,D.H. (1979) *J. Chromatogr.*, **165**, 301.
33. Lowe,C.R. and Pearson,J.C. (1984) *Methods in Enzymology*, Vol. 104, p. 97.
34. Scopes,R.K. (1986) *J. Chromatogr.*, **376**, 131.
35. Hey,Y. and Dean,P.D.G. (1983) *Biochem. J.*, **109**, 363.
36. Radcliffe,R. and Heinze,T. (1978) *Arch. Biochem. Biophys.*, **189**, 185.
37. Harris,E.L.V. and Angal,S. (eds) (1989) *Protein Purification Applications: A Practical Approach*. IRL Press, Oxford.
38. Sturgeon,C.M. and Kennedy,J.F. *Enzyme and Microbial Technology*, (regular feature).
39. Kraemer,D.M., Lehmann,K., Penneweiss,H. and Plainer,H. (1978) In *Enzyme Engineering*. Brown,G.B. and Manecke,G. and Wingared,I.B. (eds), Plenum Publishing Corp., New York, Vol. 4, p. 153.
40. Zemanová,I., Turkova,J., Capka,M., Nakhapetyan,L.A., Svec,F. and Kala,J. (1980) *Enzyme Microb. Technol.*, **3**, 229.
41. Hannibal-Friedrich,O., Chun,M. and Sernetz,M. (1980) *Biotech. Bioeng.*, **22**, 157.
42. Verheijen,J.H., Mullaart,E., Chang,G.T.G., Kluft,C. and Wijngaards,G. (1982) *Thromb. Haemostas.*, **48**, 266.
43. Solomon,B., Koppel,R., Pines,G. and Katchalski-Katzir,E. (1986) *Biotech. Bioeng.*, **28**, 1213.
44. Solomon,B., Koppel,R. and Katchalski-Katzir,E. (1984) *Biotechnology*, **2**, 709.
45. Bamberger,U., Scheuber,P.H., Sailer-Kramer,B., Bartsch,K., Hartman,A., Beck,G. and Hammer,D.K. (1986) *Proc. Natl. Acad. Sci. USA*, **83**, 7054.
46. Artmann,U. and Borchert,A. (1985) *BTF-Biotech Forum 2*, **3**, 119.
47. Laumen,K., Reimerdes,E.H. and Schneider,M. (1985) *Tetrahedron Lett.*, **26**, 407.
48. Huck,H., Schelter-Graf,A. and Schmidt,H.-L. (1984) *Bioelectrochem. Bioenerg.*, **13**, 199.
49. Bihari,V. and Buchholz,K. (1983) *Indian J. Exp. Biol.*, **21**, 27.
50. Keinan,E., Hafeli,E.K., Seth,K.K. and Lamed,R. (1986) *J. Am. Chem. Soc.*, **108**, 162.
51. Lis,H. and Sharon,N. (1973) *Annu. Rev. Biochem.*, **42**, 541.
52. Barondes,S.H. (1981) *Annu. Rev. Biochem.*, **50**, 207.
53. Olson,M.O.J. and Liener,I.E. (1967) *Biochemistry*, **6**, 105.
54. Baues,R.J. and Gray,G.R. (1977) *J. Biol. Chem.*, **252**, 57.
55. Toyoshima,S., Osawa,T. and Tonomura,A. (1970) *Biochim. Biophys. Acta*, **221**, 514.

56. Vretblad,P. (1976) *Biochim. Biophys. Acta,* **434**, 169.
57. Nicolson,G.L. and Blaustein,J. (1972) *Biochim. Biophys. Acta,* **266**, 543.
58. Le Vine,D., Kaplan,M.J. and Greenaway,P.J. (1972) *Biochem. J.,* **129**, 847.
59. Entlicher,G., Kostir,J.V. and Kocourek,J. (1970) *Biochim. Biophys. Acta,* **221**, 272.
60. Osawa,T. and Tsuji,T. (1987) *Annu. Rev. Biochem.,* **56**, 21.
61. Thomas,D.B. and Winzler,R.J. (1969) *J. Biol. Chem.,* **244**, 5943.
62. Tsuji,T., Yamamoto,K., Irimura,T. and Osawa,T. (1981) *Biochem. J.,* **195**, 691.
63. Yamamoto,K., Tsuji,T., Irimura,T. and Osawa,T. (1981) *Biochem. J.,* **195**, 701.
64. Yamashita,K., Tachibana,T. and Kobata,A. (1978) *J. Biol. Chem.,* **253**, 3862.
65. Lotan,R. and Nicolson,G.L. (1979) *Biochim. Biophys. Acta,* **559**, 329.
66. Findlay,J. (1989) In *Protein Purification Applications: A Practical Approach.* Harris,E.L.V. and Angal,S. (eds), IRL Press, Oxford.
67. Grob,P.M., Berlot,C.H. and Bothwell,M.A. (1983) *Proc. Natl. Acad. Sci. USA,* **80**, 6819.
68. Adair,W.L. and Kornfeld,S.(1974) *J. Biol. Chem.,* **249**, 4696.
69. Novick,D., Orchansky,P., Revel,M. and Rubenstein,M. (1987) *J. Biol. Chem.,* **262**, 8483.
70. Ronnstrand,L., Beckmann,M.P., Faulders,B., Ostman,A., Ek,B., Heldin,C.-H. (1987) *J. Biol. Chem.,* **262**, 2929.
71. Nilsson,S.F. and Waxdal,M.J. (1978) *Biochemistry,* **17**, 903.
72. Daniel,T.O., Tremble,P.M., Frackelton,Jr.A.R. and Williams,L.T. (1985) *Proc. Natl. Acad. Sci. USA,* **82**, 2684.
73. Dawson,J.R., Silver,J., Sheppard,L.B. and Amos,D.B. (1974) *J. Immunol.,* **112**, 1190.
74. Hayman,M.J., Skehel,J.J. and Crumpton,M.J. (1973) *FEBS Lett.,* **29**, 185.
75. Steinemann,A. and Stryer,L. (1973) *Biochemistry,* **12**, 1499.
76. Jack,G.W., Blazek,R., James,K., Boyd,J.E. and Micklem,L.R. (1987) *J. Chem. Technol.,* **39**, 45.
77. Bonnerjea,J., Oh,S., Hoare,M. and Dunnhill,P. (1986) *Bio/Technology,* **4**, 954.
78. Birch,J.R., Thompson,P.W., Lambert,K. and Boraston,R. (1985) In *Large Scale Mammalian Cell Culture.* Feder,J. and Tolbert,W.R. (eds), Academic Press, London and New York, p.
79. Birch,J.R., Lambert,K., Thompson,P.W., Kenney,A.C. and Wood,L.A. (1987) In *Large Scale Cell Culture Technology.* Lyderson,B.J. (ed.), Carl Hanser Verlag, Muncih, p. 1.
80. Janson,J.-C. (1986) *Trends Biotechnol.,* **2**, 31.
81. Hill,C.R., Birch,J.R. and Benton,C. (1986) In *Bioactive Microbial Products.* Stowell,J.D., Bailey,P.J. and Winstanley,D.J. (eds), Academic Press, London and New York, Vol. 3, p. 175.
82. Secher,D.S. and Burke,D.C. (1980) *Nature,* **235**, 446.
83. Muller,H.P., Van Tilbrug,N.H., Derks,J., Klein-Breteler,E. and Bertine,R.M. (1981) *Blood,* **58**, 1000.
84. Tarnowski,J.J. and Liptak,R.A. (1983) In *Advances in Biotechnological Processes.* Alan R.Liss Inc., New York, Vol. 2, p. 271.
85. Kenney,A.C., Lee,L.G. and Hill,C.R. (1988) In *Methods in Molecular Biology* Walker,J.M. (ed.), Humana Press Inc., Vol. 3, p. 99.
86. Hill,C.R., Kenney,A.C. and Goulding,L. (1987) *BIF-Biotech Forum,* **4**, 167.
87. Novick,D., Eshhar,Z. and Rubinstein,M. (1982) *J. Immunol.,* **129**, 2244.
88. Novick,D., Eshhar,Z., Gigi,O., Marks,Z., Revel,M. and Rubinstein,M. (1983) *J. Gen. Virol.,* **64**, 905.
89. Cobbs,C.S., Graus,P.K., Russ, ? *et al.* (1983) *Toxican,* **21**, 285.
90. Chase,H.A. (1981) *Chem. Eng. Sci.,* **39**, 1099.
91. Underwood,P.A. and Bean,P.A. (1985) *J. Immunol. Methods,* **80**, 189.
92. Gray,G.R. (1980) *Anal. Chem.,* **52**, 9R.
93. Chase,H.A. (1984) *J. Biotechnol.,* **1**, 67.
94. Hudson,L. and Hay,F.C. (1980) *Practical Immunology.* Blackwell Scientific, Oxford.
95. Kristianson,T. (1978) In *Affinity Chromatography.* Hoffman-Ostenhoff,O., Breitenbach,M., Koller,F., Kraft,D. and Scheiner,O. (eds), Pergamon, New York.
96. Chidlow,J.W., Borune,A.J. and Bailey,A.J. (1974) *FEBS Lett.,* **41**, 248.
97. Mains,R.E. and Eipper,B.A. (1976) *J. Biol. Chem.,* **251**, 4115.
98. Avrameas,S. and Ternynck,T. (1967) *Biochem. J.,* **102**, 37C.
99. Zoller,M. and Matzku,S. (1976) *J. Immunol. Methods,* **11**, 287.
100. Melchers,F. and Messer,W. (1970) *Eur. J. Biochem.,* **17**, 267.
101. Weintraub,B.D. (1970) *Biochem. Biophys. Res. Commun.,* **39**, 83.
102. Hill,R.J. (1972) *J. Immunol. Methods,* **1**, 231.
103. Anderson,K.K., Bejamin,Y., Douzov,P. and Bolny,C. (1979) *J. Immunol. Methods,* **25**, 375.

CHAPTER 6

Separation on the basis of size: gel permeation chromatography

ANNA Z.PRENETA

1. INTRODUCTION

Gel permeation chromatography (GPC) is a form of partition chromatography used for separating molecules of different sizes. GPC has been described by several other terms including gel filtration, gel exclusion chromatography, molecular sieve chromatography or even simply as gel chromatography. The basic principle of GPC is that molecules are partitioned between solvent and a stationary phase of defined porosity. The separation process is carried out using a porous gel matrix (in bead form) packed in a column and surrounded by solvent (*Figure 1a*). Consider a sample containing a mixture of molecules smaller and larger than the pores of the stationary phase matrix, as well as molecules intermediate in size. The smaller molecules can enter the matrix pores and hence move more slowly through the column, appearing as the last components in the chromatogram. The larger molecules are excluded from the stationary phase and hence elute first from the column. Molecules intermediate in size can enter the stationary phase, but spend less time within it than smaller molecules do. Thus, all the molecules are eluted in order of their decreasing size (*Figure 1b*). Some proteins show anomalous behaviour on GPC; these include proteins which are not globular in shape. Generally long thin proteins elute earlier than globular proteins of the same size. Denaturing agents such as urea or guanidine hydrochloride minimize the effects of tertiary and quaternary structures of proteins by transforming them to a random coil configuration, and hence allow a more accurate assessment of their molecular weight. However, the effective exclusion limit of a given matrix will be reduced due to the increased hydrodynamic radius (viscosity) of a random coil conformation. The separation mechanism of GPC involves both molecular shape and mass of the molecules, thus GPC can be used to determine molecular shape in solution (1) and consequently as a monitor for denaturation.

The volume of solvent between the point of injection and the peak maximum of a solute is known as the elution volume (V_e) (*Figure 1b*). This is used to characterize the behaviour of the solute molecule in gel permeation media. In the case of large molecules which are excluded from the stationary phase, the elution volume is equal to the void/exclusion volume (V_0). The elution volume of a particular molecule is dependent on the fraction of the stationary phase available to it for diffusion; this is represented by the constant K_d. Hence, $V_e = V_0 + K_d (V_s)$, where V_s represents the volume of solvent within the stationary phase. This value is difficult to determine experimentally and is therefore substituted by the term $V_t - V_0$ where V_t represents

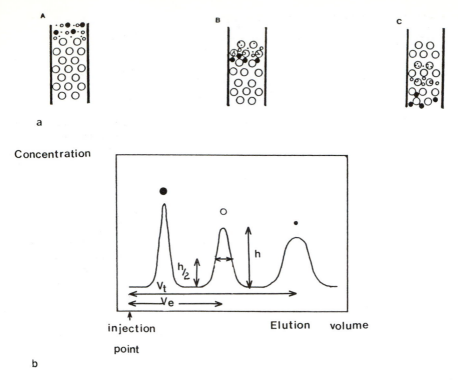

Figure 1. (a) Diagrammatic representation of the principle of GPC. ●, represents molecules larger than matrix pores; ○, represents molecules intermediate in size; ·, represents molecules smaller than matrix pores. (b) Elution profile of a mixture of three different size solutes, showing various parameters used to describe the behaviour of a solute and the matrix.

the total bed volume. Total bed volume can be measured by determining the elution volume of a small molecule (not within the fractionation range of the matrix), such as acetone. The expression therefore becomes

$$K_{av} = \frac{(V_e - V_0)}{(V_t - V_0)}$$

where K_{av} is the coefficient describing the fraction of stationary gel volume available for diffusion of a given species ($K_{av} = K_d$). Further information and theoretical models for calculating K_d and similar constants have been reviewed in ref. 2. Efficiencies of GPC columns are determined by calculating the number of theoretical plates (n) per metre (this is usually calculated using one of the latest eluting peaks) using the following equation:

$$n = 5.56 \frac{(V_e)^2}{(V_w)} \times \frac{100}{L}$$

where V_e = elution volume of sample.
V_w = peak volume at 1/2 height.
L = column length in cm.

294

Table 1. Commercially available GPC matrix types.

Product name	Matrix type	Supplier	pH range	Temperature stability (°C)	Stability to solvents
Sephadex	Dextran	Pharmacia	2	120	Subject to microbial degradation
Sephacryl	Dextran/ bisacrylamide	Pharmacia	2−11	120	Stable
Sepharose	Agarose	Pharmacia	4−9	40	Avoid urea, organic solvents and chaotropic salts
Sepharose CL	Cross-linked agarose	Pharmacia	3−14	120	Avoid strong oxidizing agents
Ultrogel A	Agarose	IBF	4−9	40	Avoid urea, organic solvents and chaotropic salts
Ultrogel AcA	Agarose/ polyacrylamide	IBF	3−10	40	Subject to microbial degradation, variable resistance to chaotropic salts
Biogel A	Agarose	Bio-Rad	4−9	40	Avoid urea, organic solvents and chaotropic salts
Biogel P	Polyacrylamide	Bio-Rad	2−10	120	Avoid strong oxidizing agents. Organic solvents cause shrinkage
Fractogel TSK	Cross-linked polyether	Merck	1−14	120	Stable

2. GEL PERMEATION MEDIA

Ideally the materials used for GPC should be inert with respect to the molecules being separated. However, interactions of packing materials with biological molecules have been observed. Partial adsorption of the protein under specific buffer conditions results in the protein eluting from the column later than would be expected for its molecular size. Most interactions of an ionic nature can be eliminated by increasing the ionic strength of the buffer, typically an ionic strength of 50 mM is sufficient to avoid ionic interactions. In cases where the interactions are of a hydrophobic nature, high ionic strength buffers should be avoided. Protein−protein interactions can also occur (e.g. specifically with subunit interactions, or non-specifically as adsorption) and again careful use of buffers may be used to minimize these interactions.

The most commonly used supports are gels consisting of cross-linked polyacrylamide, agarose, dextran or combinations of these as shown in *Table 1*.

Dextran and agarose-based matrices are biodegradable and should therefore be stored in the presence of antimicrobial agent (e.g. sodium azide; CARE—toxic, see Section 3.8).

2.1 Choice of gel matrix

Several factors should be considered when choosing the most suitable gel matrix for a particular application. These are discussed below.

2.1.1 *Fractionation range*

If all proteins in a mixture are to be separated from relatively low molecular weight

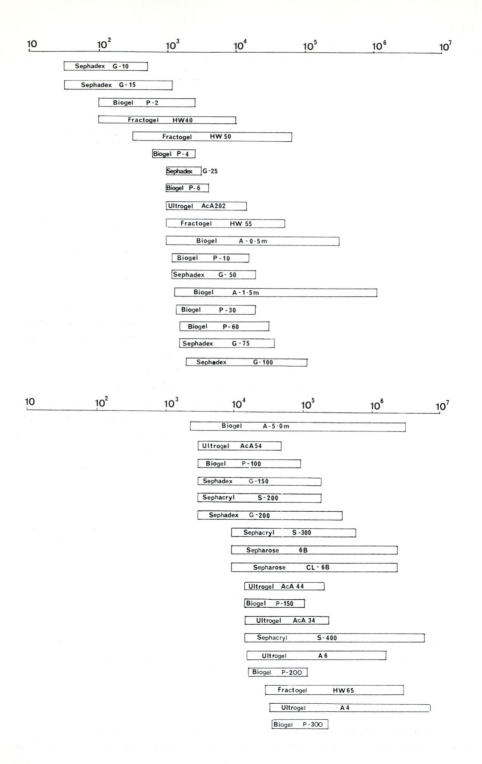

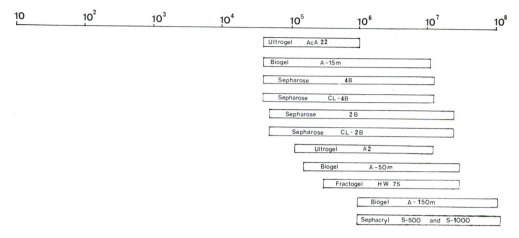

Figure 2. Fractionation ranges of commercially available gel filtration matrices.

solutes (<5000), then packings with small pores are needed; this is known as a group separation or desalting. In this instance the proteins are totally excluded from the pore matrix; suitable media include Sephadex G-25, or G-50 (Pharmacia-LKB) and Biogels P-6 or P-10 (Bio-Rad). Pre-packed Sephadex (G-50) columns (PD-10 columns) are available from Pharmacia-LKB specifically for this purpose.

For resolution of different proteins more closely related in size, a matrix of the correct fractionation range should be chosen such that the molecules do not elute in either the void or the total volume. *Figure 2* shows the fractionation ranges of many of the available matrices. If the molecular weights of the proteins to be separated are known then the gel type is chosen by examining the selectivity curves (K_{av} versus $\log_{10}$ mol. wt, see *Figure 3*) for each of the gels; this information can be found in the technical bulletins available from the gel manufacturers.

2.1.2 *Resolution*

Smaller beads (Fine and Superfine grades) usually give better resolution because molecular diffusion is lower and therefore there is less zone broadening. However, smaller particles give higher resistance to flow and are more compressible, therefore lower flow rates have to be used. Rigid and large beads allow faster flow rates and are therefore more appropriate for large-scale applications.

2.1.3 *Stability*

The final consideration is the stability of the matrix to pH, temperature and organic solvents; these should be compatible with the properties of the proteins to be separated.

3. EXPERIMENTAL

3.1 **Equipment**

The basic equipment necessary to perform a gel permeation experiment includes a column, a detector, a fraction collector and a means of controlling the flow rate, such

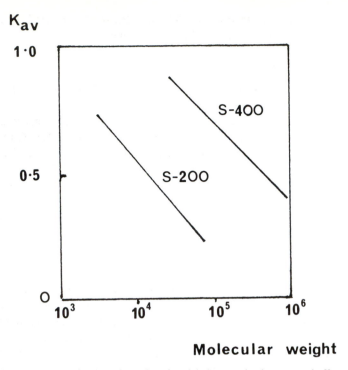

Molecular weight

Figure 3. Selectivity curves of Sephacryl matrices for globular proteins in aqueous buffer.

as a constant head reservoir or peristaltic pump. The system should be arranged with the monitor connected as close as possible to the column outlet to prevent longitudinal mixing in the tubing, similarly the connecting tubing and flow cells should have narrow dimensions to prevent mixing. A typical apparatus is shown in Chapter 4.

3.2 The column

The aim of GPC is to obtain the greatest possible resolution of the molecules of interest where

$$\text{Resolution} = \frac{\text{distance between separated zones}}{\text{zone width}}$$

The choice of column dimensions is therefore an important consideration. Generally, the longer the column the better the resolution, however, this also implies a longer separation time and greater sample dilution. Columns up to 100 cm in length and internal diameter $1-2.5$ cm ($20-40$ times longer than wider) are usually sufficient for most laboratory separations. Irregular flow is less important in long narrow columns than in short wide columns of the same volume. This is because a long thin column is unlikely to suffer so much from horizontal plane distortion. In addition, the sample occupies a greater column depth and hence any distortion that may occur has less effect. Wall effects due to surface tension drag on the sides of the column, in addition to irregular sample application, contribute to irregular flow. There are many suitable columns available from Amicon, Pharmacia-LKB and Bio-Rad for GPC at analytical to preparative scales ranging from $1-20$ cm in diameter.

3.3 Preparation of the gels

Some GPC media are obtained as dry powders (e.g. Sephadex and Biogel P), and thus need to be swollen in the eluent prior to use.

(i)   Mix the powder with excess liquid (stirring is not usually necessary, and may in fact break the particles) and then heat the gel slurry to 100°C in a water bath (ensure the matrix is stable to such treatment prior to swelling). This allows the gel to swell within a few hours and removes any dissolved air.

(ii)   After cooling to the appropriate temperature remove fines by decanting off the solution after the slurry has settled.

Gels which are supplied pre-swollen, such as Sephacryl, only require dilution in the eluent buffer before use.

3.4 Packing the column

(i)   Prior to pouring ensure that the gel consists of a slurry of approximately 75% settled gel and 25% supernatant liquid. If the gel has been swollen at an elevated temperature allow it to cool, otherwise de-aerate under vacuum. Allow the suspension to reach operating temperature before packing.

(ii)   Mount the column on a vertical stand away from sunlight and draughts, as these may result in temperature fluctuations and therefore air bubble formation, and preferably maintain it at a constant temperature to prevent formation of air bubbles. (Pharmacia-LKB K columns have water jackets to maintain constant temperature).

(iii)   Fill the dead space below the support with liquid from the outlet end to ensure no bubbles are trapped below the net, then close the outlet tubing.

(iv)   Fill the column by pouring the gel in a single step, since packing in sections often results in an unevenly packed bed. To do this a column extension tube or funnel may be required. The column should be filled by pouring the gel down a glass rod held against the wall of the column. Alternatively the column may be tilted at an angle and the slurry poured directly down the inside wall of the column; however, the column must be returned to a vertical position immediately after pouring.

(v)   Apply flow to the column as soon as possible to achieve even sedimentation. The maximum operating pressure for each gel type must not be exceeded during packing. Two to three column volumes are sufficient to stabilize the gel bed; these are usually applied at a flow rate slightly higher than that which will be used for the separation itself.

(vi)   Inspect the column visually in transmitted light for heterogeneities in the bed or air bubbles trapped during packing. As a final check apply a coloured test sample, such as Blue dextran (available from Pharmacia-LKB) and observe the progress of the coloured zone. For a well packed gel the zone should move as a straight band with little tailing. If not or if bubbles are observed repack the column. Blue dextran is also used for determining the void volume of the column since it is excluded from the matrix of most gels and so this and the packing test can be done simultaneously.

3.5 **Choice of eluent**

GPC is relatively independent of the type of eluent, except in cases where there is known to be an interaction between the matrix and protein (see Section 2). It is therefore usual to choose an eluent in which the protein is most stable. This may include the addition of various cofactors or metal ions. Alternatively if the sample is to be freeze-dried volatile buffers such as ammonium bicarbonate can be used.

3.6 **Sample size, composition and application**

High resolution in GPC depends on application of the sample in a small volume. The greater the sample volume the greater becomes the volume in which the sample is eluted; if this volume is greater than the volume separating different components they will be remixed. The sample volume must therefore be lower than the separation volume between components. Typically the sample volume should be $1-5\%$ of the total bed volume. Sample volumes below 1% do not result in improved resolution. For sample volumes greater than 5% of the total bed volume, e.g. 8%, only molecules differing in molecular size by a factor of four will be separated (3). For group separation or desalting where there is a large difference between molecular sizes, it is often possible to apply much larger sample volumes (e.g. $25-30\%$ of the total column volume) without losing the required resolution.

In GPC the partition coefficients between gel and liquid are independent of sample concentration. However, the concentration which can be used is generally limited by the viscosity of the sample relative to the eluent; this should not differ by a factor greater than two. A high viscosity results in zone instability and irregular flow pattern. This is because a dense sample runs ahead between the beads causing a diffuse leading boundary. The relative viscosity increases with increasing concentrations of the macromolecule; the protein concentration of the sample being applied to a column should be approximately $10-20$ mg ml^{-1}. With very viscous samples (e.g. redissolved ammonium sulphate precipitates), better resolution may be obtained by loading the sample on in an upward direction, since this results in a sharp leading boundary, and then reversing the column for the separation.

Care should be exercised when applying the sample to the surface of the bed so that it is not disturbed. There are several ways in which the sample may be applied.

(i)   Application to a drained bed surface. Drain the buffer from the surface of the bed without allowing the bed to run dry, then close the column outlet. Layer the sample on top of the bed with a pipette, drain the bed again and close the outlet. Take care not to disturb the surface of the gel bed. Apply a small amount of buffer and drain the bed. Finally close the column outlet again, refill the column with eluent and reconnect to the pump or buffer reservoir.

(ii)   Layering the sample under the eluent. The sample should be denser than the eluent; this may require addition of glucose, sodium chloride or sucrose. Draw the sample into a syringe connected to a piece of fine capillary tubing, close

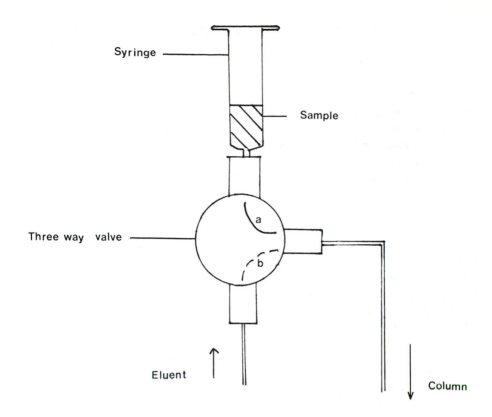

Figure 4. Diagrammatic representation of sample application by means of a three-way valve. Position a: sample loading onto column. Position b: elution buffer flowing onto column (sample isolated from column).

the column outlet and deliver the sample carefully some millimetres above the bed surface. The sample solution will therefore displace the eluent forming an even layer on top of the bed. Allow the sample to pass into the bed, fill the area above with eluent and reconnect to the buffer reservoir.

(iii) Use of a three-way valve. The easiest and most reproducible method of sample application requires the use of an adaptor, syringe and three-way valve connected as shown in *Figure 4*. Draw the sample into the syringe and place into the three-way valve. Switch the valve so that the syringe and column are connected, allow the required sample volume to drain onto the bed surface, and switch the valve back to connect the eluent buffer. It is important to minimize the dead space between the gel bed surface and adaptor to avoid mixing of the sample.

3.7 **Chromatography**

Maximum resolution is obtained using long columns and slow flow rates, but not so slow that they result in diffusion. The optimum flow rate for resolution of proteins is approximately 2 ml cm^{-2} h^{-1}; however, for Sephacryl flow rates up to 30 ml cm^{-2} h^{-1} can be used, due to the rigidity of the beads. It is often necessary to optimize the flow rate to obtain adequate resolution; gel manufacturers' recommendations offer a useful guideline.

The spread of an eluting peak depends to some extent on its elution position. The largest molecules in a mixture elute first, spending less time in the column and therefore have been subjected to less turbulence and diffusion. Smaller molecules, however, have higher diffusion coefficients and therefore elute in a broader peak (and hence in a larger volume). This emphasizes the importance of choosing a matrix with the most appropriate fractionation range for the protein to be separated. When peak tailing is observed, it is probably due to an unevenly packed bed or a badly applied sample.

It may in some cases be advantageous to perform GPC in an upward flow manner. This is particularly true for very soft gels, since these have a tendency to pack down

Table 2. Commercially available HPLC and FPLC columns.

Supplier	Column	Composition	Fractionation range for globular molecules (mol. wt)
Millipore Waters	Protein-Pak 60￼ 125￼ 300	Porous silica with hydrophilic groups	$1000 - 20\,000$ $2000 - 80\,000$ $10\,000 - 500\,000$
Toyo Soda (Anachem Beckman and Varian Associates)	TSKG2000SW TSKG3000SW TSKG4000SW	Porous silica	$500 - 60\,000$ $1000 - 300\,000$ $5000 - 1 \times 10^6$
Serva	Si60 Polyol Si100 Polyol Si300 Polyol Si500 Polyol	Silica	<5000 $<16\,000$ $6000 - 100\,000$
Synchrom. Inc.	Synchropac 100￼ 300￼ 500￼ 1000￼ 4000	Porous silica with glycerol propyl groups	Not given Not given Not given Not given Not given
Merck	LiChrosorb DIOL LiChrospher 100￼ 300￼ 500￼ 1000	As above As above	$10\,000 - 80\,000$ $10\,000 - 80\,000$ $70\,000 - 300\,000$ $250\,000 - 500\,000$ $400\,000 - 1 \times 10^6$
Pharmacia-LKB (FPLC)	Superose-12 Superose-6	Cross-linked agarose	$1000 - 30\,000$ $5000 - 5 \times 10^6$
DuPont	GF-250 Zorbax Diol	Surface modified amorphous silica	$10\,000 - 250\,000$

Absorbance

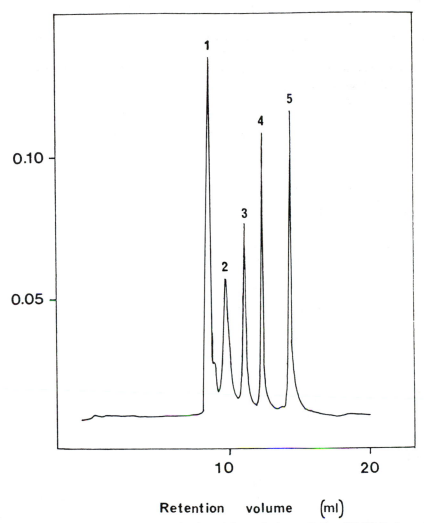

Figure 5. HPLC elution profile of Bio-Rad molecular weight standards on a DuPont GF-250 Zorbax column at a flow rate of 1 ml min^{-1} in phosphate buffer (50 mM, pH 7.6) containing NaCl (0.1 M). 1 = thyroglobulin, 670 000 mol. wt; 2 = gamma globulin, 258 000 mol. wt; 3 = ovalbumin, 44 000 mol. wt; 4 = myoglobin, 17 000 mol. wt; 5 = vitamin B$_{12}$, 1350 mol. wt.

resulting in decreased flow rates. In the case of upward flow chromatography, gravity works in the opposite direction to the flow, hence higher flow rates may be used than would otherwise be possible; in addition it may also result in improved resolution due to sharpening of bands.

3.8 Column cleaning and storage

Since GPC is generally used with relatively pure proteins, excessive cleaning is not

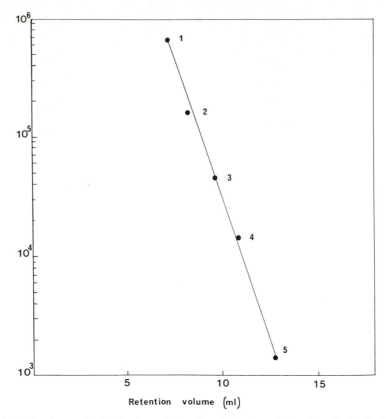

Figure 6. Calibration graph of molecular weight against elution (retention) volume for Bio-Rad molecular weight standards, using conditions described in *Figure 5*.

required. However, if necessary most gels may be cleaned by treating with 0.2 M sodium hydroxide or non-ionic detergents; the manufacturer's recommended cleaning instructions should be checked for each product. Many gels are susceptible to microbial degradation and therefore should be stored in the presence of antimicrobial agents, preferably at 4°C in the dark. One of the most commonly used antimicrobial agents is sodium azide; this is used at concentrations of 0.02−0.05% w/v. Care should be exercised when using this agent since it is very highly toxic and is incompatible with low pH. Alternatives to sodium azide include 20% ethanol. It is probably most convenient to store the gel packed in columns, particularly if the column is to be re-used in the near future.

4. HPLC AND FPLC

There are various columns commercially available which enable GPC to be performed on a HPLC system, as shown in *Table 2*. These columns are packed with microparticulate organic or silica based particles (5−10 μm in diameter). Generally HPLC GPC will give high resolution in a very short time (typically 30 min). However, only relatively small amounts (1 – 10 mg) can be loaded onto these columns and hence this is often used only as an analytical technique. Silica-based packings are unstable above pH 7.5 and therefore alkaline buffers should not be used. These columns should be stored in aqueous buffers containing antimicrobial agents such as 0.02% sodium azide.

Conventional HPLC systems are made of stainless steel, which may be corroded by the long-term use of high concentrations of chloride ions. In addition some proteins may be inactivated by metal ions leaching from the stainless steel. Hence systems based on inert materials (such as glass, teflon and titanium) have been developed (e.g. the Pharmacia-LKB FPLC, and other systems supplied by, Waters and Anachem). Gels for these systems are packed in glass columns. They are organic and have higher porosities, thus lower pressures than those for many silica gels are required. The Pharmacia-LKB Superose gels are also available in large particle sizes (e.g. 30 μm) thus allowing easy scale-up. For further information on HPLC GPC the reader is referred to ref. 4.

5. APPLICATIONS OF GPC

GPC is not often used as the first step in protein purification due to its relatively low capacity, but is often used as one of the final steps or to assess the purity of a preparation.

Other applications include group separations (desalting), fractionations (on analytical and preparative scales), and determination of molecular weights, equilibrium constants for protein binding, or complex formation (5). In addition GPC may be used for analysing aggregate formation, breakdown and denaturation of proteins. Some GPC matrices such as Sephadex G-200 can also be used for concentrating samples as described in Chapter 3.

5.1 Determination of molecular weights

GPC is a valuable alternative to SDS – PAGE for the determination of molecular weights of proteins. The elution volumes of globular proteins are mainly determined by their hydrodynamic radius which is related to their molecular weight; thus the elution volume is an approximately linear function of the logarithm of the molecular weight.

(i)   A calibration graph is usually constructed for each column to be used for the determination. Apply a mix of suitable standard proteins (in the correct fractionation range) to the column. Suitable standards for most ranges can be obtained from Bio-Rad, Sigma or Pharmacia-LKB. Determine the elution volume (ideally this should be K_d though in practice it makes little difference; see Section 1) of each standard from the chromatogram, as shown in *Figure 5*.

(ii)   Plot these values against the logarithm of the molecular weight (*Figure 6*).

(iii)   Run the protein of unknown molecular weight on the same column, and note its elution volume. Use this value to obtain the molecular weight from the calibration graph.

6. REFERENCES

1. Potschka,M. (1987) *Anal. Biochem.*, **162**, 47.
2. Yarmush,M.L., Antonsen,K. and Yarmush,D.M. (1980) *Comprehensive Biotechnol.*, **2**, 489.

3. Scopes,R. (1984) In *Protein Purification, Principles and Practice.* Cantor,C.R. (ed.), Springer-Verlag, New York, p. 151.
4. Unger,K. (1984) In *Methods in Enzymology.* Jakoby,W.B. (ed.), Academic Press, New York and London, Vol. 104, p. 154.
5. Fischer,L. (1980) *Gel Filtration Chromatography. Laboratory Techniques in Biochemistry and Molecular Biology.* Work,T.S. and Burdon,R.H. (eds), Elsevier, North Holland Biomedical Press, Oxford.

APPENDIX I
Suppliers

APV, P.O. Box 4, Manor Royal, Crawley, West Sussex, RH10 2QB, UK.

Aldrich, The Old Brickyard, New Road, Gillingham, Dorset, SP8 4JL, UK. Tel. (07476) 2211.

Amersham, Lincoln Place, Green End, Aylesbury, Bucks., HP20 2TP, UK. Tel. (0296) 395222.

Amicon, Upper Mill, Stonehouse, Gloucester, GL10 2BJ, UK. Tel. (045382) 5181.

Anachem, 20 Charles Street, Luton, Beds., LU2 0EB, UK. Tel. (0582) 456666.

Anderman & Co. Ltd, 145 London Road, Kingston upon Thames, Surrey, KT2 6NH, UK. Tel. (01) 541 0035.

Artisan Metal Products, Waltham, MA, USA.

Atlas Bioscan Ltd, Osborne House, Stockbridge Road, Chichester, PO19 2DU, UK. Tel. (0243) 773141.

BCL, Bell Lane, Lewes, East Sussex, UK. Tel. (0273) 480444.

BDH, Broom Road, Parkstone, Poole, Dorset, UK. Tel. (0202) 745520.

Beckman, Progress Road, Sands Ind. Estate, High Wycombe, Bucks., HP12 4JL, UK.

Biocatalysts Ltd, Tredforest Industrial Estate, Pontypridd, Wales, UK.

BioInvent International, S-223, 70 Lund, Sweden. Tel. (046) 46 16 85 50.

BioProbe International Inc., 2842 Walnut Avenue, Tustin, CA 92680, USA. Tel. (714) 544 4035.

BioRad Laboratories Ltd, Caxton Way, Holywell Ind. Estate, Watford, Herts., WD1 8RP, UK. Tel. (0923) 240 322.

BRL, PO Box 145, Science Park, Cambridge, CB4 4BE, UK.

Brownlea Labs. Inc., 2045 Martin Ave., Santa Clara, CA 95050, USA. Tel. (408) 727 1346.

Calbiochem, see Nova Biochem, UK.

Cambio, 34 Millington Road, Newnham, Cambridge CB3 9HP, UK. Tel. (0223) 66500.

Cecil Instruments Ltd, Milton Technical Centre, Milton, Cambridge, CB4 4AZ, UK. Tel. (0223) 420821.

Christison Scientific Equipment Ltd, Albany Road, S.H.House, East Gateshead Industrial Estate, Gateshead, NE8 3AT, UK. Tel. (0632) 77461.

Collaborative Research, PO Box 370068, Boston, MA 02241, USA.

Cooper Biomedical & ICN Immunobiologicals, Free Press House, Castle Street, High Wycombe, Bucks., HA3 6RN, UK. Tel. (0494) 443826.

Cuthbert Andrews Ltd, Watford, UK.

Dakopatts Ltd, 22 The Arcade, The Octagon, High Wycombe, Bucks., HP11 2HT, UK. Tel. (0494) 452016.

DDS Membrane Filtration, 1600 County Road, F.Hudson, WI 54016, USA. Tel. (715) 386 9371.

Desaga, Springfield Mill, Sandling Road, Maidstone, Kent, ME14 2LE, UK.

Dominick Hunter Ltd, Durham Road, Birtley, DH3 2SF, UK. Tel. (091) 4105121, see also HPLC Technology Ltd.

Drew Scientific Ltd, 12 Barley Mow Passage, London, W4 4PH, UK. Tel. (01) 995 9382.

DuPont, Wedgewood Way, Stevenage, Herts., SG1 4QN, UK. Tel. (0438) 734680.

Eastman Kodak, LRPD, Acorn Field Road, Liverpool, L33 7ZX, UK.

Edwards, Manor Royal, Crawley, Sussex.

EDT Analytical, 14 Trading Estate Road, London, NW10 7LU, UK. Tel. (01) 961 1477.

Electro-Nucleonics International Ltd, Adriann van Bergerstraat, 202−208, 4811 SW Breda, Netherlands. Tel. (7622) 2033.

Eppendorf, see Anderman.

Filtron Technology Corp., 500 Main St., PO Box 119, Clinton, MA 01510, USA. Tel. (508) 3688582.

Fisher Scientific, 711 Forbes Ave., Pittsburg, PA, USA. Tel. (412) 562 8300.

Fisons Instruments, Sussex Manor Park, Gatwick Road, Crawley, RH10 2QQ, UK. Tel. (0293) 561222.

Flow Laboratories, Woodcock Hill, Harefield Road, Rickmansworth, Herts., WD3 1PQ, UK. Tel. (0923) 774666.

Fluka, Peakdale Road, Glossop, Derby, SK13 9XE, UK.

Gallenkamp, Belton Road, Loughborough, Leics., LE11 0TR, UK.

Gelman Sciences Ltd, 10 Harrowdean Road, Brackmills, Northampton, NN4 0EZ, Tel. (0604) 765141.

Gibco UK Ltd, PO Box 35, Trident House, Renfrew Road, Paisley, PA3 4EF, Scotland, UK. Tel. (041) 889 6100.

Glen Creston Ltd, 16 Dalston Gardens, Stanmore, Middlesex, HA7 1DA, UK. Tel. (01) 226 0123.

Heraeus Equipment Ltd, Unit 9, Wates Way, Brentwood, Essex, CM15 9TB, UK. Tel. (0509) 237371.

Hoeffer Scientific Instruments, Unit 12, Croft Road Workshops, Croft Road, Newcastle under Lyme, ST5 0TH, UK. Tel. (0782) 617317.

Hoeffer (Biotech), 183A Camford Way, Luton, Beds. LU3 3ANS, UK

IBF, see Life Science Laboratories Ltd.

ICN Biomedicals Ltd, Free Press House, Castle St., High Wycombe, Bucks., HP13 6RN, UK. Tel. (0494) 443826.

Janssen, Grove, Wantage, Oxon. OX12 0DQ, UK. See also ICN.

Jencons, Cherrycourt Way Ind. Estate, Leighton Buzzard, Beds., LU7 8UA, UK.

Joyce Loebel, Marquisway, Team Valley, Gateshead, NE11 0QW, UK. Tel. (091) 482 2111.

J.T.Baker, PO Box 9, Hayes Gate House, 27 Uxbridge Road, Hayes, Middlesex, UB4 JD, UK. Tel. (01) 569 1191.

Kabivitrum, Kabi Vitrum House, Riverside Way, Uxbridge, UB8 2YF, UK.

Life Science Laboratories, Sedgewick Road, Luton, LU4 9DT, UK. Tel. (0582) 597676.

LKB, see Pharmacia−LKB.

Merck, see BDH.

Microfluidics Corp., 44 Mechanic St., Newton, MA 02164, USA. Tel. (617) 969 5452.

Microgen Inc., 23152 Verdugo Dr., Laguna Hills, CA 92653, USA. Tel. (714) 581 3880.

Miles Ltd, Stoke Court, Stoke Poges, SL2 4LY, UK. Tel. (02814) 5151.

Millipore Waters, The Boulevard, Ascot Road, Croxley Green, Watford WD1 8YW, UK. Tel. (0923) 816375.

MSE Scientific Instruments, Sussex Manor Park, Crawley, Sussex, RH10 2QQ, UK. Tel. (0293) 31100.

New Brunswick Scientific (UK) Ltd, 6 Colonial Way, Watford, WD2 4PT, UK. Tel. (0923) 223293.

Northern Media Supply, Sainsburyway, Hessle, N. Humberside, HU13 9NX, UK. Tel. (0482) 572436.

Nova Biochem, 3 Heathcote Buildings, Highfields Science Park, University Blvd., Nottingham, NG7 2QJ, Tel. (0602) 430951.

Novo Ltd, 28 Thomas Avenue, Windsor, Berks., SL1 1QP, UK.

Nucleopore Corp., 7035 Commerce Cr., Pleasanton, CA 94566, USA. Tel. (415) 463 2530.

PCI, One Fairfield Crescent, West Caldwell, NJ 07006, USA. Tel. (201) 575 7052.

Perkin-Elmer Ltd, Post Office Lane, Beaconsfield, Bucks., HP9 1QA, UK. Tel. (0494) 676161.

Perstorp Biolytica, S-223 70 Lund, Sweden. Tel. (046) 46 16 87 80. See also EDT Analytical.

Pfeiffer and Langen Dormagen, Frankenstrasse 25, D-4047 Dormagen, FRG. Tel. (02106) 52-1.

Pharmacia – LKB, Pharmacia House, Midsummer Boulevard, Milton Keynes, MK9 3HP, UK. Tel. (0908) 661101.

Phase-Separations Ltd, Deeside Industrial Park, Queensferry, Clwyd, CH5 2NU, UK. Tel. (0244) 816444.

Phillips Analytical, Building HKF, NL-5600 MD Eindhoven, The Netherlands. Tel. (040) 785213.

Pierce Europe BV, PO Box 1512, NL 3260, The Netherlands. See also Life Science Laboratories.

Rohm Pharma GmbH, Westerstadt, PO Box 4347, D-6100, Darmstadt 1, FRG. Tel. (06151) 877-0.

Romicon (Rohm & Haas Ltd), Lennig House, 2 Masons Avenue, Croydon, CR9 3NB, UK.

Russell pH Ltd, Station Road Auchtermuchty, Fife KY14 7DP, Scotland, UK. Tel. (03372) 8871.

Sarstedt, 68 Boston Road, Leicester, LE4 1AW, UK.

Sartorius, 18 Avenue Road, Belmont, Sutton, Surrey, UK.

Schleicher and Schuell GmbH, P.O. Box 246, D-3352, Einbeck, FRG.

Scientific Supplies, 618 Western Avenue, Park Royal, London, W3 OTE, UK.

Serva, see Cambridge Bioscience.

Shandon Southern Products Ltd, Chadrich Road, Astmoor, Runcorn, Cheshire, WA7 1PR, UK. Tel. (09285) 66611.

Shimadzu Scientific Instruments, 7102 Riverwood Road, Columbia, MD 21046, USA. Tel. (301) 381 1227.

Sigma, Fancy Road, Poole, Dorset, BH17 7NH, UK. Tel. (0202) 733114.

Sorvall-DuPont, see DuPont.

Sterogene Biochemicals, 136 E. Santa Clara St., Arcadia, CA 91006, USA. Tel. (818) 446 3773.

Sturge Ltd, Denison Road, Sellay, North Yorkshire, YO8 8EF, UK.

Synchrom Inc., see Anachem.

Toyo Soda, see Anachem.

Ulvac (Chemlab), Hornminster House, 129 Upminster Road, Hornchurch, Essex, KM11 3XJ, UK.

Uniscience Ltd, 12 – 14 St. Annes Crescent, London SW18 2LS, UK.

Union Carbide Corp., Old Ridgebury Road, Danbury, CT 06817, USA. Tel. (203) 7945300.

Varian Associates Ltd, 28 Manor Road, Walton on Thames, Surrey, KT12 2QF, UK. Tel. (09322) 43741.

Wako, see Cambridge Bioscience.

Watson Marlow, Smith and Nephew Pharmaceuticals Ltd, Falmouth, Cornwall, TR11 4RW, UK.

Wellcome Diagnostics, Temple Hill, Dartford, Kent, DA1 5AH, UK.

Whatman Lab Sales Ltd, Unit 1, Coldred Road, Parkwood, Maidstone, Kent, ME15 9XN, UK. Tel. (0622) 674821.

APPENDIX II
Useful recipes

SODIUM PHOSPHATE BUFFER

To make sodium phosphate buffer make the following stock solutions and mix them in the ratios given below to obtain the desired pH. The resultant buffer will be 0.5 M and can be diluted appropriately to give the desired concentration. Check the final pH.

0.5 M disodium hydrogen phosphate. Dissolve 178.05 g of $Na_2HPO_4.2H_2O$ per litre of water (add the solid slowly to the water with vigorous mixing).

0.5M sodium dihydrogen phosphate. Dissolve 156.05 g of $NaH_2PO_4.2H_2O$ per litre of water.

pH at 25°C	Na_2HPO_4	NaH_2PO_4
5.8	4.0	46.0
6.0	6.15	43.85
6.2	9.25	40.75
6.4	13.25	36.75
6.6	18.75	31.25
6.8	24.5	25.5
7.0	30.5	19.5
7.2	36.0	14.0
7.4	40.5	9.5
7.6	43.5	6.5
7.8	45.75	4.25
8.0	47.35	2.65

PHOSPHATE BUFFERED SALINE (PBS)

Dissolve the following in one litre of water and check that the final pH is 7.2. If not required the calcium chloride and magnesium chloride can be omitted.

		Final conc.
Sodium chloride (NaCl)	8.00 g	137 mM
Potassium chloride (KCl)	0.20 g	2.7 mM
Disodium hydrogen phosphate (Na_2HPO_4)	1.15 g	6.5 mM
Potassium dihydrogen phosphate (KH_2PO_4)	0.20 g	1.5 mM
Calcium chloride ($CaCl.2H_2O$)	0.13 g	0.9 mM
Magnesium chloride ($MgCl.6H_2O$)	0.10 g	0.5 mM

Tablets are commercially available for making PBS.

TRIS BUFFERED SALINE

Tris (3 g, 25 mM final concentration) is substituted for the disodium hydrogen phosphate and the potassium dihydrogen phosphate in the recipe for PBS, and the pH adjusted with hydrochloric acid.

0.5 M EDTA (ETHYLENE DIAMINE TETRACETATE)

Add 186.1 g of disodium ethylene diamine tetracetate.$2H_2O$ in approximately 800 ml of water. Adjust the pH of the solution to 8.0 with 5 M sodium hydroxide (this will take approximately 100 ml). NB This concentration of EDTA will not dissolve unless the pH is adjusted to 8.0.

100% TRICHLOROACETIC ACID.

To avoid weighing out this highly corrosive reagent add 227 ml of water to a bottle containing 500 g trichloroacetic acid.

CONCENTRATIONS OF COMMERCIALLY AVAILABLE ACIDS

	Molecular weight	*Molarity (M)*	*Concentration (g/l)*	*% by weight*	*Specific gravity*
Acetic acid	60.05	17.4	1045.0	99.5	1.05
Formic acid	46.02	23.4	1080.0	90.0	1.20
Hydrochloric acid	36.5	11.6	424.0	36.0	1.18
Nitric acid	63.02	15.99	1008.0	71.0	1.42
Perchloric acid	100.5	11.65	1172.0	70.0	1.67
Phosphoric acid	80.0	18.1	1145.0	85.0	1.70
Sulphuric acid	98.1	18.0	1766.0	96.0	1.84

INDEX